Exotic Fruits and Nuts of the New World

FSC
www.fsc.org
MIX
Paper from
responsible sources
FSC® C013604

Exotic Fruits and Nuts of the New World

Odilo Duarte

Retired Professor – Escuela Agrícola Panamericana – El Zamorano, Honduras,
Adjunct Professor, Universidad Agraria La Molina, Lima, Perú

Robert E. Paull

Tropical Plant and Soil Sciences
University of Hawaii at Manoa
Honolulu, Hawaii, USA

www.cabi.org

CABI is a trading name of CAB International

CABI
Nosworthy Way
Wallingford
Oxfordshire OX10 8DE
UK

Tel: +44 (0)1491 832111
Fax: +44 (0)1491 833508
E-mail: info@cabi.org
Website: www.cabi.org

CABI
38 Chauncy Street
Suite 1002
Boston, MA 02111
USA

Tel: +1 800 552 3083 (toll free)
E-mail: cabi-nao@cabi.org

A catalogue record for this book is available from the British Library, London, UK.

Library of Congress Cataloging-in-Publication Data

Duarte, Odilo.
Exotic fruits and nuts of the New World / Odilo Duarte, Retired Professor - Escuela Agrícola Panamericana, El Zamorano, Honduras; Adjunct Professor, Universidad Agraria, La Molina, Lima, Perú ; Robert E. Paull, Tropical Plant and Soil Sciences, University of Hawaii at Manoa, Honolulu, Hawaii, USA.
pages cm
Includes bibliographical references and index.
ISBN 978-1-78064-505-6 (hb : alk. paper) 1. Exotic plants--America. 2. Fruit--America. 3. Nuts--America. I. Paull, Robert E. II. Title.

QK109.D83 2014
581.6′2--dc23

2014020714

ISBN-13: 978 1 78064 505 6

Commissioning editor: Charlotte Hammond
Editorial assistant: Emma McCann
Production editor: Tracy Head

Typeset by SPi, Pondicherry, India.
Printed and bound by CPI Group (UK) Ltd, Croydon, CR0 4YY.

Contents

Preface

In the CABI Crop Production Science in Horticulture Series, we presented the major fruit crops found in international trade. In Volume 1 of that series, the general aspects of tropical fruit production and major tropical fruit such as banana, pineapple, papaya, mango and avocado were described. Many other tropical fruit, already well known in the tropics, are now appearing in larger temperate city markets. In Volume 2, we selected those tropical fruit crops that are being increasingly seen in overseas markets outside of the tropics. The choice of crops to present in Volume 2 was a challenge, especially as to which American tropical fruits to include. The challenge meant that many American tropical fruit and nut crops that have potential in the tropics were excluded.

In this monograph, we have attempted to cover many of the tropical fruit and nut crops that have been infrequently covered in English. These fruits and nuts with further improvement have the potential for production and commercialization in other tropical production areas. However, again a number of other fruit and nut crops often seen in local tropical American markets are not included. This absence is due both to space limitations and more often the absence of published information on their ecological requirements and responses, and production practices.

Special thanks are due to the Commissioning Editors, Sarah Hulbert and Charlotte Hammond, and Editorial Assistants, Christopher Shire and Emma McCann, at CABI for their assistance and patience during the book's development.

We would greatly appreciate receiving all comments and suggestions on this text. We can be reached at the addresses in the front of the text or via e-mail at paull@hawaii.edu or odiloduarte@yahoo.com.

In closing, we both acknowledge the continued support, assistance, and love of our wives Carla and Nancy, and of our children, which enabled us to complete this undertaking.

Odilo Duarte
Lima, Peru, 2014

Robert E. Paull
Honolulu, USA, 2014

Acknowledgments

The authors gratefully acknowledge the support, information, and ideas supplied by Jeff Anderson, Alton Bailey, Jit Baral, Lou Biad, Chris Biad, Anna Biad, Judy Bosland, Emily Bosland, Will Bosland, Emma Jean Cervantes, Danise Coon, William Daga, Jorge Escobedo, Deyuan Wang, Natalie Goldberg, Max Gonzalez, Wendy Hamilton, Steve Hanson, John Hard, Sue Hard, Jaime Iglesias, Sanjeet Kumar, Jimmy Lytle, Jo Lytle, Ariadna Monroy, Mary O'Connell, Jaebok Park, Jennifer Randall, Adrian Rodriqguez, Robert Steiner, Ousmane Sy, Betty Terrien, Nankui Tong, Pedro Undurraga, Manju Vishwakarma, Stephanie Walker, April Ulery, Everardo Zamora, and the Chile team at New Mexico State University.

1 Anacardiaceae

The Anacardiaceae includes many fruit-producing genera including *Spondias* (*S. mombin*, the hog plum, *S. dulcis*, the ambarella, *S. purpurea*, the red mombin, and others), *Mangifera* (*M. indica*, the mango, and several other species), *Pistacia* (*P. vera*, the pistachio nut, and ornamental trees) and *Schinus* (*S. molle*, the Peruvian pepper tree, and *S. terebinthifolius*, the Brazilian pepper tree). This chapter will cover the cashew, which is widely grown for both its fruit and its nut, and two *Spondias* species that are now distributed worldwide.

Cashew

Cashew, *Anacardium occidentale* L. (Anacardiaceae), is one of the important edible nuts consumed worldwide. The cashew fruit (swollen receptacle or pseudofruit) is also important and is frequently consumed fresh, and made into a juice and other products (Donadio, 1983). The Latin name means inverted (ana) heart (cardium) and the Portuguese name *cajú* comes from *akajú*, meaning yellow, in a native language. Common names include: in Arabic *habb al-biladhir*; in Bengali *hijllbadam*, *hijuli*; in Hindi *kaaju*; in Tamil *mindiri*; in Nepali *kaaju*; in Malay *gajus*, *jambugolok* and *jambu mede*; in Swahili *mbibo* and *mkanju*; in Thai *mamuang*, *yaruang*; in Chinese *yao guo* and *yao guo shu*; in French *acajou a pommes*, *noix-cajou*, *noix d'acajou*, *pomme d'acajou*; in Portuguese *cajú* and *cajueiro*; in Spanish *anacardo* or *marañón*; *jocote marañon* in Central America, *cajuil* in the Dominican Republic, *casho* or *cajú* in Peru; *merey* in Venezuela and Colombia. Synonyms include *Acajuba occidentalis* Gaertn., and *Cassuvium pomiferum* Lam.

Several species of *Anacardium* are similar to cashew, with a fleshy receptacle and a nut (Donadio *et al.*, 2002). Some of these species, such as *A. giganteum*, *A. negrense*, *A. othonianum*, *A. humile* and *A. microcarpum*, are regionally important in Brazil for their nuts and fruit. *A. microcarpum* is a species native to the north-east of Brazil, where it is called *cajui* or *caju miniaturia* ("miniature cashew") and could become important commercially. The fruit of *A. microcarpum* is about a third the size of cashew and the pseudofruit is somewhat acid. Other species found in Brazil according to Prabhakaran-Nair (2009) are: *A. nanum*, *A. corymbosum* and *A. spruceana*. On the western side of the Andes *A. excelsum* is the only species found.

Origin and distribution

Brazil is normally considered the center of origin of cashew (de Almeida *et al.*, 2003), since the largest number of varieties of the genus *Anacardium* are found in north-eastern Brazil. Johnson (1973) considers that the state of Ceará is where cashew originated. Prabhakaran-Nair (2009) indicates central Amazonia and the Planalto of Brazil as the probable places of origin. The indigenous people of Brazil consumed the nut and the swollen pedicel called the "cashew apple". They fermented the juice squeezed from the cashew apples to produce wine and roasted the nuts over a fire, thus eliminating the toxic oil from the seed coat. The trees are often found growing wild on the drier sandy soils in the central plains of Brazil and are cultivated in many parts of the Amazon rainforest (Sivakumar and Pai, 2008).

The Portuguese introduced the cashew to India in 1590, possibly through Goa, where it was grown for producing wine and brandy. Later cashew was introduced to the rest of Asia. The Portuguese also introduced the cashew to their colonies in East Africa where it became naturalized and now grows wild along the Mozambique coast. From here, it was introduced to other East African countries: Tanzania and Kenya. Cashew is now grown in tropical regions from South America to the West Indies to Florida, Africa and India.

The cashew nut entered international commerce at the beginning of the 20th century when it became a very important nut after almonds. The planted area has increased in many countries and between 1995 and 2004, world cashew nut production doubled as a result of incentives in producing countries and foreign market expansion. This expansion has slowed in the last few years. Vietnam saw a fourfold increase in the area planted to cashew and in its nut production between 1995 and 2004 (FAO, 2006). In 2004, the total area cultivated in the world was 3.09 million ha with a production of 2.27 million t (FAO, 2006). Prior to 2004, India was the largest producer of raw nuts and now is in third place with 544,000 t after Vietnam, 961,000 t and Nigeria, 594,000 t. India is still the largest processor and exporter, and the second largest consumer. India also imports around 200,000 t of raw nuts, mostly from African countries. The imported nuts are processed and India exports about 95,000 t of clean kernels. Productivity in India is improving with the use of superior varieties and better technology. Similarly, productivity in Brazil with its 691,000 ha in production and currently low yields is increasing with more technically managed orchards. The north-eastern states of Ceará, Piaui and Rio Grande do Norte account for more than 90% of Brazil's cashew production.

Ecology

Soil

The ideal soils seem to be those on flat or slightly hilly land, with light to medium texture, free from stones and aluminum toxicity, with good drainage, high in organic matter and nutrients, pH of 5.0 to 6.5 and no impervious layer in the first 100 cm, such as virgin forest soils (Crisóstomo *et al.*, 2007; Sivakumar and Pai, 2008). Soils to avoid are those that are gravelly, saline, shallow, with an underground water table deeper than 10 m or shallower than 2 m, and areas subjected to periodic flooding (da Silva, 1998). Alluvial well-drained soils normally give good production.

Cashew is frequently grown on marginal soils and also on wasteland unsuitable for other economic crops. The tree is found along sandy sea coasts, fairly steep lateritic slopes or rolling land with shallow top soils in India; alluvial soils in Sri Lanka; ferruginous soils in East and West Africa, Brazil and Madagascar; and volcanic soils in the Philippines, Indonesia and the Fiji Islands. In Brazil, especially in the north-east, the majority of cashew grows on latosols, argisols and quartz-sands that are fairly deep but poor in fertility with a pH of 4.5 to 5.5. It is very popular among poor farmers who can grow a crop of cashew without much tree management, though yields are very low. Many small farmers intercrop cashew with annual crops to obtain more income from their land.

Rainfall

Cashew grows best in a warm, moist, tropical climate with a well-defined dry season of 5–7 months that coincides with flowering and fruiting (de Almeida *et al.*, 2003), followed by a wet season of 5–7 months (1,000–2,000 mm rainfall). In regions with two dry seasons, the tree will flower twice, and with rainfall year round it will flower continuously. In areas with 500–700 mm rainfall, it performs well if it has access to an adequate underground water supply (Ohler, 1979). Cashew can also grow in places with up to 4,000 mm rainfall though a dry period is needed during flowering and fruit set. When flowering occurs during heavy rains and with humidity above 85%, flower diseases, commonly anthracnose and mildew, cause serious losses (da Silva, 1998). In most places, the plant grows without irrigation though it does respond very well to summer irrigation.

Cashew develops well in humidity between 70% and 80%, but also grows well in regions with relative humidity of 50% if there is good soil moisture or with irrigation (Crisóstomo *et al.*, 2007). Very low humidity during flowering reduces stigma receptivity or pollen viability and induces small fruit drop (da Silva, 1998).

Temperature

The tree is found growing between 27°N and 28°S, though yields are higher between 15°N and S (Crisóstomo *et al.*, 2007). Plantations can be found at altitudes of up to 1,200 m in the tropics (Vargas *et al.*, 1999). Tree growth occurs between 16 and 40°C, with an optimum of 26–28°C. Damage to young trees and flowers will occur below 7°C and above 45°C. Prolonged cool temperature does damage adult trees, although they will survive to 0°C for short periods.

Light and photoperiod

The plant prefers to grow in full sun. The optimum total sunshine is 1,285 h, or 9 h day^{-1} during the flowering and fruit set period (Sivakumar and Pai, 2008). Flowering is unaffected by day length.

Wind

Winds up to 3 m s^{-1} are generally not a problem. In areas with wind speeds greater than 7 m s^{-1} young plants have to be protected by tying them to a stake and installing windbreaks. During flowering, dehydration caused by strong winds can be a problem for fertilization and fruit set. Flower and young fruit drop as well as trees being blown down are also possible (da Silva, 1998).

General characteristics

Tree

The cashew is an attractive evergreen with smooth bark and is an erect, low-branching tree, 4–16 m tall, with a spreading canopy that can be as wide as it is tall. The lower limbs, if not pruned, sometimes touch the ground. Under sub-optimal conditions the tree will grow to 5–8 m and the stem will be tortuous, while under good conditions it is straight. In deep soils, the tree can have a deep and well-defined taproot and the lateral roots will extend beyond the drip line of the canopy. The depth of the main and lateral roots and their distribution are affected by the soil type. In good soil conditions the taproot can go down to 10 m and 82% of the root system is in the upper 30 cm of soil (de Almeida *et al.*, 2003). In Brazil, two types of cashew trees are recognized. The common cashew grows to heights of 10–12 m and can reach 14–16 m, and the canopy diameter can vary from 10 to 14 m. The other type grows to 4–6 m, with a canopy diameter of 6–8 m (de Almeida *et al.*, 2003).

The oblong-oval or obovate leaves are alternate, simple, entire and fairly large (8–20 cm by 6–12 cm). The leaves are glabrous and have a short petiole, prominent veins and come in terminal clusters (Fig. 1.1). They are normally reddish or golden when young turning into a light green color and leathery texture as they mature.

Flowers

Cashew flowers are either male or hermaphrodite (andromonoecious, perfect), like those of mango, and are borne in 15–25 cm terminal panicles (Fig. 1.1A). Each panicle can have 120–1,100 flowers (average 500). The hermaphroditic flowers are larger than the male flowers. The individual flowers are sweet-smelling and small, with usually five yellowish-green or yellowish-pink petals about 1.0–1.5 cm, five 0.5 cm sepals and ten stamens. One stamen is about 12 mm long and the other nine are about 4 mm long (Fig. 1.1C). Early flowers are mostly male with perfect flowers being generally produced about 1 month later on the panicle. One perfect flower is found for every 6–28 male flowers, depending on the genotype, climate and other factors.

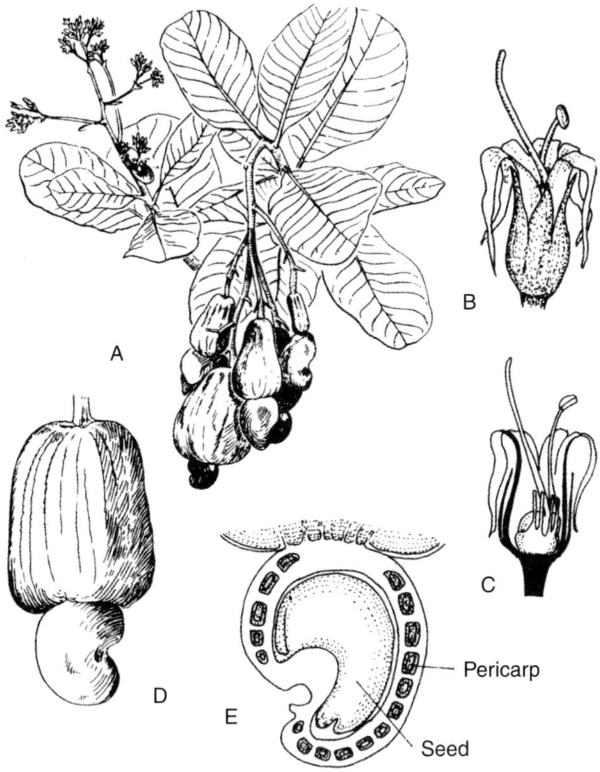

Fig. 1.1. Cashew (*Anacardium occidentale*) nut and "apple" showing (A) leaves, panicle and fruit cluster, (B) and (C) flowers, (D) fruit and swollen peduncle "receptacle" called the "apple" and (E) transverse section through the fruit showing the seed and the pericarp (used with permission from León, J. (2000) *Botánica de los Cultivos Tropicales*. Editorial Agroamérica, Instituto Interamericano de Cooperación para la Agricultura (IICA), San José, Costa Rica).

The petals turn from white or creamy-white or pale greenish with red stripes to pink or red and become recurved as the flower fully opens. The flowering period lasts 2–3 months and occurs normally during the dry season following rains, with the fruit maturing 45–75 days later.

Pollination and fruit set

Flowers normally open between 6 a.m. and 6 p.m. with a peak at noon. The stigma is receptive to pollen only on the day of anthesis, but pollen is released later, allowing both cross-fertilization and self-pollination. Studies to date have implicated both wind and a variety of insects as pollinating agents, but there is no information on their relative importance (de Almeida *et al.*, 2003). Since the flowers are scented and pollen is sticky, insects may play an important role in pollination. Beehives placed in or near the orchard do improve fruit set and yield. Ants, flies and other insects have been mentioned as pollinating insects. Since bagged flowers do not set fruit, self-fertilization might be minimal. Normally 10% of perfect flowers set fruit, with 4.4 fruit per inflorescence. Early set fruit is more prone to fruitlet drop.

Fruit

The true fruit or nut is a kidney-shaped achene (2.5–3.0 cm by 2.0–2.5 cm) with a double-walled shell (Fig. 1.1D). The shell has an outer thick exocarp, and an inner hard endocarp separated by a resinous, cellular or alveolar mesocarp filled with an oily substance called the Cashew Nut Shell Liquid (CNSL), and encloses the edible kernel that is covered by a reddish tegument (Fig. 1.1E). The nut is green at first, but becomes a gray-brown as it matures. The nut reaches its maximum size about 30–38 days after becoming visible and then starts to harden and to shrink (about 12–22%) due to loss of moisture until day 50–60, when it reaches maturity together with the "apple" or swollen pedicel (fruit stalk).

The swollen pedicel is the "cashew apple" and can be cylindrical, elongated, or pear-shaped. The "apple" starts to swell when the nut reaches its maximum size (35–38 days after the visible nut appears) and will increase about 80% in length, to 3.5–20 cm, and about 40% in width, to 3–12 cm, 15–25 days later (Donadio, 1983). At this time, the "apple" has a waxy bright yellow, red, orange or red-orange skin and a spongy, fibrous, very juicy, astringent, acid to subacid, yellow pulp. The "apple" weight (up to 650 g) is about 10–15 times that of the nut.

Cultivar development

Cytogenetics, genetics and breeding

Cashew has chromosomic polymorphism with variation in chromosome number; $2n = 42$ is considered the normal number but some types have $2n = 24$, $2n = 30$ and $2n = 40$ (Prabhakaran-Nair, 2009). Outcrossing leads to highly heterogeneous populations that limits breeding and often the only approach is to select superior types for local conditions. Current hybridization programs often use both local and introduced material to combine desired qualities of the parents. The hybrid seed is germinated and later grafted onto a mature seedling rootstock to evaluate for resistance to *Helopeltis anacardii*, a sap-sucking bug, and powdery mildew, vigor, kernel yield and quality. The selected plants are multiplied by budding or grafting for trials and evaluated over three consecutive seasons. The size and quality of the "apple" has a lower priority. In countries where the "apple" is consumed, such as Brazil and India, selection programs focus both on the "apple" and the nut. For the "apples" color, size, taste, sugar content, acidity, juice content and astringency are the typical criteria. The preferred color is red or reddish. Breeding programs exist in Brazil, India, Mozambique, Nigeria, the Philippines, Thailand and Australia.

Cultivars

Brazilian traditional cashews are of several types with the most widely known being 'Maça' (apple), 'Banana' and 'Manteiga' (butter). An aggressive breeding program was started in 1965 with both EMBRAPA and EPACE being involved and resulted in the release by EMBRAPA of the vegetatively propagated "Dwarf-Precocious" cashews in 1983, sometimes classified as *Anacardium occidentale* var. *nanum*. The first clones were 'CCP 06' and 'CCP 76' released in 1983, 'CCP 09' and 'CCP 1001' in 1987 and clone 'EPACE CL 49' in 1993. In 1996, 'EPACE MQ 10' was released for areas with irregular rainfall and frequent water deficits where other clones were not well adapted (de Almeida *et al.*, 2003). Other dwarf-precocious clones are 'CAP 11', 'CAP 15', 'CAP 22', 'EMBRAPA 50' and 'EMBRAPA 51'. All these asexually propagated clones have more compact canopies with more intensive branching, earlier flowering (6–18 months), and higher yields than the traditional type and are grown under improved management practices. These clones are normally grafted on their own seedlings.

India has more than 35 types recommended for use in different states where cashew is grown. The released hybrids include 'BPP 1', 'BPP 2' and 'BPP 8' in Andhra Pradesh; 'Goa 1' in Goa; and 'Vengurla 3', 'Vengurla 4', 'Vengurla 5', 'Vengurla 6' and 'Vengurla 7' in Maharashtra. In Kerala, the new hybrid cultivars released include: 'Priyanka', 'Amrutha', 'Kanaka', 'Akshaya', 'Dhanashree', 'Dhana' and 'Anagha'. Of these 15 hybrids 'BPP 1', 'BPP 2' and 'Vengurla 5' have small kernels (grades W400–W450; nuts per pound or 454 g) while the others have larger kernels (grades W180–W240) which is a great advantage (Prabhakaran-Nair, 2009). Varietal improvement is one component that has contributed to increased Indian production and productivity, combined with improved technology in the field. The older cultivars are 'Anakkayam 1' and 'Madakkathara 1'. In Tamil Nadu 'VRI 12' and 'VRI 13' are popular cultivars (Sivakumar and Pai, 2008). In Colombia, 'Gigante de Magdalena', 'Larga de Nazaré', 'Pequeña de Meta' and other local types such as 'Cayutín', which is resistant to humid conditions, are grown. In Panama, 'Nacional' and 'Mexicana' are widely grown. Malaysia grows 'Indicum' and 'Americanum' (Ohler, 1979). In most countries, locally selected types are grown.

Cultural practices

Propagation

SEXUAL Cashew propagated by seed normally produces low-yielding, highly variable progeny. Vegetative propagation is recommended to obtain true progeny. Seeds that are kept under dry and cool conditions remain viable for almost 2 years. Only seed with high specific gravity should be used; these seeds should sink in a 16% (w/v) sugar solution. Seeds are sown 3–5 cm deep with the convex portion pointing up and the two nut ends pointing down. Soaking the seed in water for 48 h with 15 min aeration after 24 h improves germination (Duarte *et al.*, 1991). Seed germinates in 7 to 10 days and it is epigeous.

Trees grown from seed start flowering after 3–5 years in normal type cashews while the dwarf-precocious type can start producing in 6–18 months. Full production is attained by the tenth year and the tree continues to bear economically until it is 30–35 years old, although good productions can be obtained from 50-year-old trees.

ASEXUAL To ensure a uniform plantation and better yields of uniform fruit, the ideal is to use vegetative propagation. Air layers of shoots just prior to the pre-flowering flush have been used with success, but they do not lend themselves for large-scale propagation and their establishment is poor (Sivakumar and Pai, 2008). Cuttings are sometimes difficult to root, although ringing the stem 40 days prior to removal from the parent plant has sometimes improved rooting of 1–2-year-old shoots whose stems are still light colored and somewhat flexible. Making a longitudinal splitting cut of 3 cm in the middle of the base of a sub-terminal semi-hardwood cutting with or without leaves attached and treating this wound with 8,000 ppm indolebutyric acid resulted in 56–44% rooting for with or without leaves, respectively, using a coarse sand medium in a poly-sealed chamber under 50% shade (Duarte *et al.*, 1991).

Patch budding can be done successfully (Alix and Duarte, 1999) and age of the rootstock seems to be important, with 8–10-month rootstocks being better than younger ones. Budding should be done about 1 month after flowering begins. Grafting is probably the most used and recommended method for asexual propagation of cashew. Different methods of grafting such as epicotyl grafting, softwood grafting, veneer grafting and side grafting have been tried with varying degrees of success. A successful micrografting technique has been developed using *in vitro* germinated seedlings as rootstocks and axenic shoot cultures from shoot-tip and nodal cultures as micro-scions. Approach grafting is more successful if done just prior to the pre-flowering flush of growth (Sivakumar and Pai, 2008).

Softwood grafting is the best for large-scale multiplication of cashew; the grafts can be prepared almost throughout the year with a mean graft success of about 60–70%. Higher success is achieved during the rainy season and plants can be transplanted 5–6 months after grafting. For this a 40–45-day-old seedling sown in polyethylene bag is used as rootstock; all leaves except the basal two pairs are removed. To obtain the scion a lateral shoot, pencil thick, of the mother plant with no young leaves at the tip is selected and 10–15 days before grafting the leaves cut off leaving the petioles. This scion is cut off and prepared for cleft grafting and introduced in the vertical cut made in the rootstock after cutting off the top at about 15 cm from the soil. The graft union is tied tightly and the scion wrapped with polyethylene or parafilm strips or a narrow polyethylene cap is put on top of it to avoid dehydration; these protections are removed after about 3 weeks to allow the buds to start growing. This procedure is preferably done under partial shade. Three to four weeks after grafting the bags are moved to full sun. India has been producing 10 million grafted plants per year.

Top-working is used to rejuvenate trees with poor production or low quality fruit. The trees should not be too old since the productive life of cashew is estimated to be 30–35 years, so that trees older than 20 years should not be top-worked. The tree is stumped 0.4–0.5 m high, normally done with a chainsaw in a slanting cut. After the stump sprouts are large enough, they are grafted using superior material. When 'CCP 76' is the scion, one graft per stump is necessary (Rosetti and Corrêa, 1996). The top-worked trees start fruiting from the second year. Stem end borer can be a problem in top-worked trees.

Biotechnology is being used for rapid multiplication of superior types and rootstocks. Embryo rescue to propagate hybrids, use of immature embryos, embryogenesis from somatic tissue, and micrografting are all possible.

Orchard establishment

For new plantations, the soil should be prepared as for any tree crop; this includes plowing and harrowing using machinery that is not too heavy as to cause soil compaction. Liming can be necessary to get a higher pH, to neutralize excess aluminum or to provide calcium or magnesium, and should be done a couple of months before soil preparation (de Almeida *et al.*, 2003). The quantity of lime should be sufficient to increase base saturation to 60% and the levels of exchangeable calcium (Ca) and magnesium (Mg) to a minimum of 3 and 4 mmol dm^{-3}, respectively (Crisóstomo *et al.*, 2007).

Transplanting and plant spacing

Planting is done in 60 × 60 × 60 cm holes or smaller, though the hole should be slightly larger than the root ball. In clayey soils, larger holes are better. In acid soils, calcareous dolomite should be placed in the bottom of the hole along with organic manure. Transplanting is normally done at the beginning of the rainy season or at any time if irrigation is available. In many countries seeds are either sown *in situ* or seedlings produced in poly-bags are transplanted at the onset of the rainy season.

In India, initial close plantings (3 × 2 or 4 × 3 m) are used for higher yields in the first 3–4 years. Shading of the soil during this period suppresses weeds, conserves soil moisture and provides higher initial yields. The trees will require training and pruning during these years and later thinning to the final spacing of 8–10 m after 5–6 years when canopies and root systems are intermingled with those of neighboring trees. According to Prabhakaran-Nair (2009) recommended spacing with normal type of cashew is 10 × 10 m or 10 × 8 m, thinned to 20 × 20 m or 20 × 16 m after about 10 years. Normally in India cashew is planted in squares or triangles with 8 m sides. Studies with densities from 156–2,500 plants per ha were established in Kerala and Karnataka states. A density of 625 plants per ha (4 × 4 m) during the first 11 years later thinned to 312 plants per ha spaced in triangles 8 m × 5.7 m × 5.7 m gave the highest accumulated yields.

In Brazil, normal tree types are planted at 7.5 × 7.5 m to be thinned later to 15 × 15 m, while for the dwarf clones 3.5 × 3.5, 4 × 3, 7 × 3.5 or 4 × 6 m are recommended initially in order to maximize early production, and later when the canopies overlap thin the trees to 7 × 7 or 8 × 6 m. Once established, the field needs little care. Intercropping may be done for the first few years, with cotton, groundnut, cassava or other annual crops. The young plants should be staked to avoid bending by the wind in the early years of growth.

Irrigation

Cashew trees are generally grown under rain-fed conditions in places with an annual rainfall of 600 mm or more, and 4–6 dry months. Irrigation under these conditions will significantly improve production. Studies in Brazil and elsewhere have shown increases of up to 300% in productivity with irrigation

(Crisóstomo *et al.*, 2007). Irrigation extends the harvest period and results in improved quality of the nut and "apple". The ideal method is to use the modern low pressure systems such as micro and drip irrigation that save water and labor, adapt to the topography, allow for less weed growth and can be used to fertilize. During summer, it is advisable to provide mature plants with 200 l per plant at fortnightly intervals. Other authors indicate that adult trees should receive 70–80 l day^{-1}, irrigating every 4–5 days in sandy soils and every 5–7 days in clayey soils (da Silva, 1998).

Pruning

After transplanting any shoots coming from below the graft line or close to the ground should be removed. Once the plant reaches 1 m, it should be topped to induce lateral branching. Three or four lateral branches that are growing in different directions and originate at different heights on the main stem are selected, to form the main structure of the tree. Initial training and pruning of young plants during the first 3–4 years is necessary to obtain trees with good shape. After that practically no pruning is necessary, except to remove sick, dried, weak or damaged branches or those growing in the wrong direction. Any branch overcrowding the canopy should be removed to allow for better light penetration and ventilation. Pruning should be done after harvest. All flowers during the first or the first and second year should be removed to allow for vigorous early vegetative growth of the young tree.

Fertilization

The application of fertilizers significantly improves yields. In Brazil, the dwarf-precocious clones used for improved yields need to be properly irrigated and fertilized to achieve their maximum yields. This includes applying 50 g urea 30–40 days after planting and a similar amount 30 days later. After this increasing amounts are used in the second, third and fourth year: 100, 200 and 400 g of urea; 200, 250 and 300 g of triple superphosphate; and 100, 150 and 300 g of potassium chloride, respectively. All the phosphorus and half of the urea and potassium chloride are applied at the start of the rainy season; the other half can be applied 45–60 days later in a single application or split in two parts, applied at 30–40-day intervals. A fertilization protocol is outlined in Table 1.1. In the case of common cashew the dosages used are: 70, 100, 150, 200 and 250 g of urea; 100, 150, 200, 250 and 300 g triple superphosphate; and 100, 150, 200, 250 and 300 g potassium chloride in the first, second, third, fourth and fifth year, respectively with the same distribution schedule described for the dwarf-precocious type (de Almeida *et al.*, 2003).

According to Prabhakaran-Nair (2009) fertilizer application in some parts of India is 500 g N and 125 g each of P_2O_5 and K_2O per tree annually. In the case of high-yielding varieties, response to N was still significant up to a rate of 750 g per tree. In other cases 1,000 g N, 500 g P and 500 g K are used after the sixth year with gradual increases from the first year until reaching these amounts. Fertilizers are applied into a shallow trench at the drip line when the monsoon ceases. It is recommended to apply the fertilizer in split doses during the premonsoon phase (May to June) and post-monsoon phase (September to October).

Table 1.1. Fertilization recommendations for pre-mature dwarf cashew in planting, growth and production phases in both irrigated and dryland conditions in Brazil. (From Crisóstomo *et al.*, 2007.)

		Soil P-resin (mg dm^{-3})			Soil K (mmol dm^{-3})		
	Soil N	0–12	12–30	>30	0–1.5	1.6–3.0	>3.0
Fertilization rate	(g per plant)	P_2O_5 (g per plant)			K_2O (g per plant)		
	0	200 (180)[a]	150 (140)	100 (90)	0	0	0
Age (years)							
0–1	60 (45)	0	0	0	60 (50)	40 (30)	20 (20)
1–2	80 (70)	200 (160)	150 (140)	100 (90)	100 (90)	60 (50)	40 (30)
2–3	150 (120)	250 (220)	200 (180)	120 (110)	140 (120)	100 (90)	60 (50)
3–4	200 (150)	300 (290)	250 (230)	150 (140)	180 (170)	140 (130)	80 (70)
4–5	300 (220)	300 (290)	250 (230)	150 (140)	180 (170)	140 (130)	80 (70)
Expected yield (kg ha^{-1})							
<1200	400 (300)	200 (160)	100 (80)	100 (80)	150 (120)	100 (80)	80 (80)
1200–3000	700 (520)	300 (240)	200 (160)	150 (120)	300 (240)	200 (160)	150 (120)
>3000	1000	400	300	200	450	300	200

[a]Values in parentheses refer to cultivar in dry land.
Apply 50 g of F.T.E. Br-12 per plant per year for years 2–4 and 100 g starting from year 5.
(F.T.E. Br-12 = 9% Zn, 1.8% B, 0.8% Cu, 3.0% Fe, 2.0% Mn, 0.1% Mo)

However, if a single application is used, the post-monsoon period is preferred as sufficient soil moisture is still available.

On sandy, laterite soils and on sloping land with heavy rainfall, fertilizer is applied in a circular trench 25 cm wide and 15 cm deep, at a distance of 1.5 m from the trunk. On red loamy soils with low rainfall, the fertilizer should be incorporated into the soil in a band 1.5 m wide, at a distance of 1.5 m (inner edge) to 3 m (outer edge) around each trunk (Sivakumar and Pai, 2008).

Cover cropping

Legumes are normally used to protect the surface from erosion, minimize water run-off and reduce weed growth. Legumes have the additional advantage of fixing and adding nitrogen. Crops such as *Pueraria javanica*, *Calopogonium mucunoides* and *Centrosema pubescens* are frequently used. Sowing should be done at the start of the rains in the space between trees. Before harvest, the cashew basins must be cleared of the cover crops to ensure gathering all the fallen fruit. In China, natural grass and leguminous crops are usually maintained at the time of land clearance to conserve soil, and during initial years after planting, green manure crops are also grown (Prabhakaran-Nair, 2009).

Intercropping

The purpose is to increase income in the early years in small farmers' plantations. The intercrop should be established early in the plantation to benefit from the space and sunshine in a young orchard. Field investigations have shown that the most suitable intercrop is pineapple because it helps to reduce erosion and produces an extra income. In Indonesia, groundnuts and sweet potatoes are frequently used and recently watermelon and sweet melon have been tried, as well as maize, cassava, sweet and hot peppers. In other places papaya, pomegranate, coconut and banana are popular (Prabhakaran-Nair, 2009). The key point for intercropping success is the availability of supplemental irrigation.

Pest management

DISEASES Anthracnose (*Colletotrichum gloeosporioides*) is very common in cashew plantations during the rainy season in most parts of the world. The main symptom is the appearance of white patches on branches followed by drying of twigs from the tip. It deforms leaves and can attack flowers, peduncles and young nuts. Fungicides applied preventively can be effective (Villachica *et al.*, 1996; de Almeida *et al.*, 2003). Powdery mildew (*Oidium anacardii*) dries and deforms the leaves and kills flowers. It can have a devastating effect on cashew tree yields. Powdery mildew is a significant problem in East Africa. The mildew occurs during cool, humid conditions and on succulent plant growth. Thinning the canopy by pruning, to get more air circulation and light, as well as the use of sulfur sprays is helpful (Villachica *et al.*, 1996). Other diseases of cashew include dieback and damping off in young seedlings.

INSECTS In South-east Asia, India and East Africa *Helopeltis anacardii*, a sap-sucking insect, can cause flower damage and is a major pest. A severe attack can result in up to 80% of damaged branches. It appears normally with the emergence of new flushes and panicles causing shoot dieback and inflorescence drying. Control sprays are done at the start of vegetative flushes in October–November, when panicles start to emerge in December–January and when flowering ends and fruit starts growing in January–February. The green ant (*Oecophylla smaragdina*), the meat ant (*Iridomyrmex sanguineus*), mantises (*Orthoderinae* spp. and *Mantidae* spp.), predatory bugs (*Geocoris australis*) and spiders (*Oxyopes* spp.) significantly reduce the numbers of *Helopeltis* spp. Green ants are the most abundant predatory species in cashew plantations and also significantly reduce other cashew insect pests such as the fruit spotting bug (*Amblypelta lutescens*), the mango tip borer (*Penicillaria jocosatrix*) and the leaf roller (*Anigraea ochrobasis*).

The cashew stem and root borer (*Plocaederus ferrugineus*) is another serious pest, capable of destroying the tree. Main symptoms are yellowing of leaves, drying of twigs, and the presence of holes at the base of the stem with exuding sap. To reduce the spread of infestation, it is essential to remove the dead and highly infested trees at least every 6 months.

Other insect pests include borers, thrips, mealy bugs, weevils, caterpillars and leaf miners (Sivakumar and Pai, 2008). In Brazil and Peru, the larvae of the moth *Anthistarcha binocularis* penetrate the branch terminals destroying leaves and flowers. Another insect, *Contarinia*, attacks terminal buds, deforming them and reducing the number of panicles in the tree. *Anacampsis* sp. is a moth that lays its eggs between the peduncle and the nut and the larvae penetrate the nut and feed on the kernels. *Selenothrips rubrocinctus* can attack fruits, peduncles, leaves and inflorescences reducing yields. *Marshallius* is a trunk borer and finally, *Aphis gossypii* can damage the inflorescences (de Almeida *et al.*, 2003). In Peru, the white coconut fly *Aleurodicus cocois* invades the lower part of the leaves and restricts their growth (Villachica *et al.*, 1996).

Weed control

Weeds in the plantation are normally controlled with hand tools, brush cutters or chemical weed killers. If machinery is to be used, it should not disturb the soil below 15 cm so that no roots are cut or wounded. Normally the alleys are mowed and the residues left to reduce soil temperature and erosion. The areas below the canopy are cleaned using herbicides or with hand tools. If an associated crop is to be planted it should be at least 1.0–1.20 m from the plant row and this should be done only during the first 3–4 years to avoid damaging the cashew roots when preparing the land.

Orchard protection

In areas with heavy winds, wind barriers, either natural or artificial, are needed.

Harvesting and postharvest handling

Yield

The highest yields per hectare have been reported from Vietnam (2,320 kg), followed by Tanzania (1,250 kg). In Brazil, the yields of the dwarf-precocious varieties, such as clone 'CCP 09' planted at 7 × 7 m, under adequate irrigation and fertilization, are around 4,500 kg ha^{-1} in their sixth year. The same varieties under non-irrigated conditions will yield a maximum of 1,500–1,800 kg ha^{-1}. For the common type, it is estimated that 62% of the plants produce less than 4 kg of nuts and are responsible for 30% of the total production while the other 38% produce 70%, resulting in average yields of 200 kg ha^{-1} in adult plantations (de Almeida *et al.*, 2003). The ratio of "apples" to nuts is 10:1 (w/w); for the fresh market the nut will have to go attached to the "apple", but not for processing. One thousand kg harvested nuts will convert into 200–220 kg of clean kernels. Cashew nut shell liquid (CNSL or cashew oil) amounts to 20–25% of nut weight but recovery is normally 7–12%.

Harvest

Harvest in north and north-eastern Brazil is between July and December for most varieties (Donadio, 1983), with a peak from the end of October to the end of November. However, the dwarf-precocious 'CCP 76' produces 40% of its crop from January to July and 60% from August to December while normal cashew harvest starts in August (da Silva, 1998). The nut color changes from brownish-green at fruit set to light green when the apple is one-quarter grown and later turns gray in all cultivars. The nuts can be harvested when this color is reached. Mature fruit will normally fall when the "apple" dries. Harvesting of raw nuts is done either by collection after natural drop or after shaking the tree.

The "apple" of the normal cashew type is difficult to harvest due to the height of the trees. If the "apple" abscises and falls to the ground, it is no longer marketable as a fresh fruit. Harvesting with tools damages the peduncles and neighboring unripe fruit and flowers. In Brazil, the common cashew type is not intensively cultivated for the "apple"; instead the dwarf-precocious cultivars are used as they can be harvested manually with minimal damage. The peduncle detaches from the plant with a slight twist. Only the fingers, not the entire hand, should be used and harvesters are required to cut their nails to avoid damaging the "apple". For processing, they are either harvested manually or using a pole with a bag attached.

The "apple" is non-climacteric and has to be harvested when completely mature, since it will not continue to ripen after harvest (Figueiredo *et al.*, 1999). Storage allows for water loss and the concentration of the fruit sugars; at the same time some starch converts into sugars and some sucrose can convert into glucose making the fruit 70% sweeter. Normally "apples" are marketed with the nut attached.

Postharvest treatment

The nuts are dried in the sun on mats, plastic sheets or canvas put on cement or a hardened soil surface. The nuts should be put in layers not thicker than 5 cm to provide for good air circulation and turned twice a day for uniform drying. Normally 2–3 days in full sun is necessary, or until the kernel rattles in the shell. By that time the nut should have lost 10–12% of its moisture and reached 8–10% moisture content. Too moist nuts are prone to fungal attacks while too dry nuts can crack and the CNSL spill into the kernel, as happens when overheated during drying. Sound nuts are classified according to size with the industry preferring medium and large sizes that can be processed by machines with specially shaped blades to recover the intact kernels. Nuts for processing can be stored for about 6 months without a significant reduction in quality. The kernel is extracted from the nut by mechanical shelling or by hand. The nuts are dried to 11% moisture content before processing and kernel removal.

The "apples", once harvested, are put in ventilated plastic padded harvest boxes to avoid damage and kept in the shade before transport to the packing area. At the packing area, the fruit are graded with damaged "apples" being sent for nut and juice processing and the good undamaged "apples" washed to help remove part of the field heat, then placed in a bath of citric acid or chlorine solution to reduce *Rhizopus*, *Penicillium* and *Colletotrichum* inoculum levels and then surface dried. The "apples" are then graded as to size and color and packed in single layer of 4–8 fruit in trays wrapped with PVC film with 550–800 g per tray (da Silva, 1998). At ambient temperature the "apple" lasts 24–48 h while at 5°C and 85–90% RH 10–15 days and up to 25 days at 2°C (Berry and Sargent, 2011). For processing, the "apples" can be frozen and kept in good condition for 3–4 months.

Utilization

The true fruit is the cashew nut. The kernels once extracted from the nut are classified according to size, with the largest size (160–180 per pound; W160–W180 nuts per 454 g) being the most expensive and the small size (300–320 per pound) being the most sold. There are 33 different grades, of which 26 are commercially available for domestic consumption and export. Broadly the kernels can be classified as wholes (white, scorched and dessert) and pieces (white, scorched and dessert). The kernels are mainly used as snacks, roasted or salted. Bakery, confectionery and chocolate industries use the broken kernels. The nuts can be roasted and coated with salt, sugar or honey. Inferior kernels can be used to make flour, paste, butter and other products that are very nutritious. Kernel color should be white or ivory for best quality, and they should be intact halves; pieces command a lower price. Kernels with a moisture content of 4–4.5% are vacuum-packed. Cashew kernels have the highest protein content among tree nuts and have all essential amino acids. They are low in crude fiber and high in lipids and in vitamins (Table 1.2). The ratio of saturated to non-saturated fatty acids is 4:1, close to the ideal 5:1. The nuts are also very high in magnesium.

The "apple" is very astringent due to its high tannin content that is reduced by steaming under pressure or boiling in salty water for 15 min. Cashew "apple" contains calcium, phosphorus, vitamin C, iron and fiber (Table 1.2). Currently, less than 20% of the total "apples" produced are used for human consumption. In India, the "apple" is not as widely used, while in Brazil, the "apple" is consumed fresh and the juice is widely available. The "apple" is also processed into jams,

Table 1.2. Composition of cashew nut (da Silva, 1998), cashew apple (Filgueiras *et al.*, 1999), yellow mombin (Villachica *et al.*, 1996) and red mombin fruit (Koziol and Macía, 1998) per 100 g edible portion.

	Cashew			
Proximate	Nut	Apple	Yellow mombin	Red mombin
Energy (kcal)			21.8–70	
Moisture (g)	9	84.4–88.7	72.8–88.5	65.9–86.6
Protein	20	0.10–0.16	0.6–1.4	0.09–0.26
Lipid (fat) (g)	43.5	0.05–0.50	0.1–2.1	0.03–0.17
Fiber (g)		0.4–1.0	0.6–1.2	0.2–0.6
Ash (g)	0.25	0.19–0.34	0.4–0.6	0.47–1.13
Carbohydrates (g)	25	9.08–9.75		
Reducing sugars (g)			6.7–9.4	19.1
Minerals				
Calcium (mg)	13.0–36.8	0.9–5.4	26.0–31.4	6.1–23.9
Iron (mg)	0.62–1.01	0.19–0.71	2.2–70.5	0.09–1.22
Phosphorus (mg)	16.6–15.7	6.1–21.4	27–40	31.50–55.7
Vitamins				
Thiamine (mg)	0.009–0.014	0.02–0.03	6.74–9.4	0.03–0.10
Riboflavin (mg)	0.015–0.039	0.13–0.40	0.05–0.19	0.01–0.04
Niacin (mg)	0.266–0.327	0.13–0.53	0.5	0.54–1.77
Pyridoxine (mg)			0.67	
Carotene (mg)	0.002–0.060	0.03–0.74	70–71	0.004–0.089
Ascorbic acid (mg)	90–192	147–372	11–166	26–73

juices, concentrated juices, "*cajuína*", preserved pulp, frozen pulp and candies. Cottage industries produce "apples" in syrup, in compote, jelly, "apple" syrup, jam, fruit paste and chutneys. In some places "apples" are dehydrated to a prune-like product that is very tasty. *Cajuína* is a clarified juice from pressed entire fruit that is pasteurized at 100°C for 45 minutes after bottling. In Brazil, the best varieties for juice are 'CCP 76' and 'CCP 1001' because of their high soluble solids content. The tannins in the juice bind to gelatin and form a precipitate, which can be removed by filtering. The juice is used in various countries to make wine and distilled liquors (de Almeida *et al.*, 2003).

Cashew nut shell liquid (CNSL) is an important product from the kernel shell (Fig. 1.1E) and nowadays is extracted by first steaming the shells then submerging them in hot CNSL (around 180°C) until the liquid flows out, followed by centrifugation. Alternatively, it is extracted with a low boiling point petroleum solvent. CNSL causes eye and skin irritation and is toxic and should not be inhaled or swallowed. It is 70–80% anacardic acid (C15 single unsaturated alkyl chain with a phenol attached) and 10–15% cardol. Depending on the shell weight 33–38% is extracted. It has many applications, in paints, resins, polymers, surfactants, epoxy resin curing agent and as a binder. It can also be used to prevent termite damage to wood.

The tree is used for reforestation, to prevent desertification, and as a roadside buffer tree. It was first planted in India to prevent erosion on the coast. The wood is used for carpentry, firewood and charcoal. The tree exudes a gum called cashawa, used in varnishes or in place of gum arabic. Cashew bark has about 9% tannin, which is used in tanning leather. The bark and leaves and even the shell are used medicinally. Cashew "apple" juice, without the tannin removed, is prescribed for sore throat and chronic dysentery in Cuba and Brazil. Fresh or distilled, it is a potent diuretic. The brandy is applied as a liniment to relieve the pain of rheumatism and neuralgia. The Tikuna tribe in north-west Amazonia uses the "apple" juice for influenza and for warts. The decoction of bark and leaves and the fruit juice are used against diarrhea and abdominal pains, inflammation and diabetes. Sodium anacardate destroys certain snake venoms as well as tetanus and diphtheria toxins, and the vegetative form of anaerobic bacteria. The shell oil is used to heal foot wounds and modern medicine uses cardol for its vesicant properties, as a dye for skin pigmentation, as a respiratory and circulatory analeptic and as an antagonist of barbiturates (Sivakumar and Pai, 2008).

Yellow Mombin

The yellow mombin *Spondias mombin* L. (syn. *Spondias lutea* L.) (Anacardiaceae) has several regional names: it is hog plum or yellow mombin in North America and the Caribbean islands; Spanish plum or gully plum in Jamaica; thorny hog plum in Malaysia; and Ashanti plum in Ghana. The names used for this fruit in Latin America include: *ciruela agria, ciruela amarilla, jobo, jobo jocote, ciruela de jobo, ciruela del país, hobo, hubu, jobillo, jobo de Castilla, ciruela loca, jobo de perro, jobo de puerco, jobo espinoso, jocote amarillo, jocote de chancho, jocote montañero, ciruela de monte, obo, uvu* and *ubos*. In French Guiana, it is called *prunier mombin*, in Brazil *acaiba*, *acaimiri*, *acaja*, *acajaiba*, *cajá*, *cajá mirim*, *cajazeiro* and *cajá miúdo*, and the Amazonian Indians call it *tepereba* or *tapiriba*.

Important genera and species

The genus *Spondias* was created by Linnaeus in 1753, with *S. mombin* L. as the only species. Later, this genus was expanded to include *Spondias purpurea* L., *Spondias cytherea* Sonn. and *Spondias pinnata* (L.f.) Kurz. The genus now has 17 species, of which ten occur in Asia and Oceania, and seven in the American neotropics. A new species, called *Spondias testudinis* Mitch. and Daly, was recently found in the south-west Amazon (Filgueiras and de Souza, 2008).

Origin and distribution

Yellow mombin, of unknown origin, is distributed in moist lowland tropical forests from Mexico through Central America to Peru and Brazil. In Brazil, it grows abundantly especially in the North and Northeast regions. It is also widely cultivated and naturalized in tropical Africa. In Asia, it is grown in India and Indonesia, and to a lesser extent in Malaya. In Peru, the wild populations are found in forest areas subject to periodic flooding and the fruit are food for fish when flooding occurs and wild animals during the dry season (Villachica *et al.*, 1996). Some commercial production occurs to a limited extent in Brazil but most fruit is collected from wild or from backyard trees.

Ecology

Soil

The species is not demanding of soil though it prefers alluvial soils (Mattietto and Matta, 2011). It can grow at a pH ranging from 4.5 to more than 7.0 but prefers a slightly alkaline to neutral pH. The tree tolerates flooding for several weeks.

Rainfall

The tree grows in places with an annual rainfall between 1,000 and 2,000 mm, but it can stand semiarid conditions due to the tuberous root system that stores water (Donadio *et al.*, 2002). It grows in places with a dry season of 5–6 months though leaf fall occurs.

Temperature

It is adapted to grow in humid tropical areas and even in warm subtropical areas with no frost. It does not grow satisfactorily above 1,000 m.

Light and photoperiod

This plant grows and produces better under full sun with no photoperiodic response.

Wind

The tree is regarded as fairly wind resistant.

General characteristics

Tree

Yellow mombin trees can reach 25–30 m with a canopy diameter of 8–24 m. The somewhat buttressed trunk can be 60–80 cm thick with a rough bark that in young trees bears up to 2 cm blunt-pointed spines or knobs. The lower branches tend to be whorled. The semi-deciduous or perennial leaves are compound, alternate, pinnate, 20–45 cm long, with a hairy and sometimes pinkish petiole and five to 11 sub-opposite, ovate or lanceolate pairs of pointed leaflets, 5–15 cm long, equilateral and oblique at the base (Fig. 1.2).

Flowers

The whitish male, female or bisexual flowers can occur in the same pyramidal panicle (Fig. 1.2). The terminal panicles are 15–40 cm long with almost 2,000 flowers. The panicles emerge after new vegetative growth has started. The flower has a 1–4 mm pedicel; a calyx about 5 mm wide; a round receptacle 1–4 mm long; five sepals; five 0.3 mm light yellow petals; and ten stamens, five of which are inserted in a disk and alternate with the petals and five of which are epipetalous. Two types of female flowers are reported, one with

Fig. 1.2. Yellow mombin (*Spondias mombin* L.) showing panicle leaves and fruit (from Little, E.L. and Wadsworth, F.H. (1964) *Common Trees of Puerto Rico and the Virgin Islands*. US Department of Agriculture, Agriculture Handbook #249).

visible and the other microscopic staminoids. The number of fruit that develop per panicle is highly variable both within and among plants, though is normally around 30. Flowering normally occurs after the beginning of the rainy season following a dry period. In some areas flowering can occur all year round (Mattietto and Matta, 2011).

Pollination and fruit set

The hermaphroditic and strongly protandrous flowers have a poorly developed ovary when the pollen is released, leading to cross-pollination. Wind pollination occurs as the petals curve at anthesis leaving the anthers exposed. The flowers produce copious quantities of pollen and the stigma is large with bees being frequent visitors. Flowers in north-east Brazil open between 1:30 and 6:30 p.m.

Fruit

The fruit is an ovoid to oblong drupe with a slightly flattened base (Fig. 1.2). The thin layer of yellow or orange translucent mesocarp is fleshy, juicy, sweet-sour and very aromatic and surrounds a bulky endocarp that is white and "corky". Each fruit is 3–6 cm long and up to 2.5 cm wide, and golden yellow to orange with a thin, tough skin. Fruit weight can vary from 4.8 to 37.4 g. Fruit growth is sigmoidal and the fruit takes 3–4 months to develop from anthesis; it abscises when fully ripe. The inner stone is the lignified endocarp that normally contains one to five 1.2–0.22-cm seeds; normally only one is found.

Cultivar development

The hog plum has $2n = 32$ chromosomes. There is large genetic diversity in size of the fruit, pulp color, aroma and flavor (Mattietto and Matta, 2011). Besides a few accessions at the INPA of Manaus, Brazil, no other collections are known. The major limitations for commercial cultivation are tree height, the lack of good cultivars, highly seasonal production and the perishable nature of the fruit. No clones or varieties of superior quality are recommended for cultivation though Brazil is evaluating the behavior of a number of clones. In Nicaragua, some named selections exist: 'Lapa', 'Moyo', 'Santaroseño', 'Jismoyo' and 'De cocer' (Barbeau, 1990).

Cultural practices

Propagation

SEXUAL Seed germination is low and slow. Cutting of the seed tip, chemical treatment with sulfuric acid and soaking in water increases the germination percentage. Germination will start after 60 days. Seedlings from this outcrossing species are highly variable in plant vigor, shape, physical-chemical characteristics of the fruit, yields and juvenile period. Seedlings will start to produce after 5 years.

ASEXUAL Vegetative propagation can be done using hardwood cuttings from stem or roots, air layering and grafting. Herbaceous cuttings have not been successfully rooted. Cleft and side grafting can be done and rootstocks of hog plum, *Spondias tuberosa*, *S. cytherea* or *S. mombin* can be used with up to 80% success. Propagation *in vitro* has also been tried with success.

Orchard establishment

Normal preparation procedures should be followed as for any other tree fruit crop. Recommended spacing is 10 × 10 m and tree size controlled by pruning (Villachica *et al.*, 1996).

Irrigation

No research information is available, though a short break in irrigation can induce vegetative growth and uniform flowering.

Pruning

Even asexually propagated plants grow very vigorously tending to form a large canopy. Trees should be pinched when 0.9–1.0 m tall to ensure lateral branching to reduce the speed of growth and plant size. Trees are then pruned periodically to remove undesired branches and in some cases topped to control tree height.

Fertilization

Fertilizer requirements have not yet been determined experimentally and adaptations of recommendations for other perennial fruit trees in the area are used.

Pest management

Commercial cultivation is still not significant and only the major pests and diseases have been recorded. The economic importance of these pests and diseases is unknown. In some cases, pests or pathogens only attack parts of the plant, causing superficial damage.

The major pathogens include *Glomerella cingulata*, which causes anthracnose on leaves, inflorescences and fruit. *Sphaceloma spondiadis* produces round rough-textured lesions on leaflets and fruit characterized by cream centers and light brown borders. *Botryosphaeria rhodina* causes resinosis, with the development of dark cankers that are sometimes cracked, and with abundant gum exudation that sometime kills the plant. Cercosporiosis caused by *Mycosphaerella mombin* affects the leaves and begins with small, round pit spots that become darker and coalesce, causing yellowing and leaflet fall (Filgueiras and de Souza, 2008).

Fruit flies (*Anastrepha* spp.) attack the fruit and do cause economic loss in low density planting. The leaves are attacked by leaf cutting ants. The terminal branches are attacked by larvae, and the seeds and endocarps are damaged by some weevils. Nematodes (*Meloidogyne* spp.) attack both adult plants and plantlets. The wood is easily attacked by termites (Filgueiras and de Souza, 2008).

Weed control and orchard protection

The normal procedures should be followed using hand tools, mechanical methods or herbicides. Sheltered locations are desirable.

Harvesting and postharvest handling

The plant produces normally once a year but sometimes more often, depending on climatic conditions. Adult trees in Mexico produce more than 100 kg year^{-1}. Fruit ripening starts around 100 days after anthesis and full ripening is reached at around 117 days. This climacteric fruit should be harvested after the start of chlorophyll breakdown, when it turns from light green to yellow. Green immature fruit soften and change color, but show little changes in acidity, soluble solids and starch contents, and thus are of poor eating quality (da Costa *et al.*, 1998). Fruit ripen over 4 days from the light green maturity stage (Filgueiras and de Souza, 2008).

Harvested fruit are normally transported in 20 kg containers to retail stores where they are frequently displayed in trays wrapped with 12 μm polyvinylchloride (PVC) film. Storage life at 23–25°C is about 4 days for fruit at the light green stage and less than 2 days for ripe fruit. Under refrigeration at 9–10°C the storage life can be increased to 10 days. At low temperature, the fruit develops chilling injury.

Utilization

In Brazil, the ripe fruit is eaten fresh, the edible portion being the exocarp and mesocarp, which together may represent as much as 81% of the fresh weight. Extracted pulp is frozen and sold to restaurants, hotels and snack bars and is exported for use in juices (Mattietto and Matta, 2011). It is also used in ice cream and jam. Green fruit can be eaten fresh with salt and vinegar or pickled or kept in syrup. The soluble solids can reach 11.5% in yellow fruit, though titratable acidity is also very high (around 1%) (Filgueiras *et al.*, 1999). The fruit has a good supply of ascorbic acid and minerals (Table 1.2). The major volatile components are ethyl acetate, ethyl butyrate, ethyl hexanoate, hexyl butyrate and linalool. The young leaves can be eaten as a cooked vegetable.

The yellow to yellowish-brown light and flexible wood of *S. mombin* is used to some extent for woodcraft. The bark is used to make handicrafts and its tannins are used for dyeing and tanning, and the exudate is a gum that can be used as glue. The bark and leaves contain ellagitannins with therapeutical properties (Abo *et al.*, 1999). A decoction of the bark serves as an emetic substance, a remedy for dysentery, diarrhea, gonorrhea, hemorrhoids and leucorrhea. A decoction of leaves is traditionally used in the Venezuelan Amazonia to treat malaria. In many places, it is planted as a living fence post.

Red Mombin

The red mombin, *Spondias purpurea* L. (Anacardiaceae), is a widely distributed fruit with many names. In English, it is known as red mombin, Spanish plum, scarlet plum and purple plum. In Trinidad, it is Jamaica plum, Barbados and Trinidad chili plum, and Netherlands Antilles *noba* and *makka pruim*. The most widespread name in South America and the Philippines is *ciruelo*, which means plum in Spanish (León, 1983), though in Central America and Mexico it is also widely known as *jocote* from the Nahuatl (Xocotl) or sour fruit. In some parts of South America, it is known as *ovo*; in Puerto Rico as *jobillo, jobo francés*; in El Salvador as *pitarillo*; in Colombia as *hobo colorado* and *ciruela del país*; and in Peru as *ciruela pepona, ciruela norteña, ciruela roja* and *ciruela*.

Distribution

Spondias purpurea is a well-defined and taxonomically separate species from other *Spondias* spp. (Barfod, 1987). It is native to Central America and Southern Mexico, and natural populations are found in both dry and wet areas, including a wide range of semi-deciduous forests. This aromatic fruit is highly valued for local people in the tropics. Early Spanish navigators took the red mombin to the Philippines and from there to Africa. During colonial times, it was introduced into South America, especially Ecuador and Peru (León, 1983). In Jalisco (Mexico) during the 19th century, red mombin was one of the most important fruit crops. The fruit are readily found in local markets.

The red mombin is the most cultivated species in the genus *Spondias*. There is, however, no data for world production. In Ecuador, yields of more than 4,500 t year^{-1} are reported (Macía and Barfod, 2000). During the last decades commercial plantations have been made in Mexico, Venezuela and Central America. In most of the dry or semi-dry tropical areas the tree is used as a backyard or house garden tree. In some cases, it is planted as a fence to be used as a living post.

Ecology

Soil

Red mombin prefers deep, fertile soils but it can grow in sandy, loamy, clayey and gravelly soils. A root mycorrhizal association occurs and this promotes phosphorus absorption (Macía, 2008).

Rainfall

The plant grows well in places with a prolonged dry season where defoliation is an adaptation mechanism. It can be seen in areas with an average annual precipitation varying from 300 to 1,800 mm. In dry areas, it usually sprouts and flowers at the onset of the rainy season.

Temperature

The plant is adapted to dry and semi-humid tropics up to 1,600 m, as well as subtropics. In Guatemala, it is grown from sea level to 2,100 m and it is also grown in the subtropics of the Peruvian central coast northwards. It flowers well in wet or dry climates but fruit from dry areas are of better quality. In very cool climates, it might grow but does not flower.

Light and photoperiod

The plant grows best under full sun and in warm arid areas it produces the best quality fruit. There seems to be no photoperiod response.

Wind

The spreading and low-branching habit makes the tree less prone to be damaged by wind, but the brittle branches can be a negative factor when strong winds occur.

General characteristics

Tree

In the lowland tropics, the species is a small low-branched deciduous tree or shrub (3–6 m) and in the highlands a spreading, thick-trunk tree that can reach 7–15 m. The open canopy can spread up to 15 m (León, 1983) with a trunk that can reach a diameter of 50 cm with a grey and usually smooth bark. The main branches tend to grow horizontally and all branches are fairly brittle. Cuts and bruises in the branches produce thick and transparent exudate. The compound leaves have 9–19 nearly sessile obovate to lanceolate or oblong-elliptic, alternate, 2–6 cm long by 1.25 cm wide leaflets (Fig. 1.3) that are bright red or purple becoming green when mature.

Flowers

The axillary inflorescences come in 1–10 cm long panicles with a few flowers that usually appear at older and defoliated nodes. Each panicle has male, female and bisexual flowers. The flowers have 4–5 sepals and 4–5 tiny red to purple petals that are usually 2.5–3.5 mm long at anthesis. Pollen is normally not formed because the mother cells of the micro-sporangia do not develop (Juliano, 1932) and the fruit develop parthenocarpically with no seeds present in the stone.

Flowering time varies with climate, but usually occurs during the dry season when the trees are defoliated or just as the young new leaves emerge. In areas with year-round precipitation, flowering may occur nearly every month. In dry areas, it is possible to control flowering by planning irrigation dates. Defoliating the trees with 12% urea can advance flowering by 30–40 days (Almaguer-Vargas *et al.*, 1991; Macía and Barfod, 2000).

Fruit

The parthenocarpic fruit is an oblong to obovoid, sub-globose or even pear-shaped drupe, measuring 2.5–5.0 by 1.0–3.5 cm, with a smooth and glossy peel, and looks plum-like (Fig. 1.3). The fruit appear solitary or in groups of two or three. The ripe fruit is normally dark or bright red but can be purple, orange, red-and-yellow and sometimes even yellow and can be confused with the yellow mombin. The mesocarp is fleshy and juicy, 5–7 mm thick, acid in flavor, very aromatic, yellow, and fibrous, and is attached to a fibrous and hard endocarp that can be 1.25–2.5 cm long and normally has no seeds but the vestiges of unfertilized ovules (Avilán *et al.*, 1989). The fruit takes about 115 days from anthesis to the start of ripening. The new vegetative shoots and the fruit mature at the same time during the dry season.

Fig. 1.3. Red mombin (*Spondias purpurea*) showing leaves and fruit (used with permission from León, J. (2000) *Botánica de los Cultivos Tropicales*. Editorial Agroamérica, Instituto Interamericano de Cooperación para la Agricultura (IICA), San José, Costa Rica).

Cultivar development

Red mombin has a chromosome number of $2n = 32$ (Barbeau, 1990). Little cultivar development has taken place. High species diversity is found in the Yucatan with some primitive varieties that sometimes are confused with yellow mombin. Some authors divide the non-commercial varieties into two main groups: dry-season and wet-season red mombins. Fruit of the dry-season group are smaller (2.7–3.9 by 1.9–3.0 cm), less acid and sweeter than those of the wet-season group, which are larger (3.1–4.5 by 2.4–3.5 cm) (Macía, 1997). Many cultivars bear the name of the place where they are grown or where they originated or a name that describes the fruit characteristics. Dry-season red mombin varieties include 'Tronador', 'Criollo', 'Nica', 'Morado' and 'Rojo ácido', while the wet-season mombins are 'Corona', 'Petapa' and 'Cabeza de loro' (León and Shaw, 1990).

Cultivars

In Nicaragua 'Tronador', 'Chichita', 'Guaturco' and 'Dulce' are described. In Honduras and Nicaragua the 'Tronador' type is preferred because it is crunchy. In Costa Rica, similar named 'Corona' and 'Tronador' occur as in Nicaragua and Guatemala, and are superior in taste and size with red or yellow pericarp (León, 1983). In Guatemala, the cultivar 'De corona' is similar to 'Tronador' and is considered the best. The cultivars 'Guaturco', 'Dulce' and 'Chiquita' are grown in Mexico and Venezuela (Vargas *et al.*, 1999).

Cultural practices

Propagation

The tree only infrequently produces viable seeds (Juliano, 1932) and must be vegetatively propagated via hardwood cuttings taken after the leaves have shed and before flowering. Large hardwood cuttings measuring 1–1.5 m or even longer are used. They are field planted at a depth of 30–50 cm depending on size and at the recommended spacing. The soil should be irrigated and kept moist after planting to stimulate rooting. Herbaceous cuttings have been tried with less success. Patch and chip budding have been tried successfully using buds with petioles removed. The use of rootstocks of other species such as *Spondias pinnata* is possible. Asexually propagated plants start bearing in 2–3 years (Vargas *et al.*, 1999).

Orchard establishment

Preparation should follow the normal orchard procedures. The usual planting distances when very little pruning is done are 7 × 7 m or 10 × 10 m. Some orchards that are pruned in Mexico, Venezuela and parts of Central America use 3 × 3 m to limit growth (Macía, 2008).

Irrigation

Most plantings depend entirely on rainfall, but in very dry areas such as the Peruvian coast or in certain parts of Mexico and other countries, irrigation is practiced, normally using furrows.

Pruning

Plants are pruned to form them initially, trying to keep the canopy low. Later they are topped at 2 m in Mexico, Ecuador and Central America, to keep the plant short in order to facilitate harvesting from the ground. Any water sprout, damaged or diseased branch and those growing in the wrong direction should be eliminated.

Fertilization

No specific fertilization requirements have been reported. Avilán *et al.* (1989) give a tentative program that consists of applying 15, 15 and 25 g of N, P_2O_5 and K_2O respectively per plant at planting time and 2 months later; in adult plants the dosages would be raised to 50, 50 and 75 g per plant, applied two or three times per year.

Pest management and protection

No important diseases have been recorded, but under wet conditions foliage can be affected by several diseases. The fruit is heavily attacked by fruit flies that can cause serious damage to ripe fruit (Barbeau, 1990). Aphids can attack the foliage but damage is not important. Weeding is necessary one or two times a year in dry areas while three or four times in wet regions. Wind protection is necessary in certain areas through the use of windbreaks.

Harvesting and postharvest handling

Harvesting

An adult tree can produce 40 to 50 kg year^{-1} (Barbeau, 1994; Vargas *et al.*, 1999). In an Ecuadorean Andean dry area, the average yield ranges between 2.25 and 5.0 t ha^{-1} for dry season varieties (Macía, 2008) and in Venezuela about 2.0 t ha^{-1} (Araque, 1966). Careful harvesting is necessary to avoid bruising the fruit. Fruit harvested when they start to change color (~8% TSS) attain full total solids (16.9% TSS) but do not develop full red color. When harvested at the predominantly yellow stage fruit reach the maximum soluble solid content (18–20%) and became dark red, similar to fruit that ripen attached to the tree. Titratable acidity declined from 0.55 to 0.41% citric acid during ripening (De Sousa *et al.*, 1998).

Postharvest treatment

This climacteric fruit has a short postharvest shelf life of 5–6 days at ambient temperature. Fruit can be stored for 14 days at 12.5°C and when transferred to ambient temperature of 30–32°C it still has a shelf life of 4 days. Storage below 9°C results in chilling injury.

Utilization

Red mombin is eaten green, half mature or ripe (León, 1983). Green fruit are eaten with salt and vinegar as a snack or boiled in syrup. The ripe red mombin fruit are mainly eaten fresh. The soft exocarp is easily injured and so the mesocarp is often processed into marmalade, juice, wine and liquor. The pulp is used as flavoring for ice cream. Dry fruit are eaten in a dessert made with brown sugar in Mexico and Guatemala and a fermented drink is also prepared in these countries.

The ripe fruit has a pulp yield of about 70%, soluble solid content up to 21% and 0.62% titratable acidity with a TSS/TA ratio of 34. The fruit has a good calorific density (Table 1.2) due to the high concentration of total carbohydrates (19.1%). It is a moderate source of potassium and starch, and a good source of vitamin C. The main flavor compound is 2-hexenal.

References and Further Reading

Cashew

Agtrans Research (1996) The Cashew Research and Development Program: Performance and Future Prospects for Industry Development – Background Report. *Agtrans Research*, Brisbane, Australia, pp. 1–26.

Alix, C. and Duarte, O. (1999) *Propagación de Especies Frutales Tropicales*. CURLA; PDBL; AFE/COHDEFOR; DICTA; SETCO/PROFORFITH. La Ceiba, Honduras, pp. 39–40.

Berry, A.D. and Sargent, S.A. (2011) Cashew apple and nut (*Anacardium occidentale* L.). In: Yahia, E.M. (ed.) *Postharvest Biology and Technology of Tropical and Subtropical Fruits*, Volume 2. *Acai to Citrus*. Woodhead Publishing Ltd., Cambridge, UK, pp. 414–422.

Crisóstomo, L.A., Pimentel, C.R.M., de Miranda, F.R. and de Oliveira, V.H. (2007) Cashew – dwarf variety. In: Crisóstomo, L.A. and Nuamov, A. (managing eds) and Johnston, A.E. (ed.) *Fertilizing for High Yield and Quality Tropical Fruits of Brazil*. International Potash Institute Bulletin, IPI, Horgen, Switzerland, pp. 50–69.

da Silva, V.V. (1998) *Cajú. O produtor pergunta, a Embrapa responde*. Embrapa – SPI, Fortaleza, Brazil, 220 pp.

de Almeida, J.I.L., Lopez, J.G.V. and Araujo, F.E. (2003) *Produtor de caju*. Instituto CENTEC, Fortaleza, Brazil, 76 pp.

Donadio, L. (1983) Review of South American fruits. *Proceedings of the Tropical Region American Society for Horticultural Science* 27, 21–41.

Donadio, L.C., Moro, F.V. and Servidone, A.A. (2002) *Frutas Brasileiras*. Editora Novos Talentos, Jaboticabal, São Paulo, Brazil, pp. 114–118.

Duarte, O., Nieto, J.M. and Suárez, A. (1991) Tratamientos para mejorar la germinación y el enraizamiento de estacas de marañón (*Anacardium occidentale* L.). *Proceedings of the Interamerican Society for Tropical Horticulture* 35, 9–14.

Duke, J. (1989) *Handbook of Nuts*. CRC Press, Boca Raton, Florida.

FAO, Food and Agriculture Organization of the United Nations (2006) FAOSTAT – FAO, Statistics Division, Rome. Available at: http://www.fao.org/ (accessed 29 October 2011).

FAO, Food and Agriculture Organization of the United Nations (2010) FAOSTAT – FAO, Statistics Division, Rome. Available at: http://www.fao.org/ (accessed 29 October 2011).

Figueiredo, R.W., Lajolo, F.M., Elesbao, R.E., Filgueiras, H.A.C. and Araújo, N.C.C. (1999) Alteraçoes de firmeza, pectinas e enzimas pectinoliticas durante o desenvolvimento e maturaçao de pedúnculos de cajueiro anao precoce (*Anacardium occidentale* L. var. *nanum*) CCP-76. *Proceedings of the Interamerican Society for Tropical Horticulture* 43, 82–86.

Filgueiras, H., Mosca, J., Alves, R.E. and Menezes, J (1999) Cashew apple for fresh consumption: research on harvest and postharvest technology in Brazil. *Acta Horticulturae* 485(1), 55–60.

Johnson, D.V. (1973) The botany, origin and spread of the cashew (*Anacardium occidentale* L.). *Journal of Plantation Crops* 1, 1–7.

Ohler, J.G. (1979) *Cashew*. Communication 71, Department of Agricultural Research, Koninklijk Instituut voor de Tropen, Amsterdam, 260 pp.

Prabhakaran-Nair, K.P. (2009) The agronomy and economy of some important industrial crops. *Advances in Agronomy* 101, 209–313.

Purseglove, J.W. (1987) *Tropical Crops: Dicotyledons*. Longman Scientific and Technical, Harlow, UK.

Rosengarten, F. Jr. (1984) *The Book of Edible Nuts*. Walker and Company, New York.

Rosetti, A.G. and Corrêa, M.P.F. (1996) The number of crown-grafted buds affects nut yield of early dwarf cashew (*Anacardium occidentale* L.). *Proceedings of the Interamerican Society for Tropical Horticulture* 40, 204–205.

Schaper, H., Chacko, E.K. and Blaikie, S.J. (1996) Effect of irrigation on leaf gas exchange and yield of cashew in northern Australia. *Australian Journal of Experimental Agriculture* 36, 861–868.

Sivakumar, A. and Pai, T.A. (2008) *Anacardium occidentale* Cashew. In: Janick, J. and Paull, R.E. (eds) *Encyclopedia of Fruit and Nuts*. CAB International, Wallingford, UK, pp. 8–14.

Vargas, O., Alix, C., Lobo, A.D. (authors), Duarte, O. and Sanchez, J. (technical reviewers) (1999) *Frutales y Condimentarias del Trópico Húmedo*. CURLA; PDBL; AFE/COHDEFOR; DICTA; SETCO; PROFORFITH, La Ceiba, Honduras, pp. 6–7.

Villachica, H., Carvalho, J.E.U., Muller, C.H., Diaz, S.C. and Almanza, M. (1996) Marañón. In: *Frutales y Hortalizas Promisorios de la Amazonía*. Tratado de Cooperación Amazónica, Lima, Perú, pp. 195–202.

Woodroof, J.G. (1979) *Tree Nuts: Production, Processing, Products*, 2nd edn. AVI Publishing Company, Westport, Connecticut.

Yellow mombin

Abo, K.A., Ogunleye, V.O. and Ashidi, J.S. (1999) Antimicrobial potential of *Spondias mombin* L., *Croton zambesicus* and *Zygotritonia crocea*. *Phytotherapy Research* 13, 494–497.

Barbeau, G. (1990) *Frutas Tropicales en Nicaragua*. Dirección General de Técnicas Agropecuarias, MIDINRA, Editorial Ciencias Sociales, Managua, Nicaragua, pp.112–113.

Campbell, C.W. and Sauls, J.W. (1991) *Spondias in Florida*. Fruit Crops Fact Sheet FC-63. University of Florida, Florida, pp. 1–3.

da Costa, N.P., Filgueiras, H.A.C., Alves, R.E., da Silva, A.Q. and de Oliveira, A.C. (1998) Development and maturation of yellow mombin (*Spondias mombin* L.) in Northeast Brazil. *Proceedings of the Interamerican Society for Tropical Horticulture* 42, 301–306.

da Costa, N.P., da Silva, A.Q., Alves, R.E., Filgueiras, H.A.C. and de Oliveira, A.C. (2003) Armazenamento de Cajá (*Spondias mombin*) em Refrigeração Sob Atmosfera Modificada. *Proceedings of the Interamerican Society for Tropical Horticulture* 47, 234–235.

de Souza, E.X. (2000) Effect of rootstock and grafting methods on the development of *Spondias mombin* L. plants. *Revista Brasileira de Fruticultura* 22, 286–290.

de Souza, E.X., Innecco, R. and Araujo, C.A.T. (1999) *Grafting Methods Recommended for the Production of Yellow Mombin and Other Fruit Trees of the Genus Spondias*. Comunicado Técnico 37. EMBRAPA-CNPAT, Fortaleza, Ceara, Brazil, 8 pp.

Donadio, L.C., Moro, F.V. and Servidone, A.A. (2002) *Frutas Brasileiras*. Editora Novos Talentos, Jaboticabal, São Paulo, Brazil, pp. 108–113.

Filgueiras, H.A.C., Alves, R.E., de Oliveira, A.C., Moura, C.F.H. and Araújo, N.C.C. (1999) Calidad de Frutas Nativas de Latinoamérica para Industria: Jobo (*Spondias mombin* L.). *Proceedings of the Interamerican Society for Tropical Horticulture* 43, 72–76.

Filgueiras, H.A.C., Moura, C.F.H. and Alves, R.E. (2000) Cajá (*Spondias mombin*). In: Alves, R.E., Filgueiras, H.A.C. and Moura, C.F.H. (eds) *Caracterização de Frutas Nativas da América Latina*. São Paulo State, Jaboticabal, Brazil, pp. 19–22.

Filgueiras, H.A.C. and de Souza, F.X. (2008) *Spondias mombin*, Yellow mombin. In: Janick, J. and Paull, R.E. (eds) *Encyclopedia of Fruit and Nuts*. CAB International, Wallingford, UK, pp. 31–34.

Jesus, N., Martins, A.B.G., Almeida, E.J., Junior, E.J.S. and Andrade, R.A. (2003) Propagação Vegetativa de Estacas Herbáceas de *Spondias* em Câmara de Nebulização. *Proceedings of the Interamerican Society for Tropical Horticulture* 47, 239–241.

León, J. (1983) Underutilized fruits of Central America and northern South America. *Proceedings of the Tropical Region American Society for Horticultural Science* 27, 1–20.

Lorenzi, H. (1992) *Árvores Brasileiras*. Editorial Plantarum Ltda., São Paulo, Brazil.

Mata, M.E.M.R.C., Braga, M.E.D. and Silva, M. (2003) Curvas de congelamiento de frutos de cajá (*Spondias lutea* L.) a temperaturas semi-criogénicas. *Revista Brasileira de Produtos Agroindustriais* 1, 52–62.

Mattietto, R.A. and Matta, V.M. (2011) Cajá (*Spondias mombin* L.). In: Yahia, E.M (ed.) *Postharvest Biology and Technology of Tropical and Subtropical Fruits, Volume 2. Acai to citrus*. Woodhead Publishing Ltd., Cambridge, UK, pp. 330–351.

Morton, J.F. (1987) *Fruits of Warm Climates*. Creative Resources Systems Inc., Winterville, North Carolina, pp. 245–257.
Moura, F.T., Silva, S.M., Martins, L.P., Mendoça, R.M.N., Alves, R.E. and Filgueiras, H.A.C. (2003) Evolução do Crescimento e da Maturação de Frutos de Cajazeira (*Spondias mombin* L.). *Proceedings of the Interamerican Society for Tropical Horticulture* 47, 231–233.
Oliveira, A.C. de, Filgueiras, H.A.C., Alves, R.E. and De Sousa, R.P. (1998) Ambient temperature storage of yellow mombin (*Spondias mombin* L.) in four stages of maturity. *Proceedings of the Interamerican Society for Tropical Horticulture* 42, 307–312.
Sacramento, Q.K. de and Souza, E.X. de (2000) Caja (*Spondias mombin* L.). Série Frutas Nativas 4. Fundação de Estudos e Pesquisas em Agronomia, Medicina Veterinária e Zootecnia (FUNEP), Jaboticabal, São Paulo, Brazil, 42 pp.
Villachica, H. (1996) *Frutales y Hortalizas Promisorios de la Amazonía*. Tratado de Cooperación Amazónica, Lima, Perú, 367 pp.
Villachica, H., Carvalho, J.E.U., Muller, C.H., Diaz, S.C. and Almanza, M. (1996) Ubos, *Spondias mombin* L. In: *Frutales y Hortalizas Promisorios de la Amazonía*. Tratado de Cooperación Amazónica, Lima, Perú, pp. 269–274.

Red mombin

Alix, C. and Duarte, O. (1999) *Propagación de Especies Frutales Tropicales*. CURLA; PDBL; AFE/COHDEFOR; DICTA; SETCO/PROFORFITH. La Ceiba, Honduras, p. 39.
Almaguer-Vargas, G., Espinoza-Espinoza, J.R., Martines-Bravo, A. and Amador-Gomes, J. (1991) Efecto de la defoliación química en el adelanto de la cosecha de guayabo (*Psidium guajava* L.) y ciruelo (*Spondias purpurea* L.). *Proceedings of the Interamerican Society for Tropical Horticulture* 35, 71–75.
Araque, R. (1966) *La Ciruela de Hueso*. Serie de Cultivos No. 9. Consejo de Bienestar Rural, Caracas, Venezuela.
Avilán, L., Leal, F. and Bautista, D. (1989) *Manual de Fruticultura; Cultivo y Producción*. Editorial América C.A., Caracas, Venezuela, pp. 437–441.
Barbeau, G. (1990) *Frutas Tropicales en Nicaragua*. Dirección General de Técnicas Agropecuarias, MIDINRA, Editorial Ciencias Sociales, Managua, Nicaragua, p. 113.
Barbeau, G. (1994) Plums. *Proceedings of the Third Regional Workshop on Tropical Fruits*, Grenada, pp. 131–134.
Barfod, A. (1987) Anacardiaceae. In: Harling, G. and Andersson, L. (eds) *Flora of Ecuador* 30. Arlöv, Sweden, pp. 9–49.
Cuevas, J.A. (1992) Jocote, ciruelo (*Spondias purpurea*). In: Hernández-Bermejo, J.E. and León, J. (eds) *Cultivos Marginados: Otra Perspectiva de 1492*. FAO, Rome, pp. 109–113.
De Melo, S. and Pereira, L. (2001) Carbohydrate-related changes in red mombin (*Spondias purpurea* L.) fruit. *Proceedings of the Interamerican Society for Tropical Horticulture* 45, 38–41.
De Sousa, R., Filgueiras, H.A., Alves, R., Alves, J. and Cordeiro, A. (1998) Identification of the optimum harvest stage for red mombin (*Spondias purpurea* L.). *Proceedings of the Interamerican Society for Tropical Horticulture* 42, 319–324.
Filgueiras, H.A.C., Alves, R.E., Moura, C.F.H., de Oliveira, A.C. and Araújo, N.C.C. (1999) Calidad de frutas nativas de Latinoamérica para Industria: Ciruela mexicana (*Spondias purpurea* L.). *Proceedings of the Interamerican Society for Tropical Horticulture* 43, 68–71.
Jesus, N., Martins, A.B.G., Almeida, E.J., Junior, E.J.S. and Andrade, R.A. (2003) Propagação vegetativa de estacas herbáceas de *Spondias* em câmara de nebulização. *Proceedings of the Interamerican Society for Tropical Horticulture* 47, 239–241.
Juliano, J.B. (1932) The cause of sterility of *Spondias purpurea* Linn. *Philippine Agriculturist* 21, 15–24.

Koziol, M.J. and Macía, M.J. (1998) Chemical composition, nutritional evaluation, and economic prospects of *Spondias purpurea* (Anacardiaceae). *Economic Botany* 52, 373–380.

León, J. (1983) Underutilized fruits of Central America and northern South America. *Proceedings of the Tropical Region American Society for Horticultural Science* 27, 1–20.

León, J. and Shaw, P.E. (1990) *Spondias*: the red mombin and related fruits. In: Nagy, S., Shaw, P.E. and Wardowsky, F.W. (eds) *Fruits of Tropical and Subtropical Regions*. Florida Science Source (FSS), Lake Alfred, Florida, pp. 116–126.

Macía, M.J. (1997) El 'ovo' (*Spondias purpurea* L., Anacardiaceae) un árbol frutal con posibilidades socioeconómicas en Ecuador. In: Ríos, M. and Pedersen, H.B. (eds) *Uso y Manejo de Recursos Vegetales. Memorias del Segundo Simposio Ecuatoriano de Etnobotánica y Botánica Económica*. Ediciones Abya-Yala, Quito, Ecuador, pp. 271–281.

Macía, M.J. (2008) *Spondias purpurea*, red mombin. In: Janick, J. and Paull, R.E. (eds) *Encyclopedia of Fruit and Nuts*. CAB International, Wallingford, UK, pp. 541–549.

Macía, M.J. and Barfod, A.S. (2000) Economic botany of *Spondias purpurea* (Anacardiaceae) in Ecuador. *Economic Botany* 54, 449–458.

Manzano, J.E. (1999) Comportamiento de frutos de ciruela huesito (*Spondias purpurea* L.) almacenadas a diferentes temperaturas. *Proceedings of the Interamerican Society for Tropical Horticulture* 43, 57–62.

Miller, A. and Schaal, B. (2005) Domestication of a Mesoamerican cultivated fruit tree, *Spondias purpurea*. *Proceedings of the National Academy of Sciences* 102, 12801–12806

Mohammed, M. (2011) Chili plum (*Spondias purpurea* var. *lutea*). In: Yahia, E.M. (ed.) *Postharvest Biology and Technology of Tropical and Subtropical Fruits, Volume 2. Acai to citrus.* Woodhead Publishing Ltd., Cambridge, UK, pp. 423–436.

Pimienta-Barrios, E. and Ramirez-Hernandez, B.C. (2003) Phenology, growth, and response to light of ciruela mexicana (*Spondias purpurea* L., Anacardiaceae). *Economic Botany* 57, 481–490.

Popenoe, J. (1979) The genus *Spondias* in Florida. *Proceedings of the Florida State Horticultural Society* 92, 277–279.

Ritter, S. (1993) Seasonal variation in the successful propagation of *Spondias purpurea* hardwood cuttings. *Proceedings of the Interamerican Society for Tropical Horticulture* 37, 136–137.

Silva, F.S., Silva, A.Q., Lima, P.F.U. and Silva, H. (1996) Umbuguela (*Spondias* spp.) growing in a semi-arid zone. *Proceedings of the Interamerican Society for Tropical Horticulture* 40, 206–209.

Silva, R.M., Martins, L.P., Alves, R.E. and Filgueiras, H.A.C. (2001) Carbohydrate-related changes in red mombin (*Spondias purpurea* L.) fruit. *Proceedings of the Interamerican Society for Tropical Horticulture* 45, 38–41.

Sousa de, R.P., Filgueiras, H.A.C., Alves, R.E., Costa, J.T.A. and de Oliveira, A.C. (1998) Identification of the optimum harvest stage for red mombin (*Spondias purpurea* L.). *Proceedings of the Interamerican Society for Tropical Horticulture* 42, 319–324.

Vargas, O., Alix, C., Lobo, A.D. (authors), Duarte, O. and Sanchez, J. (technical reviewers) (1999) *Frutales y Condimentarias del Trópico Húmedo*. CURLA; PDBL; AFE/COHDEFOR; DICTA; SETCO; PROFORFITH, La Ceiba, Honduras, pp. 20–21.

2 Calophyllaceae, Clusiaceae and Cactaceae

CALOPHYLLACEAE/CLUSIACEAE

Mammee Apple

The species *Mammea americana* L. formerly belonged to the family Clusiaceae (alt. Guttiferae), in the very diverse and large order Malpighiales; it has now been assigned to the family Calophyllaceae. Most of the 14 genera in the Calophyllaceae were formerly in the tribe Calophylleae of the Clusiaceae family. Calophyllaceae has 14 genera and about 595 species, most with a milky sap, and includes the genus *Mammea* with about 50 species of evergreen trees with edible one-seeded fruits.

The common names of mamey include: *mamey, mamey serrano, mamey de Cartagena, mamey de Santo Domingo, mamey dominicano, mata serrano, zapote de Santo Domingo, mamey amarillo* and *zapote mamey* (Spanish), mammee, mammee apple, South American apricot and St Domingo apricot (English), *abricot d'Amerique, abricot de Saint-Domingue, abricot des Antilles, abricot pays* and *abricotier sauvage* (French) and *abrico do Para, abrico selvagem, abricote* and *pecego de Sao Domingos* (Portuguese).

Important genera and species

Confused sometimes with mamey sapote (Sapotaceae) because of their similar external color, this fruit belongs to a different family and has other characteristics. There is an African species *Mammea africana* Sabine (syn. *Ochrocarpus africanus*), the African apricot, and several Asiatic fruits in the related family Clusiaceae mainly in the *Garcinia* genus. Among these fruits are *G. mangostana* L., the mangosteen, considered one of the finest fruits, *G. tinctoria* W.F. Wight, the gamboge, *G. livingstonei* T. Anders and *G. dulcis* Kurz, the mundu.

Origin and distribution

Mammee apple is native to the West Indies and northern South America. For some authors it apparently originated on the Caribbean islands of Jamaica, Puerto Rico or the Lesser Antilles. There are no large commercial plantations of this species although it is grown to some extent in Mexico and Central America and also in parts of South America in Colombia, Venezuela, Guyana, Suriname, French Guiana, Ecuador, Peru and northern Brazil with fruit being sold in the cities. In the USA it is considered a minor crop, and can be found growing, normally as a backyard tree, in both Florida and Hawaii. In the lowlands of Costa Rica, El Salvador and Guatemala it is grown as an ornamental, while in Nicaragua it is used as a border and windbreak tree in coffee plantations.

Ecology

Soil

The tree prefers well-drained deep soil, rich in organic matter. It can also grow in shallow, sandy soils; or in the oolitic limestone of south-eastern Florida, as well as in the volcanic soils of Nicaragua or the heavy soils of central Honduras and the alkaline soils of the Peruvian north coast.

Rainfall

This tropical or near tropical tree grows best in moist to wet climates with a mean annual rainfall of 1,500–4,000 mm. However, it produces very well in the Peruvian coast where there is practically no rain and irrigation is used year round.

Temperature

It grows in hot humid climates with average of 27–30°C. In Central America, it grows to altitudes up to 1,000 m, sometimes even 1,600 m. Growth is slow at 17–18°C and frost can kill mature trees.

Light and photoperiod

No special requirements have been reported. It flowers and produces well with no specific photoperiodic requirement.

Wind

The tree is well adapted to resist winds, and in Nicaragua it is used as a windbreak in coffee plantations (Barbeau, 1990).

General characteristics

Tree

This evergreen tree can reach 15–22 m. It has a short trunk that may attain 0.9–1.2 m in diameter with many ascending, densely foliaged branches forming an erect, oval head. The opposite, dark green leaves are glossy and leathery, broadly elliptic, 10–25 cm long and 5–12 cm wide, and attached to the branch by petioles 8–15 cm long (Fig. 2.1). The leaf base is wedge-shaped, obtuse or rounded.

Flowers

The creamish-white flowers are fragrant and 2.5–4 cm wide when fully open. They have 4–6 petals and orange stamens and/or pistils. Flowers are borne singly or in clusters of two or three on short stalks on the axils of young branches. The trees can be unisexual (dioecious), hermaphrodite or can have all three types of flowers together. In the West Indies, flowering occurs from May to October and in Trinidad and Tobago from July to September. Flowers are visited by stingless bees.

Fruit

The almost round or globose berry is sometimes slightly deformed and is attached to the plant by a thick stem. The fruit diameter ranges from 10 to 18 cm, with a weight from 0.5 to 0.8 kg, sometimes up to 2.0 kg. The texture is hard until fully ripe when it softens slightly. The thick, 4–5 mm outer peel is leathery, bitter, russet to grayish-brown and rough to the touch because of small globules of latex. Beneath the bitter skin, there is a thin, dry, whitish-yellow tissue that covers the flesh and is astringent and often bitter. The flesh can be light or golden yellow to orange, without fibers, and varies from firm and crisp to dry to tender, melting and juicy; it is fairly free from the seed except at certain points. The ripe flesh is very fragrant and, in the best varieties, has a pleasantly subacid flavor that resembles that of

Fig. 2.1. Mammee apple (*Mammea americana* L.) showing flowers, leaves and fruit (used with permission from León, J. (2000) *Botánica de los Cultivos Tropicales*. Editorial Agroamérica, Instituto Interamericano de Cooperación para la Agricultura (IICA), San José, Costa Rica).

apricot. Small fruit have normally one seed while larger fruit may have two to four seeds. The seeds are russet-brown, rough, ovoid or ellipsoid and about 5–7 cm long. The juice of the seed leaves an indelible stain. Fruit may take up to a year to mature and are ripe from July to February.

Cultivar development

Cytogenetics, genetics and breeding

No information is found on this subject. Since most of the plants are seed propagated considerable variation occurs in tree and fruit characteristics. Normally people obtain seeds from the best plants to propagate though this does not ensure that they will be true to the parent type. There are no named cultivars.

Cultural practices

Propagation

Seeds are the usual means of propagation; however, 50% can be males. The seeds are recalcitrant and should be sown soon after extraction from the fruit. Seed germination is slow, taking about 40 days. Removal of seed coats can result in 100% germination after 4 weeks while only 23% of non-peeled seeds germinate in the same time. Seed coat removal plus a 24 h soak in gibberellic acid (GA) and at a temperature of 30°C significantly improves germination (Duarte and Franciosi, 1976). Seedlings grown under light shade can reach 40–50 cm and are ready to be taken to the field 3–5 months after transplanting to nursery bags. Seedlings begin bearing after 7–8 years depending on the growing conditions.

Vegetative propagation is preferable to avoid raising male trees and to obtain earlier fruiting. Cuttings do not root easily with grafting having a higher success rate. Stems about 1 cm thick are grafted 25–30 cm from the ground. Cleft grafting can result in 100% success using terminal scions and 98% with sub-terminal scions. Splice grafting in spring is also very successful. Scions are gathered from the shoots when they stop growing and the terminal leaves are fully developed and hardened (Alix and Duarte, 1999). Grafted plants start bearing after 4–5 years.

Field preparation, transplanting and plant spacing

Normal land preparation is necessary for a commercial orchard. Nursery plants about 50–60 cm tall are usually transplanted. Spacing between 8 and 10 m is recommended.

Irrigation

Normally the tree is not irrigated when it grows in places with adequate rains. In the case of plants in windbreaks or backyards they will receive the water given to the crop or the garden.

Pruning

Normally little pruning is done. Some authors suggest a formation pruning during the first years by topping the young plants in order to obtain a lower and wider canopy.

Fertilization

No specific recommendations are available. Fertilizer rates for other fruit trees of similar size would be used.

Pest management

Mammee apple is resistant to pests and diseases. A black mildew (*Aulographum melioloides*) attacks leaves and heart rot infects older trees entering through basal scars. An unidentified root rot has been reported. Some bees like *Trigona rufricus* cut the flowers and young leaves. Wet-wood termites attack dead trunks and branches and dry-wood termites attack seasoned lumber. Since most plants are in borders or isolated, not in regular plantations, weed control is normally made together with general cleaning operations of the crop or the yard. The trees are normally used as a wind barrier.

Harvesting and postharvest handling

Yields range from 150 to 400 fruit per plant. Fruit maturity is determined by scratching the skin with the fingernail; if the color beneath is green, the fruit is not ready for harvest; if it is yellow it is ready to harvest. Sometimes fruit external color will become lighter tending to a slight yellowing as the fruit matures. Fruit is normally cut leaving a short section of the peduncle attached. Fruit will normally be firm at harvest and will take 3–4 days to soften for eating. The fruit bruise easily so they should not be allowed to fall when ripe.

The fruit is highly perishable and is very susceptible to chilling injury and should not be stored below 15°C. Harvested fruit can be left for a couple of weeks at room temperature. Fruit pulp can be frozen and kept at −30°C. Dehydration can also be used.

Utilization

The fruit may be eaten fresh or made into preserves. The pulp has a good content of vitamin A, C and B2 (Table 2.1). For processing, the peel is removed as well as the thin and bitter membrane. A compote is made by boiling strips or dices of the pulp in syrup with different spices added. Slightly under-ripe fruit that are rich in pectin are made into jelly. Wine is made from the fruit and fermented toddy is made from the sap of the tree in Brazil. In some Caribbean islands a liquor called "eau de creole" is made by the distillation of the mammee apple flowers with spirits of wine.

Mammee apple is used in folk medicine that includes treatment of scalp infections, diarrhea, digestive and eye problems. Some antibiotic principle is present in the pulp. Various extracts from the fruit, bark, leaves, or roots are toxic to moths, beetle larvae, and also to bugs, ticks and fleas. Seed extracts are toxic to fish, chicks and pigs. In Venezuela, powdered seeds are employed in the treatment of parasitic skin diseases. An infusion of the fresh or dry leaves is given in cases of intermittent fever.

Table 2.1. Composition of mammee apple, achachairú and prickly pear fruit (*Opuntia ficus-indica*) per 100 g edible portion.

Proximate	Mammee apple[a]	Achachairú[b]	Prickly pear fruit[c]
Moisture (g)	88.9	78.3	87.6
Energy (kcal)	37	–	41
Protein (g)	0.5	0.4	0.7
Lipid (fat) (g)	0.1	0.5	0.5
Fiber (g)	–	0.6	5.4
Carbohydrates (g)	9.7	14.3	14.3
Ash (g)	–	0.3	–
Minerals			
Calcium (mg)	5	–	56
Iron (mg)	0.5	–	0.3
Phosphorus (mg)	46	–	24
Vitamins			
Thiamine (mg)	0.02	–	0.014
Riboflavin (mg)	0.04	–	0.06
Niacin (mg)	0.61	–	0.46
Carotene (mg)	0.37	–	–
Ascorbic acid (mg)	2	–	14

[a]Villachica *et al.*, 1996.
[b]Ardaya, 2009.
[c]USDA Nutritional Database.

The heartwood is purple-brown or reddish and the sapwood has a lighter color. The wood is heavy, hard, fine-grained and strong; it can be easily worked and has an attractive grain that polishes well, though is highly susceptible to termites. The tree is planted as a shade and windbreak. The dark green, shiny leaves and dense foliage make it a beautiful ornamental tree.

Achachairú

The original species name of achachairú was *Rheedia laterifolia* (L.) Herzog, which was changed to *Garcinia humilis* (Vahl) C.D. Adams. Another synonym is *Mammea humilis* Vahl. In Spanish, it is called *achachairu*, *shashairu*, *ibaguazu*, *cachicheruqui* and *tapacuari*. The fruit flavor resembles that of its Asian relative the mangosteen. It has become more known and a good market exists not only in its native Bolivia but also in neighboring countries and Europe.

Important genera and species

There are several relatives of this species with edible fruit in the region that have been placed in the genus *Garcinia* (GRIN, 2006) though are less known. An exception is ocoró (*G. acuminata*) with similar volumes of fruit produced and consumed in Bolivia (Ardaya *et al.*, 1995). The bacur (*Platonia insignis*) also belongs to this family.

Origin and distribution

G. humilis (Vahl) C.D. is a native of the eastern parts of Bolivia along with other species in this genus. In the Santa Cruz de la Sierra area of Bolivia, it is a very popular fruit and most of the production originates from wild or backyard trees, although in the last 20–30 years commercial orchards have been started in the region, though no precise statistics are available. It is estimated that there are more than 1,000 ha in Bolivia. Small plantings have been made in Guatemala and some plants introduced to Honduras that are performing well at 800 m altitude. A fairly large plantation has been started in Australia and the fruit is being marketed as "achacha" in the EU.

Ecology

Soil

The tree grows in deep, alluvial, heavy or light soils, rich in organic matter, and a pH around 7, but it can also grow in stony soils with variable fertility and drainage conditions, as well as in fairly heavy soils with a pH around 5.0 (Duarte and Paull, 2008).

Rainfall

The tree grows wild in areas with 1,400–2,500 mm rainfall. Under more arid conditions, it needs to be irrigated, such as in Central America where it has grown well under 1,000 mm rainfall with a single dry season.

Temperature

In its place of origin the tree grows at altitudes that range from 150 to 600 m, preferring average temperatures of 23–25°C. It can grow in cooler climates but at a slower rate. Cold fronts, sometimes freezing temperatures, come into Bolivia from the south between June and September and can cause heavy flowering and small fruit drop.

Light

The plant grows in the wild under the canopy of taller trees and it is said that in cultivation it produces better under shade. In many cases, temporary shade is recommended during the first 3–4 years. Orchards in Honduras at 800 m elevation do well without any initial shade, with no apparent foliage damage or slowdown in growth. The plant is not seemly photoperiodic.

Wind

Wind damage is unlikely as the plant has a very compact branch and foliage system and the fruit are hidden by the leaves and branches, which are strong and flexible.

General characteristics

Tree

The tree can reach a height of 6–12 m and has a pyramidal canopy. The stem can reach a diameter of 40 cm. The leaves are slightly darker green on the upper side than the lower, opposite, measuring 15–25 × 4–7 cm, lanceolate, shiny and with entire margins and an acute or acuminate apex (Fig. 2.2). They have an obvious central vein while lateral veins are not very conspicuous and are parallel. Young leaves start pinkish-bronze or reddish, turning to yellowish-green before maturing. Young leaves are very susceptible to sun damage that can severely burn or deform them, therefore the recommendation to start the plant under temporary shade.

Flowers

The tree has about 200 hermaphrodite flowers to each male flower, although some trees produce only male flowers. The hermaphroditic flowers (17–36 mm long) have 20–34 stamens, a 3–5 mm ovary, two sepals, four petals, a pedicel of about 2 cm and appear in groups of two to five. The male flowers are 9.6–12.5 mm

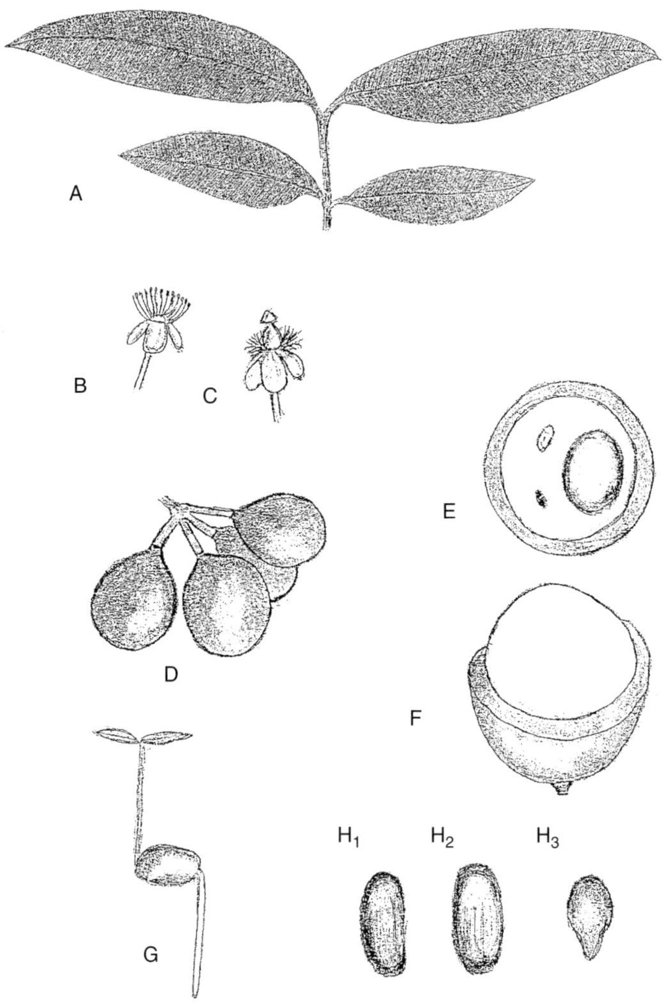

Fig. 2.2. Achachairú (*Garcinia humilis* (Vahl) C.D. Adams) showing (A) leaves, (B) male flower, (C) female flower, (D, E, F) fruit, (G) germinated seed (notice radicle), (H_1) normal seed, (H_2) polyembryonic seed and (H_3) deformed seed. (Drawing by Odilo Duarte.)

long with 26–28 stamens, can have a vestigial ovary and a pedicel of about 4 cm, and they appear in groups of 30–50 flowers (Ardaya *et al.*, 1995).

Pollination and fruit set

Anthesis occurs in the morning until noon and honey bees and other insects seem to play a role in pollination. About 57% of the flowers are self-fertile and the remaining fruit come from cross-pollination and this explains the variability found among seedlings (Ardaya *et al.*, 1995). Flowering normally occurs at the end of winter and fruit ripen at the end of spring and during most of the summer (November–February in Bolivia).

Fruit

The fruit is a berry with hard, thick skin (2 mm) that has a smooth surface. It starts out bluish-green that evolves into yellow-orange before ripening and when fully ripe it is reddish-orange. The ovoid, 4.0–5.2 cm long by 3–4 cm in diameter fruit has an average weight of 40 g of which around 40% is edible pulp, 47% is peel, and 12% are the seeds. The fruit contains three ovules of which normally two abort. After anthesis the ovary develops into a ripe fruit in about 150–160 days, depending on the climatic conditions. The fruit are not very visible since they are normally located inside the canopy. The seeds are cylindrical, brown, and large (3.0–3.4 by 1.5–2.0 cm) and polyembryonic with one to three nucellar embryos in addition to the sexual embryo. Seed number and size are related to fruit size. Seeds when cut release a yellow resin (Vasquez and Coimbra, 2002).

Cultivar development

Cytogenetics, genetics and breeding

Not much is known about this species, except that it has nucellar, but also sexual, embryos. Empirically farmers have used seeds from trees that produce more or have larger fruit, thus some selection has been carried out but this has not been systematic. During the last 20–25 years the Centro de Investigación Agrícola Tropical (CIAT) of Santa Cruz, Bolivia has been working with this crop and after evaluating seedling populations by mass selection has obtained 12 superior lines, and from them they selected the best line, which they have patented and named 'Selecto' in 2009. 'Selecto' has more edible pulp and only one viable seed, resulting in a better commercial product (Ardaya, 2009).

Cultural practices

Propagation

Seed propagation is most commonly used. Fresh seeds sown with the remains of the pulp attached germinate satisfactorily (80%), 30–40 days after sowing and sometimes to half a year. Storing the seeds at 12°C for 3 weeks retains germination percentage while at 6°C they apparently suffer chilling injury. Seeds can be stored for 12 months in a shady, cool place and lose about half of their initial germination capacity. The best procedure is to sow directly in plastic bags or to transplant spontaneous seedlings that might grow under the trees to bags. The nursery should be under 50% shade. Being polyembryonic, dividing the seed into two or three pieces can produce a plant from almost every piece.

Leafy cuttings in a plastic chamber under 60% shade showed less than 10% rooting with or without rooting hormone. Air layering has also not been successful. In Bolivia, cleft, side or crown grafts using terminal scions have given above 50% takes but growth is slow. The latex that exudates from the wounds seems to have a negative effect on the union formation. Using the same species (*G. humilis*) as a rootstock does not produce satisfactory results, while ocoró (*G. acuminata*) as a stock seems to be more successful (Ardaya *et al.*, 1995).

Field preparation, transplanting and plant spacing

The field should be prepared as for any fruit tree plantation. The plants should be set 8–10 m apart. Normally plants 50–60 cm in height are transplanted. Addition of superphosphate and aged manure to the planting hole is recommended. In Bolivia, temporary shade is provided using palm leaves arranged like a tent on top of the small plants or with pigeon peas, castor bean or similar plants sown around the planting hole, generally in advance to planting.

Irrigation

Under normal conditions in its place of origin the plants receive sufficient rain, although some years the dry period is too long and plants and yields are affected. In this case, extra irrigation would be useful.

Pruning

The tree tends to be pyramidal in shape and no formation pruning is recommended. Pruning should be limited to eliminate damaged or dead branches or those excessively close to the ground although some growers indicate that the heaviest fruit load comes in these branches so they should be left unless there is a good reason to prune them (Ardaya *et al.*, 1995).

Fertilization

Very little is known about fertilizer needs.

Pest management

Anthracnose attacking the fruit and leaves has been reported from Bolivia. The fruit, especially if very ripe, is susceptible to fruit flies. Some growers harvest before the fruit ripen to avoid possible attacks; however, the fruit is non-climacteric and early harvested fruit do not attained the best flavor. Termites (*Rhinotermitidae*) can become a problem if the plants are neglected.

Orchard protection

This tree is fairly resistant to winds because of its low canopy that touches the ground and its dense branch and foliage system.

Harvesting and postharvest handling

Trees can produce 2,000–3,000 fruit when 15 years old and 5,000–6,000 after 20 years, and a 50-year-old tree can produce up to 7,000–8,000 fruit. Each fruit weighs around 40 g (Duarte and Paull, 2008). According to Ardaya *et al.* (1995) an adult plantation can produce an estimated 19.6 t ha^{-1}.

The fruit are usually hand-picked by children climbing on the trees with bags or baskets. Only ripe fruit, determined by their skin color, are harvested. At room conditions, the fruit can last 1 week for fresh market use and 2–3 weeks for industrial use (ice cream, refreshments). The fruit will shrivel and lose firmness but will not rot, unless kept wet or if damaged. Some trials have shown that the best way to keep the fruit is by wrapping them in trays with PVC film and holding them at 12°C. This can extended postharvest life for 3 weeks, while at 6°C chilling injury develops (Duarte, 2011). Fruit can be frozen for several months without significant loss in quality.

Utilization

The fruit is normally consumed fresh, by opening it and extracting the large seed that is surrounded by a cottony juicy pulp. The pulp has a sweet subacid flavor and is low in fat and protein (Table 2.1). The peel is used to prepare refreshments by mixing it with sugar and water in a blender.

CACTACEAE

Prickly Pear

The Cactaceae family includes 125–130 genera and 1,400–1,500 species further divided into a number of tribes and subfamilies, though the taxonomy is far from settled (Nyffeler and Eggli, 2010). Cacti are found to dominate the landscapes in arid shrub lands and semideserts from the southern United States and Mexico, into north-eastern Brazil and along the eastern and western slopes of the Andes from Ecuador southward to arid areas in Argentina and Chile.

Important genera and species

The most important commercial genus is *Opuntia*, the prickly pears, with about 60 species that freely hybridize. *Opuntia* species are referred to as "tuna", nopales or paddle cactus. Only the prickly pears are now included in this genus. Cholla with their cylindrical stems have been placed in the *Cylindropuntia* instead of being a subgenus of *Opuntia*.

Common English names are Indian fig, barbary fig, cactus pear and prickly pear. Spanish names include: *nopal, cardón de México, chumbera, chumbo, chumbua, higo chumbo, higo de pala, higo México, higuera de pala, nopal de castilla, tuna de España, tuna española, tuna mansa, tuna, higo chimbo* and *tuna real*. The synonyms include *Cactus ficus-indica* L., *Opuntia ficus-barbarica* A. Berger and *Opuntia megacantha* Salm-Dyck. *Opuntia ficus-indica* has the closest affinity with the spiny *Opuntia megacantha* though other fleshy-fruited species were possibly involved.

Origin and distribution

The branched prickly pear (*O. ficus-indica* (L.) Mill.) has been long domesticated and important agriculturally in the semiarid regions of Central and South America. The selected fleshy-fruited ancestors of *O. ficus-indica* were spread through trade throughout Mesoamerica and the Caribbean and possibly into South America in pre-Columbian times (Griffith, 2004). Prickly pear, a name also applied to less common *Opuntia* species, is now spread throughout the arid and semiarid parts of the world. Its worldwide distribution started in the 15th century to Europe and North Africa for mass rearing of cochineal insects. It was then spread to the Mediterranean Basin, especially the islands, Micronesia, Australia, tropical and southern Africa, western USA, the Caribbean, temperate Asia, the Seychelles and Hawaii. This widespread distribution was due to the relative ease of vegetative propagation (Saenz-Hernandez *et al.*, 2002) and because the detached cladodes can remain alive for over 12 months and readily form new roots when placed in moist soil.

Ecology

Soil

Prickly pear tolerates a wide variety of growing conditions and soil types though growth is limited in its natural habit by poor soils and drought. It generally prefers deep sandy soils, frequently on slopes with a pH of 6 to 7.5. Clay soils that are poorly drained or become waterlogged are avoided. As with most cacti, adequate drainage is essential for good growth and it cannot withstand any prolonged waterlogging. The roots are tolerant to moderate soil salinity.

Climate

This subtropical plant is found from 0 to 2,600 m in areas with a mean annual temperature from −10 to 26°C. It is sensitive to freezing temperatures below −6°C and tolerant to high temperatures up to 65°C. In its natural habit, the mean annual rainfall is from 400 to 700 mm with a bimodal pattern. It does not do well when shaded and grows best in full sunlight.

General characteristics

Plant

The plant is large, up to 5 m tall and 3 m wide, and has a jointed succulent flattened pseudostem. The flattened stems (cladodes) are branched, green, broadly oblong to ovate to narrowly elliptic, 30–60 cm long and 6–12 cm wide (Fig. 2.3). The cladode shape has evolved to store the maximum amount of water with minimum loss (Nobel, 1994). A cross-section shows an innermost spongy tissue, with large cells to store water. During drought, water is preferentially lost from this tissue rather than from the outer photosynthetic tissue. The photosynthetic tissue is protected by a thick waxy epidermal layer that restricts water loss. *Opuntia* has Crassulacean Acid Metabolism (CAM) plants (Gibson and Nobel, 1986), with stomata that open at night for exchange of carbon dioxide and water. The carbon

Fig. 2.3. Prickly pear (*Opuntia ficus-indica* (L.) Mill.) showing plant form, cladodes with buds, fruit and flower. Used with permission of Catarina Garcìa, Madrid.

dioxide is stored as acids to be fixed into sugars during the day via photosynthesis. This approach means that water loss by transpiration is dramatically reduced and the plant has a high water use efficiency.

The cladode's meristematic unit is the areole (Gibson and Nobel, 1986), which are helically arranged on the cladodes and can develop into either branches, flowers or roots. About 7–11 areoles are present on stems per diagonal row across the mid-stem segment; they are rhombic to subcircular and may be covered with brownish multicellular hair or trichomes having a wooly appearance (Fig. 2.4). Clusters of spines in most cacti grow out of the areoles. The spines are not easily detachable but in *Opuntia* the areoles give rise to the glochids that have small, detachable barb-tipped bristles (Anderson, 2001). Glochids are small and are present along the adaxial margin of the areoles under an inconspicuous tuft of yellowish hair.

The prickly pear has rudimentary deciduous cylindrical leaves in their juvenile stages that like the cladodes are adapted to withstand water loss and drought conditions. The shallow root system is extensive, spreading just below the soil surface, and readily absorbs water, even during light rain showers.

Flowers

The cup-shaped hermaphrodite flowers are yellow to orange and about 7 cm long. The hypanthium, made up of the fused sepals, petals and stamens, is broadly cylindrical, with numerous raised wooly areoles spirally arranged and filled with glochidia.

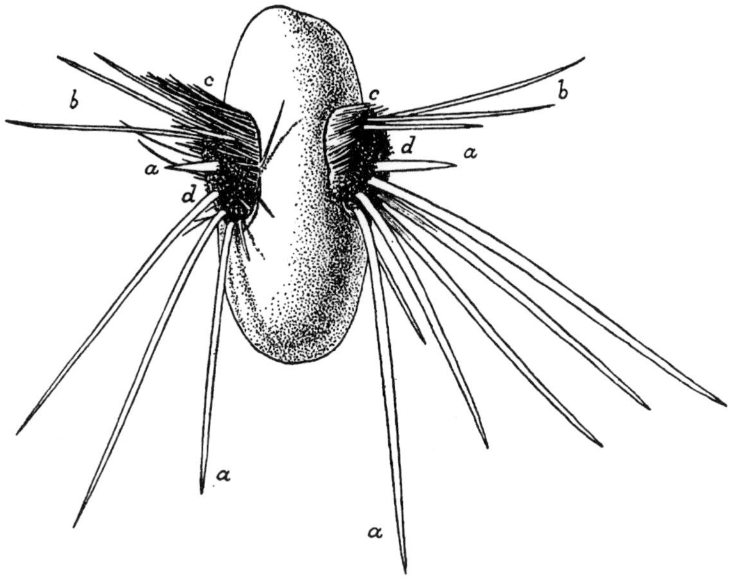

Fig. 2.4. Cacti are frequently defined by leaves that are rudimentary and function for only a few days, and the presence of specialized structures called pulvini or areoles, which are sunken buds with spines, growing points and glands. Two areoles on the edge and side of cladode showing (a) spines, (b) bristles, (c) spicules and (d) wool (from Griffiths and Thompson (1929).)

Floral buds occur on mature cladodes that are at least 6 months old (Pimienta-Barrios, 1990, 1993). Flowers normally appear in spring or early summer following a cool period. Low temperatures (15°C/5°C) favor fruit in detached cladodes, while high temperatures (35°C/25°C) favor daughter cladodes (Nobel and Castañeda, 1998). This supports an earlier finding that exposure of detached cladodes to constant temperatures of 10°C for 3 weeks led to more flower production than detached cladodes exposed to 35°C (Gutterman, 1995). Under continuous fertigation (N, P, K) increased production of floral buds in both the winter and spring crops occurred, but to a greater extent in the winter crop (Nerd *et al.*, 1991). Winter fruit command a higher price than spring fruit. The flowers are insect pollinated, but bagged flowers do set fruit and are therefore capable of self-pollination.

Most flowers open in the late morning (type A) while some open in the afternoon (type B). At night, all flowers are closed and this marks the end of anthesis in type-A flowers. Anthesis resumes in type B the following morning and the flowers finally close in the evening (Pimienta-Barrios, 1990).

Fruit

The ellipsoid fruit is a berry and weighs 100–250 g and is 5–10 cm long. A thick, fleshy skin surrounds a juicy pulp that contains many (80 to > 300) hard-coated seeds (5 mm long). Parthenocarpy has been reported in a yellow- fleshed accession from Israel though fruit quality was poor. Fruit development takes 110 days from bud differentiation and 80 days after anthesis (De la Barrera and Nobel, 2004). Drought during development can significantly reduce fruit growth.

Cultivar development

Genetics and cytogenics

Natural hybridization of *Opuntia* spp. is common, which has led to variability. The chromosome of *O. ficus-indica* is $2n = 88$ with $2n = 22$ being more common among other *Opuntia* spp., suggesting a polyploid nature that may be one of the major causes of diversity in plant size and vigor. Cultivated forms are reported to have from $2n = 66$ to $2n = 88$. All Mexican cultivars are thought to be the products of hybridization between *O. ficus-indica* and wild cactus pear forms.

Breeding

Opuntia flowers though hermaphrodite can be emasculated and controlled crosses made. *Opuntia* breeding has been described by Mondragón-Jacobo and Bordelón (1996) and Bunch (1997). Luther Burbank is claimed to have made some of the first crosses at the turn of the 20th century and the development of a spineless cactus. The long juvenile period of 6–8 years from seed has been a problem in long-term breeding programs.

Objectives of breeding have included such traits as fruit color, size and shape (oval or barrel-shaped), peel thickness and seediness, fruit sweetness with reduced cladode and fruit spines, and juvenility. Tolerance or resistance to black soft rot (*Erwinia carotovora*) and frost have been a focus of some programs.

Consumers prefer fruit with low seediness and spines. The presence of spines (glochids) is an issue; they can be removed after harvest, but glochid-free varieties or early shedding of the glochids would be more desirable.

Cultivars

Many local cultivars have been selected and named for their fruit color at maturity, particularly in Mexico ('Reyna', 'Cristalina', 'Naranjona', 'Chapeada', 'Amarilla Montesa' and 'Roja Pelona'), Chile ('Verde', 'Blanca'), Italy ('Gialla', 'Rossa' and 'Bianca'), Israel ('Ofer') and Spain ('Verdales', 'Morados', 'Sanguinos', and 'Blancos') (Tous and Ferguson, 1996). Many of the varietal comparisons do not give both yield and fruit quality in replicated field trials (Felker *et al.*, 2005). Significant differences have also been found in yield and sweetness as well as fruit firmness.

Cultural practices

Propagation

Seeds extracted from intact fruit germinate after about a week. A hot water treatment at 80°C then cooled to room temperature, before soaking in water overnight, enhances germination at 30–35°C in standard potting media with regular irrigation (Mondragón, 1999). After the first cladode reaches 5–10 cm the plant can be transferred to a nursery in a pot or bag, and transplanted to the field after two cladodes have developed.

The most frequent method of propagation is vegetative cladodes or cladode segments from selected plants. The segments should be partially dried in the shade for 1–2 weeks to allow "curing" of the cut surfaces. This "curing" improves rooting when subsequently planted in rows. Irrigation improves rooting if planting is done in the dry season. The juvenile period is from 2 to 5 years from planting when started from cuttings and varies with cultivar and growing conditions. Full production is reached in 7–8 years.

Field preparation

Field preparation is similar to that of other orchard crops.

Transplanting and spacing

Common planting densities range from 1,000 to 2,000 plants per ha, achieved with spacings of 5–7 m between rows and 1–2 m in a row.

Irrigation

Irrigation improves vegetative growth, fruit yield and fruit size. The use of drip and other similar low volume systems also allows "fertigation". When *O. ficus-indica* was irrigated in Chile, vegetative yields of 1.3 kg DM m^{-2} $year^{-1}$, which included 0.3 kg m^{-2} $year^{-1}$ as fruit, were obtained (Acevedo *et al.*, 1983).

Pruning

The plant can withstand heavy and continuous pruning. If pruned during the vegetative growing season it has greater bud emergence. When fruit are produced, the pruning should take place after harvest. For young vegetable cladode production the plant is maintained in the juvenile state by continuous pruning. When the plant is not pruned, vegetative growth continues until the cool season with flowers occurring at the beginning of spring.

Fruit thinning

This practice will improve fruit size and quality. It is recommended to leave no more than 8–12 fruits per cladode. Thinning should be done from 2 weeks before bloom to 2 weeks after fruit set (Inglese, 1995).

Out of season production (scozzolatura)

It is possible to delay harvest by forcing the plants to a second bloom. This is accomplished by removing flowers and cladodes of the spring flush at bloom time (*scozzolatura*), which results in a second bloom 30–40 days later and a fall harvest. This practice reduces by 30–40% the number of cladodes produced after a normal spring flush. The removal of flowers should be done simultaneously with good irrigation. This removal can be done before bloom, when it results in the highest reflowering, or after petal fall, when reflowering will be reduced in relation to the pre-bloom removal (Inglese, 1995).

In a trial done in Israel by Nerd *et al.* (1993), they found that if there is a high enough temperature after the normal harvest, a second harvest can be obtained by applying 120 kg N ha^{-1} and 100 mm of water. The crop will be 20–30% less but prices will more than offset this.

Fertilization

Fertilizers induce higher yields of fruits and cladodes (Mondragón, 1994; Karim *et al.*, 1997). Combining manures with synthetic fertilizers enhances fruit yield. Fertilization efficiency in semiarid environments is strongly influenced by soil moisture and is more effective with irrigation.

Pest management

Prickly pear moth (*Cactoblastis cactorum*), prickly pear cochineal insect (*Dactylopius opuntiae*) and the prickly pear weevil (*Metamasius opuntiae*) can be problems when there are no natural predators to these insects. These insects are present in the Mediterranean Basin but have no serious impact due to the presence of the predators. In other areas such as Australia and Hawaii, where prickly pear moth was introduced for biological control of cactus invasions, they cause major damage. Regular pruning of larvae-infested cladodes controls pest populations. Fruit flies can be a serious pest and limit fruit production.

Bacterial rot (*Erwinia carotovora*), snails and slugs can be serious problems in humid areas. Fungal diseases caused by *Alternaria*, *Rhizopus* and *Fusarium* often result in severe damage, especially to the cladodes, roots and fruit.

Harvesting and postharvest handling

Peel color is commonly used to determine timing for commercial harvest. Other indexes are: fruit size and fullness; abscission of the glochids; fruit firmness; and a flattening of the floral cavity or receptacle (Kader, 2000). The fruit are easily damaged during harvest and should be carefully handled; twisting fruit from the stem or cutting fruit are desirable. Cool to 5°C as soon as possible after harvest and, depending on variety, ripeness stage and harvest season, the fruit can be stored for 2–5 weeks at 5–8°C with 90–95% RH. Below 5°C the fruit will show chilling injury in 2 weeks; some varieties show injury at higher temperatures (10°C), especially summer-harvested fruit. High gloss waxes are sometimes used to maintain appearance.

World production and utilization

World production

Areas with significant fruit cultivation include Mexico, Malta, Spain, Sicily and the coasts of southern Italy, the Canary Islands, Greece, Libya, Tunisia, Morocco, Algeria, Egypt, Saudi Arabia, Yemen, Israel, Chile, Brazil, Turkey, as well as in Eritrea and Ethiopia. In the Northern Hemisphere, the flowers first appear in early April through the early summer, and the fruit ripen from June through October with varieties showing seasonal production differences (Fig. 2.5). A mature plant can yield 100–200 fruit (Knight, 1980) with yields ranging from 2,100 to 6,400 kg ha^{-1} depending upon varieties.

Utilization, including nutritional principles

Prickly pear is grown for the large sweet fruit ("tunas"), typically eaten chilled with the outer thick skin removed. The taste is similar to a juicy, sweet watermelon. The bright red/purple or white/yellowish flesh contains many tiny hard seeds that are usually swallowed. Fruit vary considerably in color, size and flavor. Fruits ripen from yellow to red. Sugar content varies from 12 to 17% and acidity is low (0.03–0.12% TA) (Kader, 2000). The fruits are an excellent source of vitamin C (10–40 mg 100 g^{-1}) (Table 2.1).

Jams and jellies are made from the fruit, with the product resembling strawberries and figs in color and flavor. Mexicans make an alcoholic drink called *colonche*, with similar drinks being produced in Sicily and Malta (Saenz, 2000).

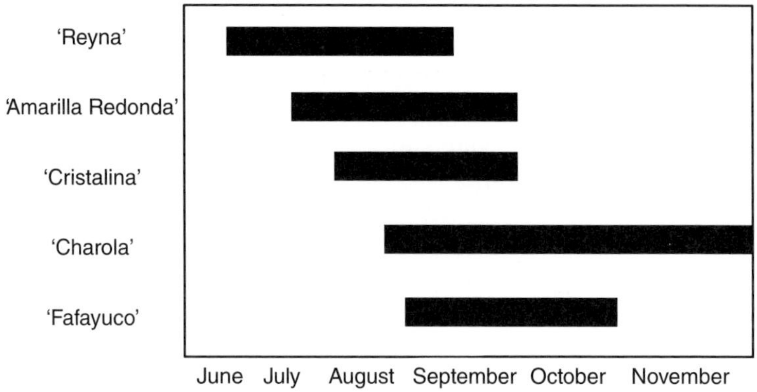

Fig. 2.5. Variation in the harvest season of different prickly pear varieties in central Mexico (redrawn from Mondragón-Jacobo, C. and Pérez-Gonzalez, S. (1996)).

The young cladodes (stem segments) are harvested before the spines have hardened, as a vegetable and as animal forage. The cladodes are sold, frequently called "nopalitos" in many local and commercial markets including the USA. During periods of drought when there is a shortage of other herbaceous plants for forage, prickly pear is a useful animal forage.

Opuntia cultivation is required for cochineal dye production from the scale insect *Dactylopius coccus*. The insect produces carminic acid, 17–24% of the dried insects' weight, that deters other insect predators. Carminic acid is extracted from the body and eggs to make carmine dye (cochineal). The plant is also used medicinally by Native Americans to treat a variety of ailments and disorders.

The *Opuntia* species are considered an invasive weed in Africa, parts of the Mediterranean Basin and Australia due to their ready ability to spread rapidly and their drought tolerance. Biological control has been obtained with the larvae of the moth *Cactoblastis cactorum* in Australia, though in other areas the moth is regarded as an invasive pest.

References and Further Reading

Mammee apple

Adams, C.D. (with contributions from Proctor, G.R., Read, R.W. and others) (1972) *Flowering Plants of Jamaica*. University of the West Indies, Mona, Jamaica.

Alix, C. and Duarte, O. (1999) *Propagación de Especies Frutales Tropicales.* CURLA; PDBL; AFE/COHDEFOR; DICTA; SETCO/PROFORFITH. La Ceiba, Honduras, p. 50.

Barbeau, G. (1990) *Frutas Tropicales en Nicaragua*. Dirección General de Técnicas Agropecuarias, MIDINRA, Editorial Ciencias Sociales, Managua, Nicaragua, pp. 175–176.

Duarte, O. and Franciosi, R. (1976) Recent advances in the propagation of some tropical and subtropical fruit species in Perú. *Acta Horticulturae* 57, 15–20.

Dunthorn, M. (2004) Cryptic dioecy in Mammea (Clusiaceae). *Plant Systematics and Evolution* 249, 192–196.

Francis, J.K. (1989) *Mammea americana* L. Southern Office – International Institute of Forestry (SO-ITF0)-SM-22. Institute of Tropical Forestry, Rio Piedras, Puerto Rico.

León, J. (1987) *Botánica de los Cultivos Tropicales*. Instituto Interamericano de Cooperación para la Agricultura, Servicio Editorial, San José, Costa Rica, 445 pp.

Martin, F.W., Campbell, C.W. and Ruberté, R.M. (1987) *Perennial Edible Fruits of the Tropics: An Inventory*. Agriculture Handbook No. 642, US Department of Agriculture, p. 30.
Morean, F. (1991) Mammee apple, an underexploited fruit in Trinidad and Tobago. *Fruits* 46, 699–702.
Morton, J. (1987) *Fruits of Warm Climates*. Creative Resource Systems, Winterville, North Carolina, pp. 304–307.
Timyan, J. (1996) *Bwa Yo: Important Trees of Haiti*. South-East Consortium for International Development, Washington, DC.
Thomas, M.B. (2008) *Mammea americana* Mamey apple. In: Janick, J. and Paull, R.E. (eds) *Encyclopedia of Fruit and Nuts*. CAB International, Wallingford, UK, pp. 267–269.
Vargas, O., Alix, C., Lobo, A.D. (authors), Duarte, O. and Sanchez, J. (technical reviewers) (1999) *Frutales y Condimentarias del Trópico Húmedo*. CURLA; PDBL; AFE/COHDEFOR; DICTA; SETCO; PROFORFITH, La Ceiba, Honduras, pp. 104–105.
Villachica, H., Carvalho, J.E.U., Muller, C.H., Diaz, S.C. and Almanza, M. (1996) Mamey. In: *Frutales y Hortalizas Promisorios de la Amazonía*. Tratado de Cooperación Amazónica, Lima, Perú, pp. 181–188.
Yahia. E.M. and Gutierrez-Orozco, F. (2011) Mamey apple (*Mammea americana* L.). In: Yahia, E.M. (ed.) *Postharvest Biology and Technology of Tropical and Subtropical Fruits*, Volume 3. *Cocona to Mango*. Woodhead Publishing Ltd., Cambridge, UK, pp. 474–481.

Achachairú

Ardaya, B.D. (2009) *El Cultivo de Achachairu* Garcinia humilis: *Manual de Recomendaciones*. Centro de Investigacíon Agricola Tropical (CIAT), Santa Cruz, Bolivia, 101 pp.
Ardaya, D., Vargas, M.A. and Kodera, Y. (1995) *El Cultivo del Achachairú*. Centro de Investigación Agrícola Tropical (CIAT) y Agencia de Cooperación Internacional del Japón (JICA), Santa Cruz de la Sierra, Bolivia.
Duarte, O. (2011) Achachairú (*Garcinia humilis* (Vahl) C.D. Adam). In: Yahia, E.M. (ed.) *Postharvest Biology and Technology of Tropical and Subtropical Fruits*, Volume 3. *Acai to Citrus*. Woodhead Publishing Ltd., Cambridge, UK, pp. 48–53.
Duarte, O. and Paull, R.E. (2008) *Rheedia laterifolia* Achachairú. In: Janick, J. and Paull, R.E. (eds) *Encyclopedia of Fruit and Nuts*. CAB International, Wallingford, UK, pp. 272–273.
Germplasm Resources Information Network (GRIN) [Web Database] (2006) US Department of Agriculture (USDA) Agricultural Research Service (ARS), National Genetic Resources Program, National Germplasm Resources Laboratory, Beltsville, Maryland. Available at: http://www.ars-grin.gov/cgi-bin/npgs/ html/splist.pl?4842 (accessed 2 August 2011).
Morton, J. (1987) *Fruits of Warm Climates*. Julia E Morton, Miami, Florida, pp. 309–310.
NotiCIAT (2009) CIAT patenta variedad de achachairú. Centro Internacional de Agricultura Tropical, Santa Cruz, Bolivia, NotiCIAT *Año* III(4), 4.
Vasquez, R. and Coimbra, S. (2002) *Frutos Silvestres Comestibles de Santa Cruz*, 2nd edn. Editorial FAN, Santa Cruz de la Sierra, Bolivia, 265 pp.

Prickly pear (*Optunia ficus-indica*)

Acevedo, E., Badilla, L. and Nobel, P.S. (1983) Water relations, diurnal acidity changes, and productivity of a cultivated cactus, *Opuntia ficus-indica*. *Plant Physiology* 72, 775–780.
Anderson, E.F. (2001) *The Cactus Family*. Timber Press, Portland, Oregon.
Bunch, R. (1997) Update on cactus pear breeding program and new products at D'Arrigo Bros. *Journal of the Professional Association for Cactus Development* 2, 60–65.
Cavagnaro, J.B. and Trione, S.O. (2007) Physiological, morphological and biochemical responses to shade of *Trichloris crinita*, a forage grass from the arid zone of Argentina. *Journal of Arid Environments* 68, 337–347.

De la Barrera, E. and Nobel, P.S. (2004) Carbon and water relations for developing fruits of *Opuntia ficus-indica* (L.) Miller, including effects of drought and gibberellic acid. *Journal of Experimental Botany* 55, 719–729.

Felker, P., Rodriguez, S. del C., Casoliba, R.M., Filippini, R., Medina, D. and Zapata, R. (2005) Comparison of *Opuntia ficus-indica* varieties of Mexican and Argentine origin for fruit yield and quality in Argentina. *Journal of Arid Environments* 60, 405–422.

Gibson, C.G. and Nobel, P.S. (1986) *The Cactus Primer*. Harvard University Press, Cambridge, Massachusetts.

Griffith, M.P. (2004) The origins of an important cactus crop, *Opuntia ficus-indica* (Cactaceae): new molecular evidence. *American Journal of Botany* 91, 1915–1921.

Griffiths, D. and Thompson, C.H. (1929) *Cacti*. United States Department of Agriculture Circular No. 66. USDA, Washington, DC, 25 pp.

Gutterman, Y. (1995) Environmental factors affecting flowering and fruit development of *Opuntia ficus-indica* cuttings during the three weeks before planting. *Israel Journal of Plant Science* 43, 151–157.

Kader, A.A. (2000) Cactus (Prickly) Pear: Recommendations for Maintaining Postharvest Quality. Produce Facts. Available at: http://postharvest.ucdavis.edu/PFfruits/CactusPear/ (accessed 11 August 2014).

Karim, M.R., Felker, P. and Bingham, R.L. (1997) Correlations between cactus pear (*Opuntia* spp.) cladode nutrient concentrations and fruit yield and quality. *Annals of Arid Zones* 37, 159–171.

Knight, R. (1980) Origin and world importance of tropical and subtropical fruit crops. In: Nagy, S. and Shaw, P.E. (eds) *Tropical and Subtropical Fruits. Composition, Properties and Uses*. AVI, Westport, Connecticut, pp. 1–120.

Mondragón, J.C. (1994) Fertilización del nopal tunero. In: *Memoria sobre Aportaciones Técnicas y Experiencias de la Producción de Tuna en Zacatecas*. Zacatecas, México.

Mondragón, J.C. (1999) Preliminary genetic studies on cactus pear (*Opuntia* spp., Cactaceae) germplasm from Central Mexico. PhD dissertation, Purdue University, West Lafayette, Indiana. Abstract available at: http://search.proquest.com/docview/304523063 (accessed 11 August 2014).

Mondragón-Jacobo, C. and Bordelón, B.P. (1996) Cactus pear (*Opuntia* spp., Cactaceae) breeding for fruit production. *Journal of the Professional Association for Cactus Development* 1, 19–35.

Mondragón-Jacobo, C. and Perez-Gonzalez, S. (1996) Native cultivars of cactus pear in México. In: Janick, J. (ed.) *Progress in New Crops*. ASHS Press, Arlington, Virginia, pp. 446–450.

Nerd, A., Karady, A. and Mizrahi, Y. (1991) Out-of-season prickly pear: fruit characteristics and effect of fertilization and short droughts on productivity. *HortScience* 26, 527–529.

Nerd, A., Mesika, R. and Mizrahi, Y. (1993) Effect of N fertilizer on autumn floral flush and cladode N in prickly pear (*Opuntia ficus-indica* (L.) Mill.). *Journal of Horticultural Science* 68, 337–342.

Nobel, P.S. (1994) *Remarkable Agaves and Cacti*. Oxford University Press, New York.

Nobel, P.S. and Castañeda, M. (1998) Seasonal, light and temperature influences on organ initiation of unrooted cladodes of the prickly pear cactus *Opuntia ficus-indica*. *Journal of the American Society for Horticultural Science* 123, 47–51.

Nyffeler, R. and Eggli, U. (2010) A farewell to dated ideas and concepts: molecular phylogenetics and a revised suprageneric classification of the family Cactaceae. *Schumannia* 6, 109–149.

Pimienta-Barrios, E. (1990) *El Nopal Tunero*. Libros de Tiempos de Ciencia. Volume 5. Universidad de Guadalajara, Jalisco, México. 246 pp.

Pimienta-Barrios, E. (1993) Vegetable cactus pear (*Opuntia* spp.). In: Williams, J.T. (ed.) *Underutilized Crops, Pulses and Vegetables*. Chapman and Hall, London, pp. 177–191.

Saenz, C. (2000) Processing technologies: an alternative for cactus pear (*Opuntia* spp.) fruits and cladodes. *Journal of Arid Environments* 46, 209–225.

Saenz-Hernandez, C., Corrales-Garcia, J. and Aquino-Perez, G. (2002) Nopalitos, mucilage, fiber, and cochineal. In: Nobel, P.S. (ed.) *Cacti: Biology and Uses*. University of California, Berkeley, California, pp. 211–234.

Tous, J. and Ferguson, L. (1996) Mediterranean fruits. In: Janick, J. (ed.) *Progress in New Crops*. ASHS Press, Arlington, Virginia, pp. 416–430.

3 Myrtaceae

The Myrtaceae or myrtle family comprises many genera of fruit trees and shrubs. Five of the most important are: *Psidium*, which includes the guava; *Eugenia*, with several species from America such as the Surinam cherry; *Syzygium* from Asia, which includes the Malay apple; *Feijoa* or *Acca*; and *Myrciaria*, which includes the jaboticabas and other fruits. Cloves, allspice and eucalyptus are also in this family. The number of genera in the family has been reported as being from 130 to 150, with more than 5,600 species in tropical and warm temperate areas of the world.

Jaboticaba

The common name for this Brazilian fruit, jaboticaba, is used for several species in the genus *Myrciaria* (Myrtaceae), especially *M. cauliflora* and *M. jaboticaba*. The common name in English, other than the Spanish jaboticaba, is Brazilian tree grape. The Portuguese names for *M. cauliflora* are: *sabara jaboticaba*, *jabuticaba sabara*, *jabuticaba de Campinas*, *guapuru*, *guaperu*, *hivapuru* and *ybapuru*. For *M. jaboticaba* (Veil.) O. Berg, the names are: *gran jaboticaba*, *jabuticabeira*, *jaboticaba de São Paulo*, *jaboticaba do mato*, *jaboticaba batuba* and *jaboticaba grauda*. Less important species, including *M. tenella* Berg, are called *jaboticaba macia*, *camboi*, *cambui*, *guayabo colorado*, *cambui preto*, *murta do campo* and *camboinzinho*, and *M. trunciflora* Berg, the long-pedicelled jaboticaba, is called *jaboticaba de cabinho*, *jaboticaba do Para* and *jaboticaba rajada*. The name jaboticaba comes from the Tupi language in reference to the fat obtained from a small tortoise, called *jabitim* or *jabuti*, that resembles the pulp of this fruit.

Important genera and species

The classification of the genus *Myrciaria* is somewhat controversial. The same common name is used for different species depending upon region (Pereira *et al.*, 2005).

Morphological and molecular markers are now being used to characterize the species. Other non-jaboticaba species with edible fruit include *M. dubia* (HBK) McVaugh, known in English as camu camu. A related species also with edible fruit is *M. floribunda* Berg, known as rumberry or guava berry. Two other species are *M. glomerata* O. Berg, yellow jaboticaba, and *M. vexator* McVaugh, called blue grape or false jaboticaba, from Panama and Venezuela.

Origin and distribution

All jaboticaba species are native to central and southern Brazil. *M. jaboticaba* grows wild in the forest around São Paulo and Rio de Janeiro, while *M. tenella* occurs in the arid zone of Bahia and the mountains of Minas Gerais in the states of São Paulo, Pernambuco and Rio Grande do Sul and also around Yaguaron, Uruguay. *M. trunciflora* is indigenous to the vicinity of Minas Gerais. *M. cauliflora* is native to the hilly region around Rio de Janeiro and Minas Gerais, Brazil, and also around Santa Cruz, Bolivia, Asuncion, Paraguay and north-eastern Argentina (Morton, 1987). Given the confusion in nomenclature there is no exact data on what species corresponds to certain areas, or if they are different species or cultivars, or hybrids.

Jaboticaba is not very well known outside its places of origin; although it has been distributed to other parts of the world, it has not acquired the popularity it has in Brazil. In 1926, it was introduced to Honduras where it has adapted well. It was also introduced to California, Hawaii and Florida where it is grown as an ornamental and for the fruit. It was taken to the Cauca Valley in Colombia and to South Africa. In Brazil, there is significant production but very few commercial orchards exist; most of the production comes from semi-wild trees and trees in home gardens, and hence very little information exists about is cultivation.

Ecology

Soil

The tree is adapted to a wide range of soils with the best being deep loamy rich soils with good drainage and slightly acid. In central Florida, jaboticaba has grown well on sandy soils and also in southern Florida on oolitic limestone. The tree is not tolerant of salty or poorly drained soil. Organic matter improves growth and production.

Rainfall

Production of jaboticaba is closely related to soil moisture. It will grow well with 1,000–1,500 mm of well-distributed rainfall. If rain is evenly distributed and temperatures are not too cold, it can flower several times during the year. The plant tolerates heavy rain up to 2,700 mm with dry periods but is not suited to the hot humid tropics where it grows well but produces little fruit. During drought, the plant stops growing and severe leaf wilting or drop and fruit abscission occur. In arid areas and during the dry season, it needs to be irrigated for good production.

Temperature

Jaboticaba is considered to be a subtropical plant but it adapts well to the tropics if not excessively warm and humid. Its area of origin is subtropical with cool winters where even short periods of frosts occur. Jaboticabas are found from 21°S in the state of Minas Gerais to Rio Grande do Sul, at 30°S, above 500 m. Trees in central Florida have survived freezing and in southern Florida have not been damaged by brief periods of −3°C. It adapts to the tropics from 0 to 1,400 m. The ideal growth temperature is 15–22°C, with a yearly average of less than 22°C.

Light and photoperiod

The plant prefers full sun exposure. Multiple flowering cycles per year suggests that photoperiod is not a factor in flower induction. Soil moisture seems to be the main factor inducing flowering.

Wind

This compact tree with very crowded branches is very resistant to strong winds having also a very strong anchored root system.

General characteristics

Tree

The two most important jaboticabas, *M. cauliflora* and *M. jaboticaba*, can reach 10–12 m and sometimes more in Brazil, while in Florida or Honduras old trees have not grown more than 5 m. Both species branch profusely from near the ground, with the branches slanting upwards and outwards. This habit gives a dense, rounded crown that can spread to 14 m giving the plant a globe form where the canopy can be wider than its height. The branches are smooth and their thin outer bark flakes off and leaves light patches. The young foliage and branchlets are hairy. The evergreen leaves are opposite and on very short, downy petioles. The leaves are lanceolate or elliptic and rounded at the base with a sharp or bluntly pointed apex, 2.5–10 cm long and 1.0–2.0 cm wide with 2–4 mm petioles (Fig. 3.1). The glossy leaves have a leathery texture and the new growth has a reddish or pinkish color that changes to pale yellowish-green before becoming dark green.

Flowers

The flowers emerge from the trunks and branches (cauliflory) on so-called floral cushions in groups of four, on very short, thick pedicels. These cushions get more numerous and produce more inflorescences when branches get thicker, with 960–1,500 flowers per m on the thick branches against 30–70 on the thin branches (Duarte, 2005d). The cushions will produce flowers during the lifetime of the plant and according to climatic conditions can flower from one to seven times per year depending on temperatures and soil moisture. Flowering can occur while there are still young fruit growing on the stem. A 50-year-old tree produced 126,000 flowers in one heavy flowering. The inflorescences push through the bark and appear as a small point surrounded by numerous bracts (Moncur, 1988).

Normally 2 weeks before the flowering process begins, the bark peels off and about 12 days before anthesis a 0.5 mm point can be detected on the surface of the branch that keeps growing until reaching the popcorn stage the day before anthesis (Duarte and Huete, 2005b). Each flower has four green sepals and four hairy white petals and an average of 52 stamens (36–68) that are about 4 mm long. The flowers emit a perfume, especially in the early hours, that attracts many insect species.

In Brazil, the major flowering peak occurs in the spring (August–September) following the first rainfall, especially in areas with dry winters, with smaller flowering peaks at other times. Under tropical conditions, trees do not flower as abundantly as in the areas where the winter is cold and dry. Flowering can be brought forward with irrigation, but the flower buds must already be developed. Heavy irrigation in the dry season can lead to several crops a year. Trees in southern Florida usually produce two crops a year (Morton, 1987) and in Honduras up to seven flowerings per year have been recorded (Duarte, 2005d).

Pollination and fruit set

Pollination can be autogamous or allogamous with the latter resulting in higher production. Fruit set from isolated flowers under fine netting is 13–24% versus 18–28% for exposed flowers, indicating that cross-pollination plays a small role in fruit set. Several insects are detected visiting the flowers including honey bees (*Apis mellifera*), the most abundant insect observed, and *Trigona* bees. Some Diptera flies also visit the flowers as well as some Chrysomelidae and Formicidae. The average fruit set of several evaluations is between 10 and 24%. Fruit set during the rainy season varies from 7 to 32% versus 7–15% during the dry season (Duarte and Huete, 2005a).

Fruit

The fruit occur singly or in clusters on short stalks and are normally not visible (Fig. 3.1), being hidden by the canopy. The round, slightly oblate, broad-pyriform or ellipsoid berries have a small disk and vestiges of the four sepals at the apex. The fruit vary in size with species and variety, ranging from 6 mm in *M. tenella* to 4–5 cm in diameter in other species. The peel is 1.5–2.0 mm thick, smooth, tough and very glossy, changing from green when immature to red-purple and maroon-purple, appearing nearly black at maturity; some varieties are red or black with red stripes or even green when mature. As the fruit matures the skin gets thinner. The skin encloses a gelatinous, juicy, translucent, all-white or rose-tinted pulp that clings firmly to the seeds. Some cultivars can have fruit with aborted sccds. Fruit weight varies from 2 to 5 g. The pedicel is 3.0–4.5 mm long and fairly hard.

Fruit growth follows a simple sigmoid curve that lasts around 2 months in *M. jaboticaba* (Barros *et al.*, 1996) while in *M. cauliflora* it lasts 27–30 days depending on temperature. During the second day after anthesis the stamens shrivel and fall the next day. On the fourth day after anthesis, petals shrivel and by day 5–6 turn brown falling the following day. On days 7–8 the ovaries are wider than their height and this reverses by day 9 and stays so until the fruit ripens. By day 15, half of the seeds are pink colored and by day 19–20 all have turned pink. By day 22–23 fibers appear on the seed surface. On days 22–24 green-colored

Fig. 3.1. Jaboticaba (*Myrciaria cauliflora*) flowers and fruit on main branches, and leaves (used with permission from León, J. (2000) *Botánica de los Cultivos Tropicales*. Editorial Agroamérica, Instituto Interamericano de Cooperación para la Agricultura (IICA), San José, Costa Rica).

fruit starts to become slightly brown and by day 27–30 all fruit is dark purple, appearing almost black, and ready to be picked. There are normally 0–4 seeds per fruit with only one seed being normally present in 35–77% of the fruit (Duarte and Huete, 2005b). The seeds are oval to nearly round measuring 3–10 mm, light brown colored, medium-hard to soft in texture and are usually swallowed with the pulp. Each seed has several embryos because of nucellar embryony.

Cultivar development

No breeding activities have been reported. The most widespread species, *M. cauliflora*, produces apomictic embryos and, for this reason, shows very little genetic variability. The zygotic species *M. jaboticaba* is more variable but is a much rarer plant. The rarer zygotic jaboticatuba is apparently monoembryonic.

There are superior selections of the three most important species that have been used for many years and due to nucellar embryony or graft propagation have kept their characteristics (Table 3.1); some might be natural hybrids. In 1990, Ahrens listed a number of Brazilian varieties, for example 'Sabara', a form of *M. cauliflora* that is the most prized and most often planted. The fruit is small, thin-skinned and sweet. The tree is of medium size, precocious and very productive. This early cultivar can bear several crops a year. The susceptibility

Table 3.1. Classification of jaboticabas (from Wiltbank *et al.*, 1983).

Type	Species	Author
Large, vigorous tree, large leaves, large sessile fruit		
'Grauda', 'Grande', 'Gigante', 'Jaboticatuba'	*M. jaboticaba* Berg	Hoehne
'Paulista', 'Mineira', 'do Matto', 'Olho du Boi', 'Acu' ('Assu')	*M. jaboticaba* Berg	Hoehne
'Branca'	*M. jaboticaba* Berg	Hoehne
'Ponhema'	*M. jaboticaba* Berg	Hoehne
'Rajada'	*M. jaboticaba* Berg	Hoehne
Smaller trees, leaves and fruit, sweeter fruit, short pedicels		
'Sabara', 'Miuda', 'Murta'	*M. cauliflora* Berg	Hoehne
Small trees, fruit with long pedicel		
'de Cabinho'	*M. trunciflora*	Hoehne and Mattos

to rust on flowers and fruit limits fruiting in the rainy season. 'Paulista' has medium-large fruit with medium-thick, leathery skin; the tree shows strong growth and is highly productive though it only bears a single crop per year and the juicy and sweet fruit appear later in the season than 'Sabara' and are resistant to rust. 'Ponhema' or 'Punhema' has large fruit with a pointed apex and somewhat leathery skin, the tree is very large and extremely productive and the large astringent fruit must be fully ripe to be eaten fresh; they are mostly used to make jelly and other preserves.

Other cultivars include 'Rajada', a medium-large fruit with green and bronze color stripes in the skin, which is thinner than that of 'Paulista', which is a mid-season producer. 'Branca' produces a medium-large green-colored fruit when mature, less sweet than 'Paulista'. 'Mineira' produces once a year a medium-large round fruit, very juicy with a sweet flavor and a dark purple medium-thick skin. 'Roxa' fruit is more red then dark purple at maturity. The names 'Miuda' and 'Murta' are probably names given to the same cultivar; they are very similar to 'Sabara'. An additional cultivar is 'Coroa', with small leaves and large fruit. Other cultivars similar to 'Paulista' are 'Grauda', 'Grande', 'Gigante' and 'Jaboticatuba' as well as 'Açu' and 'Olho de Boi'.

Propagation

SEXUAL Jaboticaba has traditionally been propagated by seed (Duarte, 1982) with the disadvantage that these plants take 8–10 years to start flowering, which is one reason for a limited number of commercial orchards (Ahrens, 1990). The seeds are polyembryonic, so that the plants arising from them are genetically equal to their mothers. This is an advantage for propagating desirable material. 'Sabara', which is the best variety, produces polyembryonic seeds and the majority of the embryos are apomictic. Each polyembryonic seed produces more than one plant and germination takes 30–40 days with an average of 1.3–1.5 plants per seed and 90% germination (Duarte, 2005b,e).

Seeds can be sown in nursery trays or beds containing 50%:50% by volume of organic matter such as shredded peat moss or decomposed sawdust and sand, or 1/3 of peat moss, 1/3 of sand and 1/3 of loamy soil. Seeds should be buried about 1 cm and sown every 1–2 cm in rows 10–15 cm apart. The use of 40–50% shade is recommended. Plant emergence will start after about a month. Seedlings grow very slowly. Once established in the pots or poly-bags the seedlings can take 6–12 months to reach a diameter of 3–6 mm at 20 cm above ground and 1–2 years to be ready for grafting or for transplanting. Once plants reach 10–15 cm they are transplanted to poly-bags or pots held under shade.

Jaboticaba seeds are recalcitrant and readily lose their viability. Clean seeds in a loosely closed container can be stored for 9 months at 12°C with 80% germination declining to 54% in a year and 5% after 18 months. Seeds stored in a tightly sealed container had only 14% germination after 1 year and 0% after 18 months (Duarte, 2005f). Jaboticaba seeds have an initial moisture content of about 50%; when this declines to 20%, germination is 80% and at 8% seed moisture no germination occurs (Fig. 3.2).

ASEXUAL Asexual propagation is used as plants flower sooner by reducing the juvenile period. Rooting of semi-hardwood cuttings under mist has not been successful while hardwood cuttings treated with rooting hormones have shown 10–14% rooting after 10 months (Duarte *et al.*, 1996a). Leafy cuttings with basal wounding can be rooted (~60%) after treating with rooting hormone in 4 months under mist (Duarte *et al.*, 1996b). Air layering has so far not been successful. Tissue culture has also been tried.

Budding is difficult because of the very hard wood and the thin bark. Conventional grafting, especially cleft grafting, has been tried successfully with 62–98% success depending upon the time of the year the scion was taken (Table 3.2). Normally 2-year-old plants have to be used as rootstocks. From grafting to transplanting normally takes 1–2 years depending on the growing conditions. Grafted plants take 4–5 years to start flowering versus 8–10 years from seed. The final grafted plant size is also smaller.

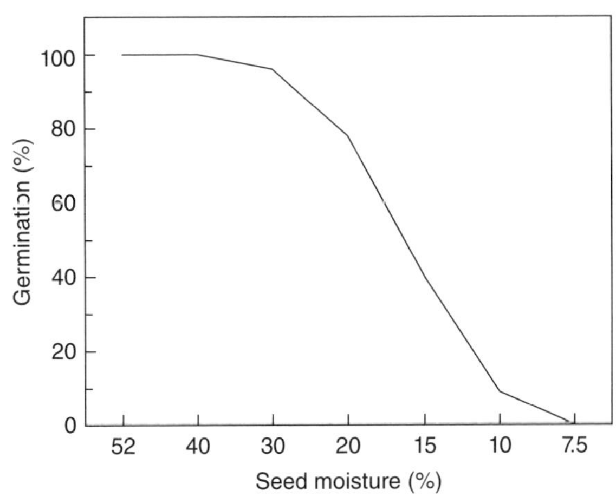

Fig. 3.2. Germination of jaboticaba (*Myrciaria cauliflora* (Mart.) Berg) seeds in relation to their internal moisture content (from Duarte, 2005e).

Table 3.2. Results of grafting methods of jaboticaba nursery trees, at different times of the year, El Zamorano, 1991–1992 (from Duarte *et al.*, 1996a).

	November	January	March	May	July	September
Cleft graft	62[a]	98[a]	70[a]	80[a]	40[a]	98[a]
Splice graft	52[b]	22[c]	52[b]	74[a]	29[a]	76[b]
Side veneer graft	29[b]	58[b]	32[c]	73[a]	21[a]	80[b]
Chip budding	4[c]	17[c]	8[d]	18[b]	4[b]	18[c]

Means followed by the same letter in a column are not significantly different by Duncan's multiple range test.

Orchard establishment

Gomes (1975) recommends plowing, harrowing and, if possible, incorporating a green manure before digging the holes. Organic matter and 200 g calcium nitrate or ammonium sulfate, plus 200 g bone meal, 100 g of triple superphosphate and 100 g potassium chloride can be added to the planting hole. Transplanting should be done preferably at the start of the rainy season, watering as soon as possible after transplanting. Plant spacing used in Brazil is 6 to 10 m depending on the growing conditions and can range from 6×4 m to 9×9 m. Initial density can be higher and later the extra trees are removed. During the first years of growth, short-cycle crops such as vegetables, beans, cassava and soybeans can grow between the trees, leaving a cleared area around the tree.

Irrigation

There is very little information available, though in arid zones or during the dry season it is recommended to irrigate two to four times per month, giving adult plants about 500–600 l of water (Anon., 1989). Irrigation promotes flowering in the dry season and is recommended in Brazil to avoid flowering in the rainy season when rust can be a problem. Irrigated trees also produce larger fruit in addition to more frequent flowering.

Pruning

Initial tree growth is slow and a seedling may take 3 years to reach 45 cm in height. It is recommended to leave the first 80 cm of the trunk free of branches and then have three primary vertical branches and later clear excess wood. Production pruning consists of eliminating damaged branches or those overcrowding the canopy to improve light and air penetration. This pruning includes removing excessively vertical interior branches or horizontal branches in older trees. Since fruit are borne on the main trunk and branches, overlapping crowns and hedging are not regarded as adversely impacting yield. Some pruning is sometimes practiced to open up the tree to allow access to the center of the tree. Tree pruning increases the size of the remaining fruit.

Fertilization

No clear guidelines are available for fertilization. Recommendations include the use of manure and slow-release fertilizers. The tree has a shallow root system and one suggestion is to dig a series of small holes around the tree and fill them with organic material. A lack of fertilizer will cause slower growth, while excess N will reduce fruit yield.

Fruit thinning

Jaboticaba flowers or fruit can be thinned to obtain larger fruit though no data has been presented to support this practice. Fruit size is expected to be dependent on stored carbohydrates and the fruit to leaf area ratio.

Pest management

DISEASES Jaboticabas do not have serious disease or pest problems in Brazil and Florida. Rust (*Puccinia psidii* Wint.) is a problem, especially during the rainy season. Trees that bloom and fruit during heavy rains are frequently affected by rust, with the variety 'Sabara' being susceptible. The initial symptoms of rust are circular spots that are at first yellow that then turn dark brown. Clearing the canopy and having greater distance between plants to improve ventilation help to reduce this problem. Other diseases reported include *Sphaceloma psidii* and *Elsinoe pitangae*, which can attack leaves. Rubelose (*Corticium salmonicolor* Berk et Br.) causes death of branches. Anthracnosis caused by *Colletotrichum gloeosporioides* Penz. can produce fruit rot. *Phyllosticta guajabae* Viegas, *Pythium debaryanum* Hasse and *Rosellinia* cause root rot, and mildew is caused by *Oidium rochae*.

INSECTS There are reports of fruit fly (*Anastrepha fraterculus* Wied.) attacks to the fruit. *Capulinia jaboticabae* Ihering attacks the branches and trunk, sucking and defoliating the plant. Aphids such as *Toxoptera auranti* Fons can attack the foliage. There are also branchlet-ringing wasps (*Trigona spinipes* Fabr.) and scales such as *Ceroplastes paranaensis* Gray. None of these insects causes serious problems. Fruit-eating birds and mammals like racoons can be a big problem, eating mature fruit. Bagging is sometimes used to protect the fruit.

Weed control

Weeds need to be controlled during the early years of tree growth by the use of hoes, machetes or weed killers. After 3 to 4 years, the tree canopy produces enough shade to suppress further weed growth under the tree.

Harvesting and postharvest handling

Yields

Adult plants can produce 30 kg year^{-1} with a total of 7–12 t ha^{-1}. In the Cauca Valley of Colombia, Sanchez *et al.* (1985) reported that the production on an 8-year-old tree was about 25 kg, while a 50-year-old isolated tree in Honduras produced 170 kg during the main harvest season.

Harvesting

Harvesting of fruit from grafted or cutting propagated trees will start between 5 and 6 years and between 8 and 10 years for seedling plants when the tree is 2–3 m tall.

In the subtropics, harvest is normally done during the spring and early summer months. In the tropics with adequate moisture and favorable temperatures there can be several harvests per year. The fruit is non-climacteric thus it is harvested at full maturity to obtain the best flavor when the fruit is dark purple or almost black. Fruit maturation is not uniform on a tree and since the fruit are so packed together on the tree, it is preferable to wait 1 or 2 days extra until all fruit are dark so that no unripe fruit are harvested.

Several methods are used for harvesting; one is to rub the fruit off the tree branch or pick them individually. Plant rubbing can be done with a broom or a stick so the fruit will drop to the ground. Putting a canvas under the canopy can help in gathering the fallen fruit. The use of a pipe with an end like a funnel and a hose at the other end can be practical with the rim of the funnel rubbing the fruit, which falls into it and rolls down the hose. The branches can also be shaken and ripe fruit will drop normally. Although the skin is tough careful handling is still recommended.

Postharvest treatment

The fruit in Brazil is now packed in cardboard boxes or cushioned trays containing 4–5 kg (Ahrens, 1990), replacing the 20 l wooden boxes (Teixeira *et al.*, 2011). Fruit can also be bulk packed with no dividers. The thick peel does protect the fruit from physical damage but not from dehydration. At room temperature, the fruit will last 2–3 days and then some fruit will develop a sour flavor in addition to gloss loss. Fruit can be stored at 13–15°C and 90–95% RH with a postharvest life of 4–5 days. Coating the fruit with wax and/or wrapping in plastic increases the storage life to 21 days when held at 12°C plus 2 days at room temperature (Duarte *et al.*, 1996c). Fruit stored at less than 12°C develop chilling injury. The fruit is also very prone to absorb odors from the storage environment. Mature fruit can be successfully frozen.

Utilization

Jaboticaba is eaten mainly fresh out of hand soon after harvest for its sweet, slightly acidic and faintly spicy flavor. Squeezing the fruit between the thumb and forefinger causes the skin to rupture and the pulp to slip out for eating. The skin is high in tannins and normally not consumed. The majority of people swallow the seeds with the pulp. Some fruit can be mildly to disagreeably resinous and astringent. The fruit has a moderate level of ascorbic acid (Table 3.3). The pulp of mature 'Sabara' has a total soluble solids content that varies from 11 to 18% and a pH from 2.9 to 3.7 (Oliveira *et al.*, 2003). The sugars found are glucose, fructose and some sucrose with citric and oxalic acids.

The pulp containing the seed can be mixed with sugar and water in a blender for a few seconds to make a good tasting juice. It is also used to prepare jellies, jams and nectars, and domestic liqueurs, wines and even vinegar. The fruit is 8–15% seeds; 25–35% peel and 50–65% pulp (Duarte, 2005c).

The high tannin-containing sun-dried skins are traditionally used to make an astringent decoction as a treatment for hemoptysis, asthma and diarrhea, and

Table 3.3. Composition of fruits of jaboticaba (Morton, 1987), camu camu, araza (Villachica *et al.*, 1996), pitanga (Morton, 1987) and feijoa (Vissor and Burrows, 1983) per 100 g edible portion.

Proximate	Jaboticaba	Camu camu	Araza	Pitanga	Feijoa
Moisture (g)	87.1	94	90	85.4–90.7	
Energy (kcal)	45.7	17	39.8	43–51	43
Protein (g)	0.11	0.5	6.0–10.9	0.84–1.01	0.8
Lipid (fat) (g)	0.01		0.5–3.8	0.4–0.88	0.3
Fiber(g)	0.08	0.6	5.5–6.5	0.34–0.6	4.3
Carbohydrates (g)	12.58	4.7	70.0–80.6	7.9–12.5	
Ash (g)	0.2	0.2	3.4	0.34–0.5	
Minerals					
Calcium (mg)	6.3	27	0.16–0.22	9	5
Magnesium (mg)			0.08–0.12	0.2	6
Potassium (mg)			1.83–2.47		130
Iron (mg)	0.49	0.5			0.09
Phosphorus (mg)	9.2	17	0.09	11	12
Vitamins					
Thiamine (mg)	0.02	0.01		0.03	0
Riboflavin (mg)	0.02	0.04		0.04	0.05
Niacin (mg)	0.21	0.06		0.03	1.85
Carotene (mg)			7.75		30
Ascorbic acid (mg)	22.7	2290	7.7–74.0	20–30	27

gargled for chronic inflammation of the tonsils. It is a popular ornamental tree common in parks and gardens throughout Rio de Janeiro. In Florida, it is planted as a backyard or roadside tree. Trunks from old trees are used for railroad building because of its hard wood. Another use of jaboticaba is to obtain dyes from the purple anthocyanin in the fruit peel (Trevisan *et al.*, 1972).

Camu Camu

Myrciaria dubia (HBK) McVaugh is another species with edible fruit in the Myrtaceae. It was also classified as *M. paraensis* Berg (McVaugh, 1958, 1963). Other synonyms are: *M. spruceana*, *M. divaricata* and *Psidium dubium*. The plant is known in English as camu camu, in Portuguese as *araza de agua* and *cacari*, while in Spanish it is *camo camo* and *camu camu*. The fruit is one of the richest sources of vitamin C.

Important genera and species

A closely related species with edible fruit is *M. floribunda* Berg (synonyms *M. protracta* O. Berg, *Eugenia floribunda* West ex. Willd.). This species is known as rumberry or guavaberry and is found throughout the Caribbean and in Central and northern South America. Some authors consider this as the arboreal camu camu that lives in more inland areas where flooding is rare while the bushy type lives close to the rivers where it is flooded during almost half the year and is the true camu camu. This arboreal type has not been fully classified and may not be the same species as it differs in fruit color and size, seed number, color, size and form, and ascorbic acid content. Two other species are *M. glomerata* O. Berg (Basionym: *Plinia glomerata* (O. Berg) Amshoff), the yellow jaboticaba and guayabito of Minas Gerais and Rio de Janeiro, and *M. vexator* McVaugh, called blue grape or false jaboticaba, from Panama and Venezuela.

Origin and distribution

Camu camu is concentrated along the Ucayali and Amazon river basins in Peruvian territory where it grows along the river banks, around lagoons and smaller tributaries of these rivers. It is also found in the lower Peruvian river basins of the Marañón and Napo rivers and further down the Amazon basin as well as the upper Orinoco basin but in fewer numbers (Villachica, 1996). The species is found in northern and western South America in Bolivia, Brazil, Colombia, Ecuador, Peru and Venezuela. It has not been spread widely outside of South America. Cultivation has started in Peru, which has become the main producer and exporter of frozen pulp and other products.

Ecology

Soil

The tree is found in shallow lagoons and on Amazonian river banks where the lower parts of the plant can be underwater for several months during the rainy season (Cavalcante, 1979). Under these flooded conditions, silt from eroded Andean soils is deposited from the rivers. These silt deposits are very fertile. The tree also adapts well to drier non-flooded soils with both good and poor drainage. It tolerates poor acid soils but yields are lower (Villachica *et al.*, 1996).

Rainfall

This small bushy tree (3–5 m) withstands flooding and requires a continuous water supply. It performs well in places with 1,500–4,000 mm rainfall, but it can grow under less rainfall with irrigation. It grows successfully in Honduras with 5–6 months of 1,000 mm rainfall supplemented with irrigation during the dry season.

Temperature

The tree grows naturally in hot, damp tropical climates where average temperatures are 25°C or higher, with average maximums of 28–35°C, and average minimums of 20°C. It will survive low temperatures above freezing and produce well in areas up to 800 m in the tropics where temperatures in winter can drop below 15°C.

Light and photoperiod

Camu camu needs full sunlight in exposed areas along rivers and may not grow well under shade. It can flower during different months of the year in the tropics indicating that day length is not a major factor in flowering.

Wind

The small size of the plant and the low canopy make it fairly wind resistant. Strong winds are not common in the area where it grows.

General characteristics

Tree

A small tree or shrub from 1 to 8 m, it starts branching at the base of the trunk forming a vase-shaped canopy. The trunk and limbs are hairless, light brown or reddish with bark peeling in large pieces periodically like many species in this family. The roots are deep with many root hairs. The opposite leaves are ovate-elliptical to lanceolate (Fig. 3.3), 4–12 cm long by 1.5–4.0 cm wide, with a pointed apex and rounded base, entire border and not very prominent veins and a 3–8 cm petiole (Calzada-Benza, 1980).

Flowers

Flowers occur on small axillary inflorescences containing four fragrant, subsessile flowers with four white glandular petals measuring 4 mm (Calzada-Benza,

Fig. 3.3. Camu camu (*Myrciaria dubia*) showing leaves, flower and fruit (used with permission from León, J. (2000) *Botánica de los Cultivos Tropicales*. Editorial Agroamérica, Instituto Interamericano de Cooperación para la Agricultura (IICA), San José, Costa Rica).

1980). The flower pedicel is 1–1.5 mm long and each node can have 12 flowers. The stamens are numerous (120–130), about 6–10 mm long and the ovary is inferior (Fig. 3.3).

The flowers are hermaphrodite and anthesis occurs early in the morning with flowers being receptive for about 4 h. After pollination the stamens shrivel and the corolla dries and drops the next day. The flowers are protogynous with styles appearing first, followed several hours later by the stamens when normally the stigma of the same flower is no longer receptive. Pollen from other flowers on the same plant can pollinate up to 91% of neighboring flowers in a tree (Villachica *et al.*, 1996). The camu camu has a facultative allogamy and exhibits no genetic mechanisms of incompatibility (Peters and Vásquez, 1987). Plant size, as established by their trunk diameter, has a direct relationship to the number of flowers and fruit produced; plants with a trunk diameter of 12–13 cm produce around 10,000 flowers.

Flowering starts when the plant reaches a certain size, around the second or third year in the field, and can occur more than once a year in cycles. Normally flowers develop and open first in the distal part of the branches and the upper part of the plant and later in the proximal and lower portions, so that a plant can have flowers and fruit at different stages of development simultaneously.

Flowering in the wild occurs when the Amazonian river waters diminish exposing most of the lower branches and leaves to the sunlight. This takes place around September and October with fruit maturing between December and February. In non-flooded orchards in the same area there are two flowering seasons, one in September–October and the other in March–April with fruit maturing 3–4 months later (Villachica, 1996).

Pollination and fruit set

Pollination can be by wind but most of it is done by small bees visiting the fragrant flowers for nectar. In the Peruvian Amazon, pollinating bees *Melipona fuscopilara* and *Trigona portica* are involved (Peters and Vásquez, 1987). The pollination percentage is 42–47%, resulting in 36–43% fruit set after natural fruit drop in wild trees.

Fruit

The fruit is a small, round, 1–3 cm berry (Fig. 3.3), weighing 6.3–8.8 g, with a pink to red to wine-red color when ripe (Cavalcante, 1979). It has one to four kidney-shaped, flattened seeds, 8–15 mm by 5.5–11 mm and 0.7–2.8 g, covered by a mesh of fibers (Calzada-Benza, 1980). The pulp is soft greenish translucent and very sour due to the high concentration of citric and ascorbic acids and low sugar content.

Cultivar development

The largest diversity is found in the Peruvian Amazon. Superior types have been collected and evaluated under cultivation and proper care. The research branch of the Peru Ministry of Agriculture (INIA) has a germplasm collection of selected superior camu camu types. These plants have been installed in flooded and non-flooded or conventional fields. The selection criteria included yields of more than 25 kg and between 2,000 and 3,000 fruit per plant to obtain fruit greater than 8.5 g. These selections have been propagated by seeds and others clonally, and the best have been propagated commercially. There are no named cultivars. There is still ample room for doing more selection and evaluation.

Cultural practices

Propagation

SEXUAL This is the traditional method of propagation. Fresh seeds have an initial viability above 90%. The seeds are washed, or allowed to stand in a plastic bag for a couple of days until the remaining pulp ferments and then washed. Seeds can be stored at 10°C after partial drying in the shade. After soaking for 24 h in water seeds are sown in moist peat moss or decomposed sawdust or by putting them in a single layer between moist newspapers or absorbent paper layers. Hypogeal germination takes place in 15–20 days and the germinating seeds with emerging roots are sown in a nursery bed or trays or better trays with cells until they have five pairs of leaves or are 10 cm. The small seedlings are then transplanted to a field bed at 10 × 10 cm and after about 6 months they are about 70 cm tall and ready to be transplanted to the field. If grafting or budding is intended the plants will go to another field to be grafted (budded), where they are planted at 40 cm between plants and 60 cm between rows. Ideally plants should be raised in nursery bags or pots instead of nursery beds to reduce handling costs, and increase efficiency and success.

ASEXUAL Cross-pollination can lead to diversity and asexual propagation is desirable. Cuttings have been tried successfully in Brazil, where terminal leafy cuttings treated with rooting hormone under mist responded with about 32% rooting. Grafting or budding is carried out when the stems are 6–9 mm diameter at 30 cm from the ground. Chip budding is very effective (Enciso, 1992) and in 6–8 months the plants are ready to be transplanted.

Orchard establishment

The normal procedures will have to be followed taking the soil to a tilth, and if possible sowing a cover crop like *Arachis pintoi* to reduce the weed problem. Under irrigation conditions, the land will have to be prepared for the system to be used. In degraded soils with good drainage 300 g dolomite and phosphate rock should be added to the bottom of the planting hole.

Transplanting is normally done with bare rooted plants. Distances recommended are from 2 to 4 m between rows and 2 to 4 m between plants, resulting in 833 plants ha^{-1} at 4 × 4 m (Villachica *et al.*, 1996). At the higher densities, used to increase early income, trees are thinned to 2 × 4 m and later to 4 × 4 m.

Irrigation

Under normal conditions this is not done. If an area has a long dry period furrow, flooding, sprinkler and drip irrigation can be used to ensure that an equivalent of about 1,500 mm rainfall is applied.

Pruning

Pruning has not been studied in detail. Initially the plant tends to grow vertically so that eliminating the tip when it reaches 1 m height will induce lateral branching, slow down vertical growth and a better-formed plant will be obtained. Branches are selected according to orientation and vertical spacing. During the life of the plant all diseased, damaged or broken branches as well as those going in a wrong direction should be eliminated.

Fertilization

It is recommended that 160, 60 and 160 kg ha^{-1} of N, P_2O_5 and K_2O, respectively, be applied to adult plants per year (Villachica *et al.*, 1996). This is applied in two or three yearly applications, one just before flowering, the second after fruit set and a third could be done after the harvest. The plant is susceptible to phosphorus and potassium deficiencies in poor soils.

Pest management

DISEASES There are no reports of important diseases of camu camu. There is a dieback whose cause has not been identified.

INSECTS Insect problems have so far not become serious since the crop is not being planted extensively. The potential insect problems in the Peruvian Amazon for trees are mealybugs like *Dysmicoccus brevipes* (Hemiptera) that attack foliage and branches or the plant base in the nursery. Yellow scale (*Ceroplastes flosculoides*) is another potential problem that, if not controlled in time, can result in limb deaths similar to those produced by red scale (*Austrotachardiella sexcordata*). Black scale (*Parasaissetia nigra*) can damage plants if not controlled in time, by covering and sucking on the twigs and leaves. The camu camu weevil (*Conotrachelus dubiae*) lays eggs in the fruit and the larvae damage them. The camu camu gall fly (*Dasineura* sp.) produces leaf galls but is of little importance. Aphids like *Aphis gossypii* Glover can be a problem sometimes by damaging young shoots and new leaves. The limb cutter (*Ecthoea quadricornis* Olivier) female lays eggs on limbs and later cuts them leaving a cut similar to a sharpened pencil tip. The jumping louse (*Tuthillia cognata*) produces leaf deformation. Finally, the limb borers *Laemosaccus* sp. and *Xylosandrus compactus* can cause serious damage if not controlled. The chinch *Edessa* has been related to twig dieback and round fruit spots but it is not a big problem at this time.

Weed control

Weeds will be a problem in rainy locations so that cover crops or green manures are beneficial. The area around the plant should be as clean as possible to avoid competition. Cut weed can be used as mulch around the plants to reduce weed problems.

Harvesting and postharvest handling

Yields

Expected yields can be around 15–18 t ha^{-1} $year^{-1}$ for adult orchards. Superior adult plants can yield up to 25–30 kg each, so that the potential yield could be 20 t and, more realistically, around 12 t ha^{-1} in the case of small farmers.

Harvesting

The camu camu fruit is harvested when it starts changing color from green to red. Sometimes completely colored fruit are harvested to ensure a nice pink-colored pulp, though the fruit has a shorter postharvest life. Wild plants are harvested from boats in Amazonia between December and March. People stand in the boat and hand-pick the fruit that is put into baskets or bags. In non-flooded lands, the main harvest period is from November to May with fruit also found at other times of the year. Harvest during peak season is done every 4–5 days and every 8–10 days during the low season. Harvested fruit should be put in containers with not more than 5 kg each to avoid fruit crushing. The fruit is washed and allowed to dry before processing.

Postharvest treatment

The recommended storage conditions for ripe fruit are 12°C and 90% RH. These storage conditions can be combined with passive modified atmospheres using low-density polyethylene film that develops about 8% CO_2 and 10% O_2. Fruit that is going to be processed can be held at 5°C for 10–15 days (Hernández *et al.*, 2011). Ripe fruit stored under modified atmosphere (PVC film) at 20°C and 68% RH had lower acidity than half-ripe fruit, lower ascorbic acid, higher soluble solids (8.2 vs. 6.9) and higher anthocyanins (Silva and Andrade, 1996). Frozen pulp held at −5 to −10°C can be stored for some months (Villachica *et al.*, 1996).

Utilization

The red/purple cherry-like fruit have a very high vitamin C content of 2–3% of fresh weight in half colored fruit; this declines slightly as the fruit ripens (Table 3.3). Predominantly green fruit have 1,910 mg 100 g^{-1}, increasing to 2,061 mg 100 g^{-1} when fruit is predominantly purple (Alves *et al.*, 2002). The concentration of ascorbic acid can be as high as 4,000 mg but normally it is around 2,000–3,000 mg 100 g^{-1} in the pulp and even higher in the peel (about 5%). These vitamin C concentrations are the highest of all known fruits. The fruit also has a high concentration of phenolic compounds that impart astringency, which declines significantly as the fruit ripen but is still significant in ripe fruit. The citric acid content is about 2.6%. The pulp amounts to about 50% of fruit weight and the peel is a very small percentage.

The main use of the fruit is as a source of vitamin C, with natural vitamin C commanding a higher price than its synthetic counterpart. This has led to an expansion of acreage in the Peruvian Amazonia in order to export frozen pulp.

In the local markets, fruit is sold to make lemonade-like refreshments because of its acidity. It is also used in ice creams, marmalade and vinegar. Its acidity means that it needs to be mixed with other fruit pulps to prepare jams or marmalades. The fruit is also used as food in fish farms in the Amazon region.

Araza

Araza (*Eugenia stipitata* McVaugh), also known as *Arazá-buey* in Peru and *Araca-boi* in Brazil, is a relatively new fruit. It has potential as a fruit crop because it is very precocious, can produce on very poor soils, and the fruit, though acid in taste, has an appealing flavor and has a very good potential for processing. According to McVaugh (1956, 1958) there are two subspecies: one is ssp. *stipitata*, which is a larger tree (12–15 m) with large leaf blades (3.5–9.5 by 8–18 cm), 100–150 stamens and 70–180 g fruit; the other is ssp. *sororia*, which is almost a shrub (2–4 m) with dense foliage, abundant branches, smaller leaves (2.5–4.5 by 6.5–13 cm), 75 stamens, fruit that range from 150 to 300 g, sometimes up to 800 g, and is found in lower Ucayali, Huallaga and Marañón river basins. Chávez and Clement (1984) suggest that the ssp. *sororia* is a domesticated form of ssp. *stipitata*.

Important genera and species

Several other *Eugenia* species, most of them originated in Brazil, have edible fruits: *Eugenia uvalha* the ubaia, *Eugenia aggregata* the cherry of Rio Grande, *Eugenia dombeyi* the grumichana, and *Eugenia luschnathiana* the pitomba. Another member of this genus, described in a separate entry in this chapter, is *Eugenia uniflora*, the pitanga or Surinam cherry, a native to Surinam, Guyana, French Guiana, Uruguay and the south of Brazil (Crane, 2008).

Origin and distribution

Araza originated and was domesticated by the native peoples in the Peruvian part of the Amazon basin close to Brazil, mainly in the lower Ucayali River basin. In the last 30–40 years some research has been done and the plant has been distributed to Amazonian and subtropical areas of Brazil, Ecuador, Colombia, Bolivia and Central America (Gentil and Clement, 1996). Outside the region it is still fairly unknown. It is commercially produced in Brazil, Colombia, Ecuador, Peru and Costa Rica. The lack of market information is probably the major reason that araza production has not expanded more rapidly.

Ecology

Soil

The plant prefers well-drained loamy soils, rich in nutrients and organic matter. It tolerates well-drained poor clayey oxisols and ultisols or very poor and acid soils.

It responds well to nitrogen applications under poor soil conditions. It also adapts well to soils with high aluminum saturation.

Rainfall

In the wild it occurs in areas with an annual rainfall of 1,700–3,200 mm, but it can grow in the Atlantic coast of Costa Rica with almost 4,000 mm or in places with 1,500 mm rain. It tolerates moderate drought and will also grow with irrigation in areas with little rainfall or in arid zones such as the Peruvian coast.

Temperature

Araza prefers the hot humid tropics at altitudes below 700–800 m and wild populations are found below 400 m. In its area of origin, the average temperature ranges from 25 to 28°C (Gentil and Clement, 1996). The minimum average temperature for growth is 18°C and the maximum 30°C. Below 17°C, growth stops and resumes when temperatures rise. It does not tolerate frost.

Light and photoperiod

The plant prefers sunny locations for best performance. Apparently there is no photoperiodic influence as it flowers and produces fruit under various day lengths.

Wind

Little is known about its response to wind; however, the tree's structure and small size would help to make it fairly tolerant to winds.

General characteristics

Tree

Araza is a small tree or large bush that can go from 3 to 5 m in height (Donadio *et al.*, 2002). Branching starts close to the ground with smooth hairless branches. The canopy is rounded. Leaves are simple, opposite, sub-sessile, green in color, slightly pubescent or smooth, and elliptical in form, 8–12 cm long and 3–6 cm wide (Fig. 3.4). Veins are prominent only on the underside (Vargas *et al.*, 1999).

Flowers

Flowers appear either solitary or grouped in racemes of 4–8 flowers. The flowers have four green-yellowish sepals, four white petals and around 100 stamens. From flower button to anthesis takes approximately 3 weeks.

Pollination and fruit set

Flowers open early in the morning between 4 a.m. and 9 a.m. Up to 1,770 flowers per plant have been reported of which only 2.3% set fruit in Peru (Villachica *et al.*, 1996), while another study indicated 25% (Gentil and Clement, 1996). Flowering can occur once or several times a year or even continuously, depending on temperature and soil moisture conditions.

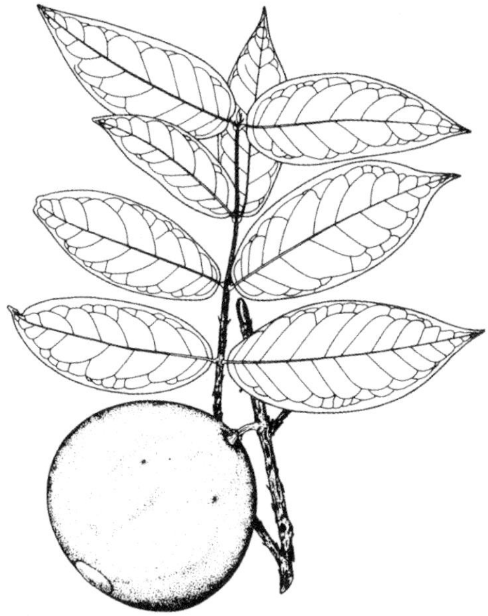

Fig. 3.4. Araza (*Eugenia stipitata*) showing leaves and fruit (used with permission from León, J. (2000) *Botánica de los Cultivos Tropicales*. Editorial Agroamérica, Instituto Interamericano de Cooperación para la Agricultura (IICA), San José, Costa Rica).

Self-pollination is possible, but most pollination is dependent on insects such as honey bees (*Apis mellifera*) and other bees (*Eulaema bombiformis*, *Eulaema mocsaru*, *Melipona lateralis*, *Melipona pseudocentris* and *Megalopta* sp.) (Villachica *et al.*, 1996).

Fruit

The fruit is a round or oval berry 2–12 cm wide by 2–10 cm, slightly compressed at the poles (Fig. 3.4). Fruit weight can vary from 50 to 600 g (Cavalcante, 1974). The immature fruit is a dull green and turns yellow then yellow-orange as it ripens. The peel is very thin and can be smooth or velvety to the touch. The pulp is acid, very aromatic and soft when ripe. It contains 1–20 seeds, with an average of ~10. The seeds are 1–3 cm long and slightly compressed and seed weight varies from 0.7 to 4.3 g (Villachica *et al.*, 1996). The largest seeds are similar to cocoa beans and have a moisture content of around 50% when the fruit matures.

Fruit growth follows a sigmoid curve and the fruit matures in 80–90 days (Galvis and Hernández, 1993). The main harvest season in Belém (Brazil) is February to May, but in other places there can be three or four harvest seasons per year or production can be almost continuous.

Cultivar development

Genetic variation is seen in field collections in Peru and Brazil with more than 20 identified ecotypes in the Peruvian Amazon. No specific breeding has been reported. Selections have been made by natives from wild trees and the research branch of the Peruvian Ministry of Agriculture's Instituto Nacional de Innovación Agraria (INIA) has collected germplasm from these ecotypes at several research stations.

There are no identified cultivars. INIA has selected some superior types that produce more than 10 kg fruit per plant, 4 years after planting. These selections are seed propagated.

Cultural practices

Propagation

SEXUAL Seed from highly productive plants is the most used propagation method. The seeds are recalcitrant and drying seeds for 5 days in the shade reduces germination from 100% to 70% (Gentil and Clement, 1996). Seeds should be extracted from mature sound fruit, cleaned and washed to remove the pulp remaining around them and then immediately planted. Fermenting the seeds with pulp for 3–4 days in a closed plastic bag makes pulp removal easier.

Superficially dried seeds can be stored up to 2 months in a moist medium or in a closed plastic bag above 15°C (Villachica *et al.*, 1996). Germination can take from 1.5 to 4 months and be completed after 9 months. The delay is due to the relatively impermeable seed coat that is difficult to remove. If the seed coat is removed, germination starts in 21 days and is completed in 4 months. Initial seedling growth is slow and it will take about 3 months to reach 7–10 cm and almost a year to attain 25–30 cm (Vargas *et al.*, 1999). Transplanting into poly-bags occurs when plants are 10–15 cm and up to a year is needed for the plants to attain sufficient size to be transplanted into the field.

ASEXUAL Cleft grafting has been tried in Brazil, using seedlings as rootstocks, and flowering occurred 8–12 months after transplanting. This offers little advantage over seeds that start fruiting in 11–12 months. Grafting has the advantages of retaining the characteristics of superior material.

Orchard establishment

In the humid tropics where this plant is normally cultivated the best practice is to clear the land and prepare the planting holes leaving a good clean area around them to avoid competition. The addition of 2 kg decomposed manure and/or 4 kg compost and 100 g Sulpomag (potassium magnesium sulfate) to the planting hole is recommended.

Plants are transplanted when they reach 30–50 cm in the poly-bags. The initial spacing recommended is 3 × 3 m (Vargas *et al.*, 1996) and is adjusted to 3 × 6 m by tree removal when the plants get larger and more space is needed between the rows (Gentil and Clement, 1996). Later the orchard can be thinned to 6 × 6 m to avoid further canopy competition for light and nutrients. Intercrops can also be used such as cassava, rice, cowpea, pineapple or other annuals during the first 2 years.

Irrigation

Irrigation is normally not needed unless it is planted in a low rainfall or arid area.

Pruning

The trees should be pruned in the nursery to leave a single stem. After transplanting, formation pruning should promote the formation of three or four main branches. Normal pruning involves the elimination of damaged or diseased material as well as branches growing in the wrong direction or in a wrong position.

Fertilization

Potassium seems to be an effective fertilizer for this crop and in acid soils phosphorus and nitrogen have produced good responses (Villachica *et al.*, 1996). The use of composted manure in the planting hole is recommended in previously used land. The recommendation is to put 10 kg composted animal manure in the planting hole, followed by chemical fertilization of 60 g N, 120 g P_2O_5 and 120 g K_2O, with a quarter applied during the first year, half in the second, three-quarters in the third and the full dose in the following years (Gentil and Clement, 1996).

Pest management

DISEASES According to Villachica *et al.* (1996) rust caused by *Puccinia psidii* can be a problem in certain areas of Brazil and Costa Rica. In Honduras, a progressive death of branches occurs and the causal agent seems to be a soil-borne fungus (Vargas *et al.*, 1999). Fernández-Trujillo *et al.* (2011) mention several postharvest fungus problems including anthracnose (*Cylindrocladium scoparium*) that causes fruit rot and is difficult to control. *Curvularia* sp. is another necrotrophic fungus that is present mainly after harvest and is difficult to control.

INSECTS Fruit flies, especially *Anastrepha oblicua* and *A. striata*, can be a serious problem. The Mediterranean fruit fly (*Ceratitis capitata*) has been detected in Costa Rica. McPhail traps should be used to capture flies and infested fruit should be picked up and buried at least 50 cm deep. A Curculionidae beetle (*Atractomerus inmigrans*) attacks araza seeds and can damage the pulp; the female lays eggs in green fruit and the larvae penetrate the seed, consume it and pupate there, making control difficult. *Plectrophoroides impressicollis* is a weevil that feeds on young foliage and flowers during the day and is favored by the planting of green manure crops where it hides, so these crops should be avoided if this pest is present (Couturier *et al.*, 1997). *Trigona branneri* stingless wasps feed on the peel, pulp and sometimes the seeds, causing damage (Villachica *et al.*, 1996). In young plants, leaf cutting ants (*Atta*) can be a problem.

Weed control

During the first and second years, araza can be grown with an associated crop like cassava, cowpea, corn and others. This will help to control most weeds between the rows. Plastic or organic matter mulch can be used.

Orchard protection

In the areas where araza is grown, strong winds rarely occur. It is a short plant with flexible branches that do not break easily. Foliage will probably be damaged. Wind barriers would be advisable in exposed locations.

Harvesting and postharvest handling

Yields

Fruit harvest can start after 18 months, and fruit yield become significant in the third year and plateau by the fifth year. Adult trees can produce between 10 and 35 kg fruit per year. In Peru, an experimental planting at 3 × 3 m produced 2.5, 9.5, 9.8, 21.5 and 40.6 t ha^{-1} from the second to the sixth year after transplanting (Villachica *et al.*, 1996). Normally lower yields should be expected from a commercial planting, probably between 10 and 20 t ha^{-1}. In Peru, Pinedo *et al.* (1981) reported 28.7 t ha^{-1} in the eighth year with individual plants yielding from 20 to 35 kg $year^{-1}$. With adequate soil moisture and warm temperatures fruit may be available all year round.

Harvesting

Fruit for the fresh fruit market should be harvested at the mature green stage just as fruit start to turn yellow. Green fruit do not ripen (Galvis and Hernández, 1993). Harvesting later than the turning stage reduces postharvest life though the fruit is more suited to processing. It takes only 2–3 days to ripen from the turning stage to the soft, full-yellow stage. Ripe fruit are too soft and difficult to transport and market. The peel is very tender and careful handling is necessary to avoid injury. Due to the rapid ripening, it is necessary to harvest every other day.

Postharvest treatment

The fruit is climacteric and ripens very quickly after harvest, emitting a strong and delicate aroma. Fruit are graded to size, stage of ripeness and to remove damaged fruit (Fernández-Trujillo *et al.*, 2011). If kept cool the postharvest life can be extended to 5 days and up to 10 days in a refrigerator. Modified atmosphere and keeping the fruit at 10°C prevents anthracnose development in harvested fruit (Fernández-Trujillo *et al.*, 2011).

Utilization

Araza pulp is very acidic and aromatic with a good content of vitamins A, B1, C, calcium, phosphorus and magnesium (Table 3.3). The peel constitutes about 8% of the fruit, the seeds about 22% and the rest is pulp (~70%). Araza's juiciness, acidity and very attractive flavor make it an excellent juice. The juice also has great potential for flavoring ice cream, sherbets and yogurt. The pulp is mainly used for making jams or jellies since it has a good content of pectic acid. Araza jams are very tasty and the addition of 90% sugar and 12% pectin of the total pulp weight is recommended. The pulp can be stored frozen, though at −12°C ascorbic acid and total carotenoids decline with storage (Andrade and Caldas, 1996). Dried pulp is also possible to prepare using artificial drying. A final potential use is the extraction of aroma from the fruit rind, since when fully ripe it has a delightful penetrating aroma. The extraction of these essential oils and other components might yield a new aroma for the perfume and cosmetic industry (Gentil and Clement, 1996).

Pitanga

Eugenia uniflora L. is named Surinam cherry, pitanga, Brazilian cherry and Florida cherry in English. Pitanga, the name given by the Tupi Indians, is used in Brazil. The most common Spanish names include: *cereza de Surinam*, *cereza de Cayena* or *pitanga*. Other Spanish names include: *guinda* in El Salvador; *arrayán* in Argentina; *cereza cuadrada* in Colombia; and *pedanga* or *pendanga* in Venezuela. In French, it is *pitanga*, *cerise caree* and *cerisier de Suriname*. In Guadelupe and Martinique *cerese a cotes* and *cerises-cotes* are used with *cerise de Cayenne*, *cerise de pays* and *cerise caree* in French Guiana. In Surinam, the name is *Surinamishe kersh* or *zoete kers*. In Portuguese, it is *ginja* or *pitanga*, in Indonesia and Malaysia *ceremai belanda* and in Thai *mayom-farang*. The Hindi name in India is Brazilian cherry and in Sri Lanka *goraka-jambo*.

Many synonyms exist for *Eugenia uniflora* L. in the older literature and include *Stenocalyx michelii* Berg, *Stenocalyx lucidus* O. Berg, *Eugenia costata* Cambess., *Myrtus brasiliana* L., *Eugenia michelii* Lam. and *Plinia rubra* Veil.

Several other *Eugenia* species have edible small fruit of limited commercial importance and are outlined above for araza.

Origin and distribution

Pitanga is indigenous to Brazil and occurs in many Brazilian states and grows wild along stream banks and at forest edges (Lorenzi, 1992). It has been expanded as far as French Guiana, Guyana, Surinam, southern Brazil and Uruguay and is grown in many parts of tropical and subtropical America. It is not very well known in Africa or Asia, although it is found in Sri Lanka, China, South-east Asia and South Africa. In the USA, it is grown in California, Hawaii and Florida, mainly as a living fence or hedge, or as a garden or backyard plant. Except for Pernambuco in Brazil, where more than 300 hectares of orchards exist (Vizzotto *et al.*, 2011), and smaller areas in other states, it is not grown elsewhere commercially for fruit production.

Ecology

Soil

Pitanga is well adapted to most soil types including heavy clay, soft limestone, sand and sandy loam. It does better if the soils are well drained, although it can stand waterlogging for some time. It does not tolerate saline soils or waters.

Rainfall

Pitanga will grow well in medium to high rainfall areas. It is tolerant to long dry seasons because of its deep root system and will grow well with irrigation in arid areas such as the Peruvian coast.

Temperature

The plants thrive from sea level to 1,800 m in Guatemala. Young plants are damaged by temperatures of −2.2°C, while adult plants can withstand temperatures as low as −6°C (Crane, 2008).

Light and photoperiod

The plant grows better under full sun and apparently there is not much photoperiodic influence.

Wind

Specific information is not available, but given its small size, slender branches and deep root system it should stand fairly strong winds.

General characteristics

Tree

This evergreen shrub or a small multi-trunked tree can reach 7–8 m and has slender branches and a spreading growth habit. Leaves are simple, opposite, ovate to ovate-lanceolate with a rounded base and short obtuse-acuminated to sharp pointed apex, and 2–6 cm long by 1–3 cm wide (Fig. 3.5). The petiole is short and the leaf color, when young, is reddish or bronze and later becomes glossy and deep green (Villachica *et al.*, 1996). In cold weather leaves turn red.

Flowers, pollination and fruit set

The flowers occur on long stalks, borne in axillary inflorescences singly or in groups of two to four. The small flowers have four white, persistent sepals and four delicate, recurved, white petals. There are 50–60 prominent white stamens with yellow anthers. Flowers appear on the basal parts of the current season's shoots or on the previous season's growth (Verheij and Coronel, 1992). Flowering may extend over 6–8 weeks. The tree can have multiple crops in a year, though usually only two, or it produces more or less continuously depending on climate and soil moisture. In Brazil, plants flower in September and are harvested in October then bloom again in December and are harvested in January. In Florida and the Bahamas, pitanga may be harvested from March to April through to June and again in September through to November. In California, fruit are harvested in late summer (Crane, 2008). In areas with cold winters, as occurs on the Peruvian coast, it normally flowers once a year at the end or in the middle of the winter and fruit mature by spring. No detailed information is available on pollination and fruit set.

Fig. 3.5. Pitanga (*Eugenia uniflora*) showing leaves and fruit (used with permission from León, J. (2000) *Botánica de los Cultivos Tropicales*. Editorial Agroamérica, Instituto Interamericano de Cooperación para la Agricultura (IICA), San José, Costa Rica).

Fruit

The fruit is a 2–4 cm wide berry with six to ten ribs and is flattened at the poles (Fig. 3.5). When mature, it turns from green to yellow then orange and finally a dark red to wine-red or dark purplish-maroon when ripe. The peel is very thin and delicate. The pulp has the same color as the peel, is aromatic, juicy and sweet or sour-sweet, and often has a resinous flavor that is not so appealing. The pulp represents about 60–65% of the total weight. The fruit can have from one to six seeds with about 40% of them having only one seed. The time from anthesis to ripening is around 40 days and overripe fruit will drop fairly soon after the full ripe stage (Villachica *et al.*, 1996). Birds and mammals disperse the seeds.

Cultivar development

Pitanga has a chromosome number of $2n = 22$ (Barbeau, 1990). There is a large genetic diversity as shown by fruit color (from red to almost black), flavor (from very sweet to sweet-sour to resinous), size (1.5–5.0 cm diameter), number and size of seeds, external fruit ribs and frost tolerance (Villachica *et al.*, 1996). Germplasm collections have been established in some agricultural research institutions in Brazil, at INPA in Manaus and at the Dirección Federal de Viçosa in Minas Gerais, in Peru at INIA in Iquitos, in Queensland Australia at the Tropical Fruit Research Station and in the USA at the Subtropical Research Station in Miami. No active program of breeding, selection and evaluation exists (Crane, 2008).

Only one named cultivar called 'Tropicana' exists in Brazil (Vizzotto *et al.*, 2011). In other parts of the world no defined cultivars have been reported. Two distinct types of pitanga are reported, those with bright red fruit and the other with dark purplish-maroon (nearly black) fruit (Morton, 1987; Villachica *et al.*, 1996). In Brazil, three types are recognized: the orange pitanga or *pitanga-laranja*; the red pitanga or *pitanga-vermelha*; and the purple pitanga or *pitanga-roxa*. All three types are found in the wild but the orange and red pitangas are rarer (Lorenzi *et al.*, 2006). The darker fruit are reported to be sweeter and less resinous.

Cultural practices

Propagation

SEXUAL This plant is normally propagated by seeds. The ideal is to take seeds from completely mature fruit, wash and clean to remove the pulp then sow them as soon as possible. Seeds are recalcitrant and will remain viable for a month or less depending on the rate of dehydration; they will not germinate when their moisture content drops below 18% (Villachica *et al.*, 1996). Germination is hypogeal and emergence of the plants will occur after about 20–25 days (Vargas *et al.*, 1999). The ideal is to sow in bags or pots and the plants should be ready in 3 or 4 months when they reach 20–25 cm. Seedling pitangas begin flowering and fruiting 2–4 years after planting.

ASEXUAL Cuttings and suckers can be rooted and air layering is also successful (Vargas *et al.*, 1999). Veneer or cleft grafting onto pitanga seedlings has also been used, as well as patch budding (Villachica *et al.*, 1996).

Orchard establishment

Normal land preparation is recommended. Pitanga may be planted when the seedlings are 6–10 months old or 30–40 cm tall. The distances used are 3–4 × 3–4 m or 5–6 × 5–6 m apart if planted as a conventional orchard or more commonly in hedgerows leaving 1–2 m between plants and 4–6 m between rows (Villachica *et al.*, 1996; Crane, 2008).

Irrigation

Irrigation is normally performed as part of the garden maintenance when planted as a hedge or in a garden. Watering improves growth, production and fruit size.

Pruning

To reduce the time to reach fruit production, the plants should not be pruned until production begins. When used as a hedge the plant is pruned like a wall, leaving foliage from the ground level to the top. Early topping induces low branching and this is continued until the hedge fills; afterwards periodic light pruning is performed to keep the hedge height and width. As a garden specimen the plant can be pruned as a tree leaving four to five limbs with a single trunk or pruned as a bush with multiple stems (Crane, 2008).

Fertilization

No specific fertilizer program has been developed for pitanga although it has been observed that periodic fertilization improves plant growth and production.

Pest management

DISEASES Although relatively disease-free, a number of diseases have been reported to attack pitanga including several leaf spots (*Cercospora eugeniae*, *Helminthosporium* sp., and *Phyllosticta eugeniae*), shoot and root dieback (*Rhizoctonia solani*, *Clitocybe tabescens*) and fruit rot (*Colletotrichum gloeosporioides*).

INSECTS Pitanga fruit are susceptible to fruit fly infestation (*Anastrepha* sp. and *Ceratitis* sp.). The most important problem in Brazil is the beetle *Costalimaita ferruginea* that attacks the foliage and can cause up to 40% defoliation (Villachica *et al.*, 1996). The larvae of *Oiketicus kirbyi* can damage the plants and occasionally an unidentified root-feeding larva can be a problem. Plants are also attacked by scales and caterpillars (Verheij and Coronel, 1992).

Weed control

In gardens or backyard orchards or fences, this will be done as a garden maintenance. In commercial orchards, care has to be taken to keep the area in the vicinity of the plant clean.

Orchard protection

Since it is not grown in commercial plantings but more frequently as a border or fence, wind protection is not provided. Its short size will make it fairly tolerant of strong winds although the leaves will be damaged or torn.

Harvesting and postharvest handling

Yield

Fruit yields generally range from 2.5 to 3.6 kg per plant per year (Villachica *et al.*, 1996; Crane, 2008). Some reports exist that claim 11 kg have been obtained from large untrimmed plants.

Harvesting and postharvest treatment

This non-climacteric fruit should be picked when fully ripe (dark red or nearly black) otherwise they may have a resinous flavor. Green mature fruit, if harvested, do not ripen fully and are of poor quality. The fruit reaches maximum total soluble solids, total titratable acidity and reducing sugars about 50 days after anthesis when fully ripe. Red fruit show higher respiratory activity and are lower in total titratable acidity than the purple type. Fruit rapidly lose water after harvest and should not be exposed to the sun. It takes about 7 days from the mature green to full ripe stage (dos Santos *et al.*, 2001).

Pitanga is a very fragile fruit and is easily damaged by abrasion and impact injuries, even with careful hand-picking. Harvest as a soft ripe fruit leads to a very short postharvest life at ambient temperatures. It can be stored at 8–10°C (Vizzotto *et al.*, 2011).

Utilization

Pitanga fruit pulp is acid to tangy sweet, sometimes mildly bitter or resinous. It is a good source of carotene and ascorbic acid (Table 3.3). Pitanga has both commercial potential as a processed fruit and as an ornamental in the landscape where it is commonly used as a small tree, shrub or hedge. The fruit is eaten fresh soon after picking although it is recommended that the resinous seeds be removed and the pulp chilled for several hours during which most of the objectionable resinous flavor dissipates thus improving flavor.

The fruit can be utilized fresh in fruit salads and processed into juices, ice cream, marmalade, jams, jellies, sauces, syrup and wine (Villachica *et al.*, 1996). It is also used for making chutneys, relish and even distilled liquor in the home. The leaves are rich in several essential oils and they are spread over the floor of homes in Brazil to repel flies with the pungent oil released when walked upon. Leaf infusions are taken against fever and stomach ache. The bark is rich in tannins that can be used in the leather industry (Morton, 1987).

Feijoa

Feijoa *Acca sellowiana* (O. Berg) Burret belongs to the section Myrtoideae of the family Myrtaceae, distinguished by their multilocular, fleshy berries (Thorp, 2008). Feijoa was first collected by Fredrich Sellow, in southern Brazil near Uruguay, and named *Orthostemon sellowianus*. Later the genus was changed to *Feijoa sellowiana* and then included in genus *Acca*, because of the similarities with the two *Acca* species (Burret, 1941). The two other *Acca* species are from the Peruvian Andes: *Acca lanuginosa* (Ruiz & Pavon ex G. Don) McVaugh, from the tropical montane forests (above 3,000 m), with a purple many-seeded fruit; and *Acca macrostema* (Ruiz et Pavon ex G. Don) McVaugh, with a similar range as *A. lanuginosa* but with small (1 cm diameter) dark burgundy red skin and fleshy fruit.

Origin and distribution

Native to subtropical southern Brazil, Uruguay, the higher parts of western Paraguay and the north-east of Argentina (Fischer, 2003), *Acca sellowiana* (Berg) Burret, feijoa, has two distinct populations based upon seed size (5–9 mm^2 versus 2–3.4 mm^2) that grow on the forest margin or as an understory tree in south-eastern

Brazil, and the southernmost states of Brazil and in Uruguay, respectively (Thorp and Bieleski, 2003). The synonyms include *Orthostemon obovatus*, *F. sellowiana*, *F. obovata*, and *F. schenkiana*. The common names for feijoa include *goiabeira-serrana*, *goiabeira do mato* and *goiaba do campo* in Brazil, *guayabo del pais* and *guayabo* in Uruguay and in English pineapple guava, Brazilian guava, fig guava, and guavasteen (Thorp, 2008).

In Brazil and Uruguay, apart from a few small plantings, the plant is not cultivated on a commercial scale, although some fruit are collected from the wild or from seedling trees in home gardens. Commercial development of feijoa has mainly occurred outside these countries, starting in Europe with France and Italy, then Israel, the USSR, California and New Zealand. Commercial development is more advanced in New Zealand where approximately 200 ha of commercial feijoa orchards have been planted producing 950 tonnes of fruit for fresh consumption and 200 tonnes for processing. Recent planting in Colombia has expanded their total area to more than 550 ha.

Ecology

Soils

Feijoa is found on nutrient-poor and acid soils not tolerated by many crops. Best yields are obtained on soils that are reasonably well drained and mildly acid (pH 6.0 to 6.5) (Thorp, 2008). Trees can survive in highly acid soils but are somewhat intolerant of alkaline soils. In California, feijoa plants growing in a soil pH above 7.0 commonly show severe leaf chlorosis due to iron deficiency, although in the Peruvian coast in sandy alkaline soils with alkaline water it does well without organic matter addition. The tree is somewhat tolerant of salinity though it slows growth and reduces yield. It can grow well in sandy soils with additions of organic matter and in heavy soils with proper drainage and better if planted on raised beds.

Rainfall

It is drought resistant and prefers areas with low humidity. The plant needs about 700–1,200 mm rain and can tolerate up to 2,000 mm if there are high sunlight levels and the relative humidity is low, to avoid fungal diseases. Dry periods with plenty of sunlight favor fruit set and minimal *Botrytis*. If there is a bimodal rain pattern, two harvests per year can be expected.

Temperature

This evergreen warm temperate/cool subtropical shrub does not tolerate too cold winters. Damage occurs below −3°C in summer and −8°C in winter (Stanley and Warrington, 1984). Frost damages fruit within a few hours and is shown as water soaked flesh that turns brown. Plants require 100 to 200 h of chilling at about 7°C to maximize flowering (Sharpe *et al.*, 1993). Pollination and fruit set are disrupted by hot (> 32°C) and dry conditions in spring. Root growth starts in spring when soil temperatures reach 8–10°C and continues until autumn when soil temperatures drop below 8°C (Thorp, 2008).

In Colombia, it is planted at 1,800–3,000 m altitude, with the range 2,100–2,600 m apparently being the best. It should be grown in the tropics in areas where temperatures are between 13 and 21°C with an average of 16°C (Fischer, 2003).

Light and photoperiod

The plant is tolerant of partial shade but prefers plenty of light so locations with more than 1,500 h of sunshine are the ideal if not combined with high temperatures or drought. In Colombia in areas with less than 1,500 h of sunlight, the other climatic conditions compensate for low sunlight and plant geometry has to be adapted for efficient light interception by forming the canopy and pruning practices.

Wind

Feijoas are very wind resistant, including to salty winds, and they are used in many cases in wind barriers, but for proper yields the orchard should be protected against strong winds. Mild winds favor drying out of the plant parts and lower fungal problems (Fischer, 2003).

General characteristics

Tree

Feijoa is a small tree or shrub up to 4 m high. New shoot growth on young trees is mainly from terminal buds, but with age terminal bud death is more common with growth occurring from lateral and sub-terminal buds leading to the mature compact, multi-stemmed spherical tree shape (Thorp, 2008). Vegetative budbreak occurs in early spring and the resulting flush of vegetative growth ceases in midsummer, about 4 months later. A second, minor growth flush of mainly lateral shoots will sometimes occur on vigorous branches, especially on young trees, in late summer. The young unbranched growth has whitish or rusty wooly hairs up to 1 mm long. Young twigs are moderate to densely wooly, becoming hairless with age. The young bark is smooth, light reddish-brown, becoming gray to light brown and flaky with age. Mature feijoa trees have a shallow, fibrous root system that generally extends horizontally to the edge of the drip line, and vertically to 0.5–1 m depth, depending on soil type.

The leaves are obovate to elliptic (3–7 cm long, 1.5–4 cm wide) and 1.5–2.3 times as long as wide (Fig. 3.6), densely wooly below, and wooly to hairless along the mid-vein above. The blade is glabrous to thinly pubescent above with an acute to rounded apex, less often abruptly acuminate or emarginate. The leaf blade's base is acute, cuneate, or rounded. The petiole has a shallow channel with the mid-vein impressed slightly above and prominent below. The lateral veins occur in 6–10 pairs and are slightly raised or impressed above, prominent below, and the secondary veins are often also prominent; marginal veins arch between the laterals, about equaling the laterals in prominence. The blades dry to a gray-green to reddish-brown, lustrous to dull above.

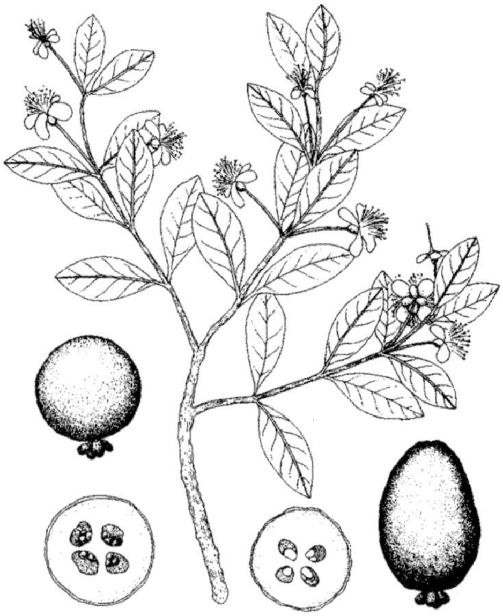

Fig. 3.6. Feijoa (*Acca sellowiana*) showing single flowers, leaves and two different shaped fruit and their transverse sections (used with permission from FAO Plant Production and Protection Series No. 26, Neglected Crops: 1492 from a different perspective, 1994 (A001/2013)).

Flower

The peduncle bears a single hermaphrodite flower and is densely wooly (0.5–3 cm long, 1–1.5 mm wide) and somewhat flattened with narrowly triangular to linear bracteoles (2.5–3 mm long, 0.5–1 mm wide). The calyx lobes are circular, oblong, or oblong-ovate (4–9 mm long, 4–8 mm wide) and wooly over most of the surface and reddish within. The inner calyx lobes are longer than the outer. The four to six fleshy petals are also almost circular to elliptic (~1.5 cm long), hairy or essentially so, glandular, and reddish to pink (Fig. 3.6). The disk is densely pubescent and 4–5 mm across. There are 60–70 stiff stamens each 1.3–2.4 cm long and dark red with ~1 mm long anthers. The style is hairless (1.7–2.7 cm long) and dark red with a 3–4 locular ovary. The septa are commonly fused for the proximal third and unfused above, the placentation thus appearing parietal on protruding T-shaped placentae distally. Each locule has 36–60 ovaries.

The juvenile period is about 4 years. The flowers are borne towards the base of the current season's shoot growth in the axils of leaf bracts. The leaf bracts form opposite each other in the over-wintering apical buds just before the emergence of new shoots. Flower buds develop along with the flush of leaf and shoot growth about 70 days after vegetative bud-break in New Zealand. Flowering continues for 4–6 weeks with each flower remaining at full bloom for 1–2 days (Thorp, 2008).

Pollination and fruit set

Feijoa is generally self-incompatible even though pollen may germinate and the pollen tubes grow to the base of the style; the barrier to self-fertilization occurs in the ovary. Some varieties are self-compatible but can also outcross. To promote outcrossing, the stigma becomes receptive to pollen germination well before the pollen on that same flower becomes ripe. Two different cultivars need to be interplanted to promote cross-pollination with a success rate of 60–90%. No nectar is produced and the flowers are not particularly attractive to bees. Pollination in New Zealand is generally carried out by birds like *Turdus merula* and *Acridotheres tristis* attracted to the brightly colored flowers and feeding on the fleshy sweet and juicy petals (Ducroquet *et al.*, 2000). In the Colombian highlands the main pollination agents are birds (*Turdus fuscater*); the advantage of this and other birds is that by eating the petals they help to control *Botrytis*, which more than offsets the possible damage they could later do to the fruit. Honey bees (*Apis mellifera*) are also mentioned (Quintero, 2003). In Brazil, pollination is also performed by birds like *Turdus* spp., *Thraupis* spp. and *Tangara* spp. Bumble bees (*Xylocopa*) also participate and are effective while normal bees are not very effective due to their small body size that does not make contact with the stigma (Manica, 2002; Donadio *et al.*, 2002). The sugar content of the petals increases just before and during flower opening, and is sweetest when the stigma is receptive to pollen germination (Stewart and Craig, 1989).

Fruit

The fruit ranges in shape from ellipsoid to ovoid, sometimes spheroid and ~2.5–10 cm long and 2.5–6 cm wide (Fig. 3.6), and is green to dark green. The ovary expands to form the flesh or pericarp with numerous small seeds embedded in a translucent pulp inside the ovary cavity. The flower sepals remain forming the calyx at the distal end of the fruit. The fruit is subacid, with oil glands just below the epidermis. The pericarp is sclerous becoming soft when ripe with clusters of stone cells that are more concentrated in the exocarp. A gelatinous substance forms around the seeds and emits a spicy odor when ripe. There are numerous seeds, or most seeds are aborted, which are more or less round, flattened and 1.9–3.4 mm long (Thorp, 2008).

Fruit growth is initially slow, with the first period of expansion beginning at the end of the main period of shoot growth in January (summer) in New Zealand, about 100 days after pollination (Thorp, 2008). Fruit growth then accelerates reaching its maximum over the last 40 days when fruit quickly reaches full size before harvest. Similarly, the total sugar level remains low until 90–100 days after flowering, but then increases rapidly during the last 30 or so days of fruit development. Flowering and fruit maturity can be 8 weeks earlier in warmer climates than in colder ones. In the same orchard, fruit maturity can vary by up to 4 weeks from one season to the next, depending on summer temperatures (Thorp, 2008). In humid, high altitude regions near the equator, for example in Colombia in regions above 1,600 m, feijoas may flower and fruit twice in one year (Ovalle and Quintero, 1992).

Cultivar development

According to Manica (2002) there are two groups of feijoa based on their geographic origin. One is the "Brazil" group that was collected in southern Brazil at altitudes of 1,200–1,600 m where frosts frequently occur; the fruit of this group have large seeds (0.45–0.6 g per 100 seeds) and the leaf underside is light green with little pubescence. The second is the "Uruguay" group, collected in that country; it has fruit with much smaller seeds (0.2 g per 100 seeds) and the leaf underside is white with a dense pubescence.

Edouard André imported feijoa to Europe from Uruguay in the 1890s (Andre, 1898) and these are the parents of nearly all feijoas that have been propagated outside South America (Thorp, 2008). The conclusion is supported by the major genotypes domesticated outside Brazil and Uruguay that have a very narrow genetic base and probably come from a common ancestor (Dettori and Palombi, 2000). Recent interest from Brazilian researchers in one of their native fruits has resulted in a significant feijoa breeding program in Santa Catarina and in the conservation of wild germplasm with its wider genetic diversity (Ducroquet *et al.*, 2000; Nodari *et al.*, 1997). Comprehensive collections of feijoa germplasm, including accessions from Brazil and Uruguay, are now established *ex situ* in Sao Joaquim (Brazil) and New Zealand. Colombia has also done some work in this direction.

Cultivars

Selection and breeding of high quality, large fruited feijoa cultivars are well advanced in New Zealand (Thorp, 2008). These selections came from materials that were first introduced to New Zealand as seed in 1908. More recent released New Zealand cultivars include 'Unique', 'Gemini', 'Apollo' and 'Opal Star' (Thorp, 2008); a more complete list of cultivars is given by Thorp and Bieleski (2002). Protected New Zealand varieties include: 'Anatoki', 'Kaitery', 'Pounamu', 'Wiki tu' and 'Opal Star'. Colombia imported 1,600 accesssions and selected 'Unique', 'Triumph', 'Mammoth', 'Apollo', 'Gemini', 'Niza', 'Tibarosa' and 'Rio Negro'. Additionally they have six new clones: 'Marion', '15-1', '9-3', 'UN', '1' and 'Caldas Colombia' (Miranda, 2003). Other cultivars are '8-4'and '41'. In California, there are some improved materials such as 'Edenvale Supreme', 'Nazemetz' and 'Trask', while in Australia they have selected 'Coolidge'and 'Choiceana' (Morley-Bunker, 2003). In Spain the cultivars 'Alpe', 'Castro-viejo' and 'Vilagarcía' are grown in Galicia. Israel grows 'Slor'. In Florida they use 'Coolidge'and 'Mammoth' (Manica 2002).

'Unique' is an early maturing fruit which is small to medium in size (up to 120 g) with rough, green to light green skin. The flesh is relatively smooth, soft and juicy. Flavor is mildly aromatic and subacid. The flowers are self-fertile. Postharvest storage life is poor because of rapid browning of the very soft flesh immediately around the seed. 'Gemini' fruit are small to medium in size (up to 120 g) and ovoid, with smooth dark green skin. Fruit mature early in the season and have a good storage life. 'Apollo' has very large fruit (up to 260 g) with moderately rough, green to light green skin. They have excellent eating quality with mildly aromatic and sweet flavor, smooth soft texture, and are very juicy. The fruit mature in mid-season and store well though are easily bruised. 'Opal Star' has medium to large fruit (up to 160 g) with smooth dark green skin. The flesh is smooth and juicy, and extends into the centre of the fruit. The fruit mature late in the season with good storage characteristics (Thorp, 2008).

Cultural practices

Propagation

SEXUAL Plants are easily propagated from seed though the progeny vary considerably due to cross-pollination. Seed extracted from ripe fruit is usually sown immediately to produce seedling rootstocks. Seeds can be washed, air-dried and stored for 2–3 years in sealed containers at 5°C. Seed is easily germinated in standard potting mix.

ASEXUAL Trees are normally clonally propagated, either from cuttings or grafted onto rootstocks, to obtain uniform quality and yields. Seedling rootstocks are ready for grafting when about 1 year old and the stems are 5–10 mm in diameter. Good results are obtained with either whip-and-tongue or cleft grafts using scion wood having the same diameter as the rootstock (Thorp, 2008).

For propagation by cuttings, evenly sized shoots are taken during late summer, when the current season's growth has almost stopped and the wood is beginning to harden. Three-node cuttings are prepared with all but the top two leaves removed. The base of the cutting is scored and then dipped in rooting hormone using a concentration suitable for semi-hardwood cuttings, before placement in propagation beds with bottom heat at 25°C and overhead misting. Root formation generally takes place within 8–10 weeks (Thorp, 2008). Another system tried in Colombia is to cut mother plants when their stem is about 2.0 cm diameter at the base to a 2–4 cm stump that will resprout and when these sprouts are about 5 cm they are detached and taken to root in a protected area (Miranda, 2003). Protocols for *in vitro* propagation have been developed (Oltramari *et al.*, 2000).

Orchard establishment

Cross-pollination is vital to achieving good fruit set and large fruit (Patterson, 1989) and is achieved by planting alternating rows of two different cultivars. In New Zealand, only 'Unique' is sufficiently self-fertile to be planted in single-cultivar blocks.

Trees are planted about 3–3.5 m apart in the row, with rows 4.5–5 m apart. Canopies will meet along the rows after 5 years with around 500–650 trees per ha. In Colombia, many orchards are planted at 4.0 m between rows and 2.5 m between plants in the row. Frequently the orchard is interplanted with banana passionfruit or *curuba* (*Passiflora mollisima*) that starts producing during the first year and lives 8–10 years and poses no real competition for the developing trees. In this case, row spacings of 5.0 m are used with feijoa spaced at 3.0 m between plants; the banana passionfruit trellis runs in the middle of the rows and plants are set at 5.0 m (Quintero, 2003).

Pruning

'Opal Star' naturally produces several lateral branches to give a compact tree shape, while other cultivars ('Apollo') have more vigorous growth with less sub-terminal branching. Cultivars with less sub-terminal branching require detailed training and pruning to achieve a compact, fruiting tree. The objective is a single-stemmed tree about 2.5 m high, with a strong and compact branching habit and thus optimum fruiting potential. At planting, the tree typically is 75 cm tall with a single strong leader and some lateral branching above 50 cm. Often, the branching pattern that develops includes one or two vigorous shoots with three or four less vigorous side shoots. More vigorous shoots are removed and well-spaced branches are selected on the main stem. The very fine branches on young trees are not removed as in most cases this wood type produces the next season's fruiting wood and crop. Mature trees are subjected to 'skirting' to remove 50–75 cm of the canopy base, especially those growing downwards from the parent branch. Tree height is controlled by two or three heavy pruning cuts each winter, towards the tree centre, on 3- to 4-year-old wood, removing the tallest or most vigorous branches limiting the height of the mature tree to about 2.5 m (Thorp, 2008). This height control allows the fruit to be hand-harvested from the ground. As with most orchard pruning schemes branch thinning is practiced to keep the canopy reasonably open. In the case of feijoa, this also allows birds to enter the canopy for pollination, and also to enhance penetration of light and orchard sprays into the tree. In Colombia the tree is shaped into a conic form and is no higher than 2.5 m (Quintero, 2003).

Fertilization

The feijoa is a slow-growing tree, with a relatively low requirement for nitrogen in relation to potassium and phosphorus. The slow growth is better served by a slow-release fertilizer than a highly soluble one (Thorp, 2008). The typical leaf and soil analyses ranges for feijoa in New Zealand are given in Table 3.4. Fertilization recommendations are to apply 25–30 kg ha^{-1} of N in the first year increasing the yearly amount gradually until reaching 120 kg

Table 3.4. Typical leaf nutrient analysis for healthy, fully expanded feijoa leaves, taken in late summer from the middle section of the current season's growth, and suggested levels for soil analyses in New Zealand (Thorp and Bieleski, 2002).

Leaf nutrients		Soil nutrients	
Macroelement	Leaf nutrient range (%)	Soil property	Suggested level
Nitrogen (N)	1.53–1.84	pH	6.0–6.5
Potassium (K)	1.03–1.86	Phosphorus	40–60 µg/ml
Phosphorus (P)	0.12–0.13	Potassium	1.0–1.2 me/100 g
Calcium (Ca)	0.99–1.7		4–5% BS
Magnesium (Mg)	0.15–0.94	Calcium	9–15 me/100 g
Sulfur (S)	0.10–0.16		55–65% BS
Sodium (Na)	0.02–0.08	Magnesium	1.7–2.5 me/100 g
Microelement	(ppm)		8–10% BS
Iron (Fe)	48–145	Cation exchange capacity	High > 25 me/100 g
Manganese (Mn)	102–202		Medium 12–25 me/100 g
Boron (B)	38–121		Low < 12 me/100 g
Zinc (Zn)	14–28		
Copper (Cu)	2–8		

after 10 years. For phosphorus, 40 kg ha^{-1} is recommended for the first year increasing to 80 kg ha^{-1} and for K 20 kg ha^{-1} during the first year reaching 100 kg ha^{-1} in adult plants (Morley-Bunker, 2003).

Irrigation

There is little detailed information available on the water requirements of feijoas. Feijoa orchards in New Zealand generally obtain sufficient moisture from rainfall for good tree and fruit growth (Thorp, 2008). In very dry summers and areas with low rainfall, irrigation is essential to obtain good fruit quality and yield. The shallow, compact root systems are easily watered with either drip irrigation or microsprinklers.

Pest management

In feijoa's natural habitat in Brazil and Uruguay there are several indigenous insect pests and diseases (Table 3.5). In commercial production outside its natural habitat, it is not particularly susceptible to either pests or diseases though a number have been reported.

DISEASES Anthracnose (*Colletotrichum gloeosporioides*) fruit rot has been identified as an important disease (Nodari *et al.*, 1997). *Phytophthora* root rot is becoming a problem in Colombia. *Botrytis* rot affects the flowers and can wipe out the whole harvest. *Alternaria* can affect stems and leaves.

INSECTS The most important insect pest is the South American fruit fly, *Anastrepha fraterculus* (Ducroquet *et al.*, 2000). Other insect pests include leafroller caterpillars that damage the surface of the fruit as well as leaves, looper caterpillars that damage flower buds and young leaves, beetles, armored scale insects, Chinese wax scale, mealy bugs, thrips and mites. In Colombia,

Table 3.5. Important pests and diseases of feijoa recorded in Brazil and Uruguay (Thorp, 2008).

Organism	Common name
Anastrepha fraterculus	South American fruit fly
Aphis gossypii	Aphid
Chrysomphalus ficus	Scale
Conotrachelus spp.	Coleoptera
Dorcadocerus barbatus	Long-horned beetle
Liothrips spp.	Thrips
Megalopyge urens	Flannel moth
Oiketicus kirbyi	Bagmoth
Phrasterothrips conducens	Thrips
Timocratica albella	Stenomid moth
Ulotingis nitor	Lace bug
Colletotrichum gloeosporioides	Anthracnose
Pestalotia psidii	Stem end collapse
Sphaceloma psidii	Leaf russet fungus

several frugivorous bats can be a problem and can be partially controlled by leaving torches in the field, using nets or burning sulfur to produce repellent smoke (Benavides and Mora, 2003).

Weed control

During the initial years when plants are small it is important to keep an area around the trunk as clean as possible; later the shade under the canopy should control weed growth. Feijoas are tolerant to some herbicides and mechanical weeding can be done. Epiphytic algae (ex. *Cephaleuros virescens*), lichens and mosses can be a problem in Colombia and occur under high humidity conditions often due to high planting densities; pruning and the use of copper fungicides (ex. Bordeaux mixture) can control this problem (Blanco, 2003).

Orchard protection

In areas with strong winds protection has to be provided for the orchard; in this case live shelter-belts are preferred as the trees can provide nesting sites for the birds required for pollination (Thorp, 2008).

Harvesting and postharvest handling

Yields

An orchard of feijoa trees can yield around 4–5 t ha^{-1} when young (2–4 years) and 20–25 t ha^{-1} when adult (Morley-Bunker, 2003) or approximately 30–40 kg of fruit per tree (Thorp, 2008). Reducing crop load, by selective thinning of fruitlets (small and misshapen fruit) soon after flowering, increases remaining fruit size. In addition, fruitlet thinning improves harvest efficiency and reduces the cull percentage by removal of most of the reject fruit at an early stage.

Harvest

Fruit are harvested before natural fruit drop, by gently pulling sideways on the fruit and harvesting those fruit that easily detach. Alternatively, harvesting nets to catch fallen fruit can be used for varieties with firmer flesh. Bruising during harvest or postharvest handling can result in serious internal damage to fruit during storage even though there are no apparent external signs at harvest or packing (Thorp and Klein, 1987). Damage from bruising is more visible in fruit with soft flesh, e.g. 'Apollo', than it is with fruit that have relatively firm flesh at harvest, e.g. 'Triumph'. Fruit of some cultivars are at eating ripeness at the time of natural fruit drop (e.g. 'Apollo' and 'Unique') while 'Triumph' and 'Gemini' fruit need to be further ripened off the tree.

Postharvest handling

The fruit show a typical climacteric pattern of ripening and ripen from the inside out (Harman, 1987). Harvested fruit are generally graded for size and shape, packed into plastic pocket packs and placed inside polyethylene-lined single-layer trays (Thorp, 2008). Recommended industry practice is to store feijoa fruit at 4°C. At this temperature fruit of the main commercial cultivars ('Apollo', 'Gemini',

'Opal Star') have a commercial storage life of approximately 4 weeks, with 5 days of subsequent shelf life at 20°C (Thorp and Klein, 1987). Though after this storage duration there are no external symptoms to indicate the end of acceptable storage life, the fruit will rapidly become unacceptable in terms of flavor and pulp color.

Feijoa fruit may suffer chilling injury when stored at 0°C. The severity of damage is cultivar-dependent and related to length of time in storage, as even long periods at 4°C may still induce off-flavors related to chilling injury. The initial external symptom of chilling injury is the appearance of sunken tissue at the stem end of the fruit. Internal browning occurs in the vascular bundles from the stem end to the calyx. Even mild chilling injury can result in the early development of off-flavors. Controlled atmosphere (CA) storage can be used to extend the storage life of feijoa fruit with low oxygen and zero carbon dioxide (Thorp and Bieleski, 2002).

Utilization

Feijoa is a delicious fruit best eaten fresh. When cut open the fruit reveal a honey-colored flesh and translucent seed pulp that is sweet, juicy and highly aromatic. The fruit is an excellent source of vitamin C and dietary fiber (Table 3.3). The resinous skin is used in certain recipes such as in chutneys. Processed feijoa products include frozen desserts, juice, wine or juice products. Freeze-dried feijoa chips are also used in breakfast cereal mixes.

An infusion is prepared from the leaves and administered in traditional medicine as a somewhat astringent drink to cure dysentery and cholera, especially in children (Thorp, 2008). Homeopathic pharmacies sell "feijoa tea" for this purpose. The more potent infusion, however, is prepared from the dried skins of feijoa fruit.

The tree, because of its attractive evergreen foliage, is planted as an ornamental. It is shaped as a hedge and used as a windbreak. The wood is hard and brittle.

References and Further Reading

Jaboticaba

Ackerman, E.D. (1978) Jaboticaba nutrition experiment. *Proceedings of the Florida State Horticultural Society* 91, 187–189.

Ackerman, E.D. (1979) Tropical fruit tree potting media experiment. *Proceedings of the Florida State Horticultural Society* 92, 273–275.

Ahrens, M.J. (1990) Jaboticaba. In: Nagy, S., Shaw, P.E. and Wardowski, W.F. (eds) *Fruits of Tropical and Subtropical Origin, Composition, Properties and Uses*. Florida Science Source Inc., pp. 266–275.

Alexander, D. McE., Scholefeld, P.B. and Fordsham, A. (1982) *Some Tree Fruits for Tropical Australia*. CSIRO, Australia.

Alix, C. and Duarte, O. (1999) *Propagación de Especies Frutales Tropicales*. CURLA; PDBL; AFE/COHDEFOR; DICTA; SETCO/PROFORFITH. La Ceiba, Honduras, pp. 66–67.

Andersen, O. and Gomes, F.R. (1975). Propagación vegetativa de jaboticaba (*Myrciaria* sp.). *Anais do III Congreso Brasileiro de Fruticultura* 2, 423–427.

Anonymous (1989) Agora, a jabuticabeira precoce. *TodaFruta* 30, 15–17. Brasil.

Argles, G.K. (1976) *Myrciaria cauliflora* and related species - Jaboticabas. In: *The Propagation of Tropical Fruit Trees*. Horticultural Review No. 4. Commonwealth Bureau of Horticulture and Plantation Crops.

Aroeira, J.S. (1962) Dormencia e conservacao de sementes de algunas plantas fruteiras. *Experientiae* 2, 541–609.

Bailey, L.H. (1929) *The Standard Cyclopedia of Horticulture*. Macmillan, New York.

Barros, R.S., Finger, F.L. and Magalhaes, M.M. (1996) Changes in non-structural carbohydrates in developing fruit of *Myrciaria jaboticaba. Scientia Horticulturae* 66, 209–215.

Brunini, M.A., Oliveira, A.L. de, Salandini, C.A.R. and Bazzo, E.R. (2004) Influencia de embalagens e temperatura no armazenamento de jabuticabas (*Myrciaria jaboticaba* (Veil) Berg) cv. 'Sabara'. *Ciencia e Tecnologia de Alimentos* 24, 378–383.

Calzada-Benza, J. (1990) *143 Frutales Nativos*. Librería El Estudiante. Lima, Perú.

Campbell, C.W. (1977) Cultivation of tropical fruits of the Myrtaceae in Southerrn Florida. *Proceedings of the Tropical Region of the American Society for Horticultural Science* 21, 3–7.

Campbell, C.W. (1988) Tropical fruits produced and marketed in Florida. *HortScience* 23(2), 247.

Chandler, W.H. (1958) *Evergreen Orchards*, 2nd edn. Lea & Febiger, Philadelphia, Pennsylvania.

Donadio, L.C. (1983) A review of South American fruits. *Proc. Tropical Region Amer. Soc. Hort. Sci.* 27(A), 21–41.

Duarte, O. (1982) Propagation methods for tropical and subtropical fruits. *Proceedings of the 21st International Horticultural Congress*, Hamburg, Germany, 1982, pp. 415–424.

Duarte, O. (1992) *La Jaboticaba; origen, distribución, cultivo y usos*. Escuela Agrícola Panamericana – El Zamorano, Honduras.

Duarte, O. (1995) Vermehrung, Blueten- und Fruchtentwicklung sowie Lagerung von Jaboticaba (*Myrciaria cauliflora* (Mart.) Berg). Doctoral Dissertation, Technical University, Berlin.

Duarte, O. (2005a) Efecto del ácido giberélico en la germinación de la semilla de jaboticaba (*Myrciaria cauliflora* (Mart.) Berg). *Proceedings of the Interamerican Society for Tropical Horticulture* 49, 152–154.

Duarte, O. (2005b) Tratamientos para una mejor germinación de la jaboticaba (*Myrciaria cauliflora* (Mart.) Berg) en vivero. *Proceedings of the Interamerican Society for Tropical Horticulture* 49, 155–156.

Duarte, O. (2005c) Cambios en la composición química durante el desarrollo y almacenaje y características físicas de los frutos maduros de jaboticaba (*Myrciaria cauliflora* (Mart.) Berg). *Proceedings of the Interamerican Society for Tropical Horticulture* 49, 160–161.

Duarte, O. (2005d) Comportamiento reproductivo de la jaboticaba (*Myrciaria cauliflora* (Mart.) Berg). *Proceedings of the Interamerican Society for Tropical Horticulture* 49, 162–165.

Duarte, O. (2005e) Determinación del punto de inicio y finalización de la capacidad de germinación de jaboticaba (*Myrciaria cauliflora* (Mart.) Berg). *Proceedings of the Interamerican Society for Tropical Horticulture* 49, 168–170.

Duarte, O. (2005f) Ensayos para la conservación de semillas de jaboticaba (*Myrciaria cauliflora* (Mart.) Berg). *Proceedings of the Interamerican Society for Tropical Horticulture* 49, 157–159.

Duarte, O. (2007) Maduración de los frutos de jaboticaba (*Myrciaria cauliflora* (Mart.) Berg). *Proceedings of the Interamerican Society for Tropical Horticulture* 51, 53–55.

Duarte, O. and Huete, M. (2005a) Estudios de la polinización de la jaboticaba (*Myrciaria cauliflora* (Mart.) Berg). *Proceedings of the Interamerican Society for Tropical Horticulture* 49, 166–167.

Duarte, O. and Huete, M. (2005b) Fenología de la floración y fructificación de la jaboticaba (*Myrciaria cauliflora* (Mart.) Berg). *Proceedings of the Interamerican Society for Tropical Horticulture* 49, 171–174.

Duarte, O., Lüdders, P. and Huete, M. (1996a) Propagation of jaboticaba (*Myrciaria cauliflora* (Mart.) Berg) by air layers, hardwood cuttings and grafts. *Proceedings of the Interamerican Society for Tropical Horticulture* 40, 61–64.
Duarte, O., Lüdders, P. and Huete, M. (1996b) Propagation of jaboticaba (*Myrciaria cauliflora* (Mart.) Berg) by terminal leafy cuttings. *Proceedings of the Interamerican Society for Tropical Horticulture* 40, 57–60.
Duarte, O., Lüdders, P. and Huete, M. (1996c) Extending storage life of jaboticaba (*Myrciaria cauliflora* (Mart.) Berg) fruits. *Proceedings of the Interamerican Society for Tropical Horticulture* 40, 65–68.
Fouque, A. (1972) Especes fruitieres d'Amerique tropicale II. *Fruits* 27(2), 120–139.
Gomes, R.P. (1975) The Jaboticaba. In: *Fruticultura Brasileña*. Biblioteca Rural, Livraría Nobel S.A., 448 pp.
Gutierrez, L.E., Cezar W.P. Jr., Ferrari, S.E. and Guimaraes, G.L. (1976) Carbohidratos soluveis em frutos. I. Roma, Manga, Banana, Jabuticaba, Limao, Abacaxi, Laranja e Cabelhudinha. *Anais da Escola Superior Agricultura "Luiz de Queiroz"* XXXIII: 167–173.
Kennard, W.C. and Winters, H.F. (1960). *Some Fruits and Nuts for the Tropics*. Miscellaneous Publication No. 801, USDA, Washington, DC.
Knight, R.J. (1988) Miscellaneous tropical fruits grown and marketed in Florida. *Proceedings of the Interamerican Society for Tropical Horticulture* 32, 34–41.
León, J. (1987) *Botánica de los Cultivos Tropicales*. Instituto Interamericano de Cooperación para la Agricultura. San José, Costa Rica, 445 pp.
Lino, J. (1985) *Fruticultura Geral*. Univ. Federal de Vicosa, Brasil.
Litz, R.E. (1984) *In vitro* somatic embryogenesis from callus of jaboticaba (*Myrciaria cauliflora*). *HortScience* 19(1), 62–64.
Litz, R.E. (1984) Tissue culture for improvement of tropical fruits. *Florida Agricultural Research* 3(1), 26–28.
Lüdders, P. and Duarte, O. (1997) Jaboticaba - eine (sub) tropische Obstart mit Zukunft? *Erwerbsobstbau* 39, 72–73.
Machado, N.P., Coutinho, E.F. and Caetano, E.R. (2007) Plastic packages and cold storage on the postharvest preservation of jabuticaba fruits. *Revista Brasileira de Fruticultura* 29, 166–168.
Martin, F.W., Campbell, C.W. and Ruberté, R.M. (1987) *Perennial Edible Fruits of the Tropics: An Inventory*. Agriculture Handbook No. 642, USDA, Washington, DC.
Mattos, J.R. (1983) *Fruteiras Nativas do Brasil - Jaboticabeiras*. Instituto de Pesquisas de Recursos Naturais Renovaveis, Porto Alegre, Brasil, 76 pp.
McGregor, B.M. (1987) *Tropical Products Transport Handbook*. Agricultural Handbook No. 668, USDA, Washington, DC.
Mohlenbrock, R. (1979) *You Can Grow Tropical Fruit Trees*. Great Outdoors Publishing, St. Petersburg, Florida.
Moncur, M.W. (1988) *Floral Development of Tropical and Subtropical Fruit and Nut Species*. Natural Resources Series No. 8. CSIRO, Australia.
Morton, J.F. (1987) *Fruits for Warm Climates*. Media Incorporated, Greensboro, North Carolina, pp. 371–374.
Mowry, H., Toy, L.R. and Wolfe, H.L. (1953) *Miscellaneous Tropical and Subtropical Florida Fruits*. Bulletin 156, Agr. Ext. Serv., University of Florida, Florida.
Ochse, J.J. and Reark, J.B. (1950) The propagation of subtropical fruit plants by cuttings. *Proceedings of the Florida State Horticultural Society* 63, 248–251.
Ogden, M.A. and Campbell, C.W. (1982) Intergeneric and interspecific rootstock trials for jaboticaba (*Myrciaria cauliflora* (Mart.) Berg). *Proceedings of the Florida State Horticultural Society* 95, 119–121.
Oliveira, A.L de, Brunini, M.A., Salandini, C.A.R. and Bazzo, R.R. (2003) Caracterização tecnológica de jabuticabas 'Sabara' provenientes de diferentes regiões de cultivo. *Revista Brasileira de Fruticultura* 25, 397–400.

Paull, R.E. (2008) *Myrciaria* spp. Jaboticaba. In: Janick, J. and Paull, R.E. (eds) *Encyclopedia of Fruit and Nuts*. CAB International, Wallingford, pp. 536–539.

Pereira, M., Oliveira, A.L. de, Pereira, R.E. de A., Sena, J.A.D., Costa, J.R.V. da, Almeida, M. de and Goncalves, A.N. (2005) Morphologic and molecular characterization of *Myrciaria* spp. species. *Revista Brasileira de Fruticultura* 27, 507–510.

Phillips, R.L. and Goldweber, S. (1980) *The Jaboticaba*. Fruit Crops Fact Sheet No. 39. Cooperative Extension Service, Institute of Food and Agriculture Science, University of Florida, Florida.

Popenoe, W. (1974) *Manual of Tropical and Subtropical Fruits*, Facsimile of the 1920 edition. Macmillan, New York.

Sanchez, L.A., Torres, R., Salazar, R. and Ríos, D. (1985) Comportamientos de cuarenta frutales tropicales no explotados comercialmente en Colombia. *Revista I.C.A.* 20, 1–12.

Soares, N.B., Pommer, C.V., Sarmento, B.M. de M., Ribeiro, I.J.A., Araujo, A.P., Jung-Mendacolli, S. and Pereira, R.A. (2001) *Jaboticaba, Instruçoes de cultivo*. Cinco Continentes Editora Ltda, Porto Alegre, P.S. Brazil, 33 pp.

Soubihe Sobrinho, J. and Gourgel, J.T.A. (1952) Características das sementes de Myrtaceae frutiferas. *Revista de Agricultura, Piracicaba* 27, 83–89.

Sturrock, D. (1972) *Fruits for Southern Florida*. Horticulture Books Inc., Stuart, Florida.

Teixeira, G.H. de A., Durigan, M.F.B. and Durigan, J.F. (2011) Jaboticaba (*Myrciaria cauliflora* (Mart.) O. Berg. Myrtaceae). In: Yahia, E.M. (ed.) *Postharvest Biology and Technology of Tropical and Subtropical Fruits, Volume 3. Cocona to mango*. Woodhead Publishing Ltd., Cambridge, UK, pp. 246–274.

Traub, H.P. (1939) Polyembryony in *Myrciaria cauliflora*. *Botanical Gazette* 101, 233–234.

Trevisan, L.M., Bobbio, F.O. and Bobbio, P.A. (1972) Carbohydrates, organic acids and anthocianins of *Myrciaria jaboticaba* Berg. *Journal of Food Science* 37, 818–819.

Vargas, O., Alix, C., Lobo, A.D. (authors), Duarte, O. and Sanchez, J. (technical reviewers) (1999) *Frutales y Condimentarias del Trópico Húmedo*. CURLA; PDBL; AFE/COHDEFOR; DICTA; SETCO; PROFORFITH, La Ceiba, Honduras, pp 204–205.

Wiltbank, W.J., Chalfun, N.N.J. and Andersen, O. (1983) The Jaboticaba in Brazil. *Proc. Trop. Region Amer. Soc. Hort. Sci.* 27(A), 57–69.

Camu camu

Alves, R.E., Filgueiras, H.A.C., Moura, C.F.H., Araujo, N.C.C. and Almeida, A.S. (2002) Camu-camu (*Myrciaria dubia* McVaugh): A rich natural source of vitamin C. *Proceedings of the Interamerican Society for Tropical Horticulture* 46, 11–13.

Calzada-Benza, J. (1980) *143 Frutales Nativos*. Librería El Estudiante, Lima, Peru, pp. 75–80.

Cavalcante, P.B. (1979) *Frutas Comestiveis da Amazonía III.* Instituto Nacional de Pesquisas da Amazonia, Museo Paraense Emilio Goeldi, Publicaciones Avulsas No. 33, Belém, Pará, Brasil, p. 39.

Chavez, W. (1993) Camu camu. In: Clay, C.W. and Clement, C.R. (eds) *Selected Species and Strategies to Enhance Income Generation from Amazonian Forest*. FAO: Misc./93/6. Working Paper, FAO, Rome, pp. 139–146.

Couturier, G., Tanchiva, E., Cárdenas, R., Gonzales, J. and Inga, H. (1994) *Los Insectos Plaga del Camu Camu (Myrciaria dubia HBK) y del arazá (Eugenia stipitata McVaugh). Identificación y Control*. Serie Informe Técnico No. 25. INIA, Lima, Perú.

Enciso, R. (1992) *Propagación del Camu Camu (Myrciaria dubia) por Injerto*. Informe Técnico No. 18. Programa de Investigación en Cultivos Tropicales. INIA, Lima, Peru, 17 pp.

Hernández, M.S., Carrillo, M., Barrera, J. and Fernández-Trujillo, J.P. (2011) Camu-camu (*Myrciaria dubia* Kunth McVaugh). In: Yahia, E.M. (ed.) *Postharvest Biology and Technology*

of Tropical and Subtropical Fruits, Volume 3. Acai to citrus. Woodhead Publishing Ltd., Cambridge, UK, pp. 352–373.
McVaugh, R. (1958) Flora of Peru, Myrtaceae I. Field Museum of Natural History. *Botanical Series* 13(2), 736–737.
McVaugh, R (1963) Tropical American Myrtaceae II. Field Museum of Natural History. *Botanical Series* 29(8), 395–532.
Peters, C.M. and Vásquez, A. (1987) Estudios Ecológicos de Camu Camu (*Myrciaria dubia* L.) I. Producción de frutos en poblaciones naturales. *Acta Amazónica* 16/17, 161–173. Mimeographic copy.
Scaloppi, E.J. Jr., Geraldo, A., Andrade, R. and Jesus, N. (2003) Clonagem do camu-camu em Jaboticabal, SP. *Proceedings of the Interamerican Society for Tropical Horticulture* 47, 134–136.
Silva, C.T.C. and Andrade, J.S. (1996) Postharvest modifications in camu-camu fruit (*Myrciaria dubia* McVaugh) in response to the stage of maturation and modified atmosphere. *Proceedings of the Interamerican Society for Tropical Horticulture* 40, 90–92.
Villachica, H. (1996) *El Cultivo del Camu Camu (*Myrciaria dubia *H.B.K. McVaugh) en la Amazonía Peruana*. Tratado de Cooperación Amazónica, Lima, Perú, 95 pp.
Villachica, H., Carvalho, J.E.U., Muller, C.H., Diaz, S.C. and Almanza, M. (1996) Camu camu In: *Frutales y Hortalizas Promisorios de la Amazonía*. Tratado de Cooperación Amazónica, Lima, Perú, pp.76–83.
Zapata, S.M. and Dufour, J.P. (1993) Camu-camu *Myrciaria dubia* (HBK) McVaugh: Chemical composition of fruit. *Journal of the Science of Food and Agriculture* 61, 349–351.

Araza

Andrade, J.S. and Caldas, M.L.M. (1996) Quality of Araça-boi (*Eugenia stipitata* McVaugh) pulp during freezing. *Proceedings of the Interamerican Society for Tropical Horticulture* 40, 80–82.
Cavalcante, P.B. (1974) *Frutas Comestiveis da Amazonía II*. Instituto Nacional de Pesquisas da Amazonia, Museo Paraense Emilio Goeldi, Publicaciones Avulsas No. 27, Belém, Pará, Brazil, p. 26.
Chávez F., W.B. and Clement, C.R. (1984) Consideraçóes sobre o araçá-boi (*Eugenia stipitata* McVaugh, Myrtaceae) na Amazonia Brasileira. *Anais Congresso Brasileiro de Fruticultura* 7, 167–177.
Couturier, G., Tanchiya, E., Gonzales, J., Cárdenas, R. and Inga, H. (1997) Observations preliminaires sur les insectes nuisibles à l´araza (*Eugenia stipitata* McVaugh, Myrtaceae) nouvelle cultura fruitière en Amazonie. *Fruits* 51(1), 229–239.
Crane, J.H. (2008) *Eugenia uniflora*, pitanga and *Eugenia uvalha*, ubaia. In: Janick, J. and Paull, R.E. (eds) *Encyclopedia of Fruit and Nuts*. CAB International, Wallingford, UK, pp. 534–536.
Donadio, L.C., Moro, F.V. and Servidone, A.A. (2002) *Frutas Brasileiras*. Editora Novos Talentos, Jaboticabal, São Paulo, Brazil, pp. 57–59.
Fernández-Trujillo, J.P., Hernández, M.S., Carrillo, M. and Barrera, J. (2011) Arazá (*Eugenia stipitata* McVaugh). In: Yahia, E.M. (ed.) *Postharvest Biology and Technology of Tropical and Subtropical Fruits, Volume 3. Acai to citrus*. Woodhead Publishing Ltd., Cambridge, UK, pp. 98–124.
Galvis, V.A. and Hernández, M.S. (1993) Análisis del crecimiento del fruto y determinación de cosecha del arazá (*Eugenia stipitata*). *Colombia Amazónica* 6(2), 107–121.
Gentil, D.F.O. and Clement, C.R. (1996) The araza (*Eugenia stipitata*): Results and research directions. *Proceedings of the Interamerican Society for Tropical Horticulture* 40, 83–89.
McVaugh, R. (1956) Tropical American Myrtaceae I. *Fieldiana Botany* 29(3), 145–228.
McVaugh, R. (1958) Flora of Peru, Myrtaceae I. Field Museum of Natural History. *Botanical Series* 13, 736–737.

Pinedo, M., Ramírez, F. and Blasco, M. (1981) *Notas Preliminares Sobre el Arazá (*Eugenia stipitata*) Frutal Nativo de la Amazonía Peruana*. Lima, Ministerio de Agricultura y Alimentación/ I.I.C.A., Publicación Miscelánea No. 229, 58 pp.

Suquilanda, M.B. (1995) *Arazá, Manual para la Producción Orgánica*. Fundación para el Desarrollo Agropecuario, Quito, Ecuador.

Vargas, O., Alix, C., Lobo, A.D. (authors), Duarte, O. and Sanchez, J. (technical reviewers) (1999) *Frutales y Condimentarias del Trópico Húmedo*. CURLA; PDBL; AFE/COHDEFOR; DICTA; SETCO; PROFORFITH, La Ceiba, Honduras, pp. 198–199.

Villachica, H., Carvalho, J.E.U., Muller, C.H., Diaz, S.C. and Almanza, M. (1996) Arazá. In: *Frutales y Hortalizas Promisorios de la Amazonía*. Tratado de Cooperación Amazónica, Lima, Perú, pp. 25–32.

Pitanga

Barbeau, G. (1990) *Frutas Tropicales en Nicaragua*. Dirección General de Técnicas Agropecuarias, MIDINRA, Editorial Ciencias Sociales, Managua, Nicaragua, pp. 225–226.

Crane, J.H. (2008) *Eugenia uniflora*, Pitanga; *Eugenia uvalha*, Ubaia. In: Janick, J. and Paull, R.E. (eds) *Encyclopedia of Fruit and Nuts*. CAB International, Wallingford, UK, pp. 534–536.

dos Santos, A.F., Silva, S., Mendoça, R.N., Alves, R.E. and Filgueiras, H.A.C. (2001) Maturation and ripening changes in Surinam cherry (*Eugenia uniflora* L.) fruit. *Proceedings of the Interamerican Society for Tropical Horticulture* 45, 45–47.

dos Santos, A.F., Silva, S., Mendoça, R.N., Silva, M.S., Alves, R.E. and Filgueiras, H.A.C. (2002) Alterações fisiológicas durante a maturação de pitanga (*Eugenia uniflora* L.). *Proceedings of the Interamerican Society for Tropical Horticulture* 46, 52–54.

Lorenzi, H. (1992) *Árvores Brasileiras*. Editora Plantarum S.A., São Paulo, Brazil.

Lorenzi, H., Bacher, L., Lacerda, M. and Sartori, S. (2006) *Frutas Brasileiras e Exóticas Cultivadas*. Editora Plantarum S.A., São Paulo, Brazil.

Martin, F.W., Campbell, C.W. and Ruberté, R.M. (1987) *Perennial Edible Fruits of the Tropics: An Inventory*. Agriculture Handbook No. 642, US Department of Agriculture (USDA), Agriculture Research Service (ARS), US Government Printing Office, Washington, DC, pp. 40–41.

Morton, J.F. (1987) *Fruits of Warm Climates*. Creative Resource Systems, Winterville, North Carolina, pp. 386–388.

Vargas, O., Alix, C., Lobo, A.D. (authors), Duarte, O. and Sanchez, J. (technical reviewers) (1999) *Frutales y Condimentarias del Trópico Húmedo*. CURLA; PDBL; AFE/COHDEFOR; DICTA; SETCO; PROFORFITH, La Ceiba, Honduras, pp. 200–201.

Verheij, E.W.M. and Coronel, R.E. (eds) (1992) *Edible Fruits and Nuts. Plant Resources of South East Asia No.2*. PROSEA Foundation, Bogor, Indonesia, pp. 165–167.

Villachica, H., Carvalho, J.E.U., Muller, C.H., Diaz, S.C. and Almanza, M. (1996) Pitanga. In: *Frutales y Hortalizas Promisorios de la Amazonía*. Tratado de Cooperación Amazónica, Lima, Perú, pp. 227–232.

Vizzotto, M., Cabral, L. and Lopes, A.S. (2011) Pitanga (*Eugenia uniflora* L.). In: Yahia, E.M. (ed.) *Postharvest Biology and Technology of Tropical and Subtropical Fruits, Volume 4. Mangosteen to white sapote*. Woodhead Publishing Ltd., Cambridge, UK, pp. 272–286.

Feijoa

André, E. (1898) A new fruiting tree - *Feijoa sellowiana*. *Revue Horticole* 70, 264–265.

Benavides, M. and Mora, H. (2003) Problemática del Complejo Moscas de la Fruta (Diptera: Tephritidae) y otras Plagas de Importancia Económica en el Cultivo de la Feijoa. In: Fischer, G.,

Miranda, D., Cayón, G. and Mazorra, M. (eds) *Cultivo, poscosecha y exportación de Feijoa* (*Acca sellowiana* Berg). Universidad Nacional de Colombia, Bogotá, Colombia, pp. 73–85.
Blanco, J.O. (2003) Manejo de Enfermedades de la Feijoa. In: Fischer, G., Miranda, D., Cayón, G. and Mazorra, M. (eds) *Cultivo, poscosecha y exportación de Feijoa* (*Acca sellowiana* Berg). Universidad Nacional de Colombia, Bogotá, Colombia, pp. 87–93.
Burret, M. (1941) Myrtaceen studien II. *Repertorium Specierum Novarum Regni Vegetabilis* 50, 59.
Dettori, M.T. and Palombi, M.A. (2000) Identification of *Feijoa sellowiana* Berg accessions by RAPD markers. *Scientia Horticulturae* 86, 279–290.
Donadio, L.C., Moro, F.V. and Servidone, A.A. (2002) *Frutas Brasileiras*. Editora Novos Talentos, Jaboticabal, São Paulo, Brazil, pp. 154–157.
Ducroquet, J.P., Hickel, E.R. and Nodari, R.O. (2000) *Goiabeira-Serrana* (*Feijoa sellowiana*). Funep (Série Frutas Nativas, 5), Jaboticabal, Brazil.
Fischer, G. (2003) Ecofisiología, Crecimiento y Desarrollo de la Feijoa. In: Fischer, G., Miranda, D., Cayón, G. and Mazorra, M.(eds) *Cultivo, poscosecha y exportación de Feijoa* (*Acca sellowiana* Berg). Universidad Nacional de Colombia, Bogotá, Colombia, pp. 9–26.
Harman, J.E. (1987) Feijoa fruit: growth and chemical composition during development. *New Zealand Journal of Experimental Agriculture* 15, 209–215.
Manica, I. (2002) *Frutas Nativas Silvestres e Exóticas 2*. Cinco Continentes Editora Ltda, Porto Alegre, RS, Brazil, pp. 11–140.
Miranda, D. (2003) Sistemas de propagación y establecimiento del cultivo. In: Fischer, G., Miranda, D., Cayón, G. and Mazorra, M. (eds) *Cultivo, poscosecha y exportación de Feijoa* (*Acca sellowiana* Berg). Universidad Nacional de Colombia, Bogotá, Colombia, pp. 27–48.
Morley-Bunker, M. (2003) Feijoas. In: Jackson, D.I. and Looney, N.E. (eds) *Temperate and Subtropical Fruit Production*, CAB International, Wallingford, UK, pp. 301–304.
Nodari, R.O., Ducroquet, J.P., Guerra, M.P. and Meler, K. (1997) Genetic variability of *Feijoa sellowiana* germplasm. *Acta Horticulturae* 452, 41–46.
Oltramari, A.C., Vesco, L.L., Pedrotti, E.L., Ducroquet, J.P., Nodari, R.O. and Guerra, M.P. (2000) Protocol for micropropagation of feijoa (*Acca sellowiana* (Berg) Burret). *Ciência Rural* 30(1), 61–68.
Ovalle, G. and Quintero, O. (1992) La feijoa (*Feijoa sellowiana* Berg) fruta promisoria para Colombia. *Acta Horticulturae* 310, 239–248.
Patterson, K.J. (1989) Effects of hand pollination on fruit set and fruit quality of the feijoa (*Acca sellowiana*). *Acta Horticulturae* 240, 197–200.
Quintero, O. (2003) Selección de cultivares, manejo del cultivo y regulación de cosechas de feijoa. In: Fischer, G., Miranda, D., Cayón, G. and Mazorra, M. (eds) *Cultivo, poscosecha y exportación de Feijoa* (*Acca sellowiana Berg*). Universidad Nacional de Colombia, Bogotá, Colombia, pp. 49–71.
Sharpe, R.H., Sherman, W.B. and Millor, E.P. (1993) Feijoa history and improvement. *Proceedings of the Florida State Horticultural Society* 106, 134–139.
Stanley, C.J. and Warrington, I.J. (1984) Seasonal frost tolerance of feijoa (*Feijoa sellowiana*). *New Zealand Journal of Experimental Agriculture* 12, 315–317.
Stewart, A. and Craig, J.L. (1989) Factors affecting pollinator effectiveness in *Feijoa sellowiana*. *New Zealand Journal of Crop and Horticultural Science* 17, 145–154.
Thorp, G. (2008) Feijoa *Acca sellowiana* (Berg) Burret, Myrtaceae. In: Janick, J. and Paull, R.E. (eds) *Encyclopedia of Fruit and Nuts*. CAB International, Wallingford, UK, pp. 526–533.
Thorp, T.G. and Bieleski, R.L. (2002) *Feijoas: Origins, Cultivation and Uses*. David Bateman Ltd., Auckland, New Zealand, 87pp.
Thorp, T.G. and Klein, J.D. (1987) Export feijoas: post-harvest handling and storage techniques to maintain optimum fruit quality. *The Orchardist of New Zealand*, June, 164–166.
Vissor, F.R. and Burrows, J.K. (1983) *Composition of New Zealand Foods. 1. Characteristic fruit and vegetables*. DSIR Bulletin 235, Wellington, 1–35.

4 Sapotaceae

The pantropical family Sapotaceae has approximately 65 genera and includes about 800 species of evergreen trees and shrubs. The name sapote is believed to be derived from the Aztec "tzapotl" used for soft sweet fruits. A number of species in different genera produce edible fruit: *Chrysophyllum cainito* L. – caimito, star apple; *Manilkara zapota* (L) von Royen – chiku, chicozapote, sapodilla; *Pouteria sapota* (Jacq.) H.E. Moore and Stearn – mamey sapote; *Pouteria campechiana* (HBK) Beahni – canistel; *Pouteria obovata* HBK – lucmo or lúcumo; *Pouteria caimito* (Ruiz and Pav.) Radlk – abiu or caimo; *Pouteria viridis* Pittier – green sapote or *injerto*. The black sapote (*Diospyros digyna* Jacq.) belongs to the family Ebenaceae and is therefore not a true sapote.

Star Apple

Chrysophyllum cainito L. (Sapotaceae), the star apple, has also been named *Achras caimito* Ruiz & Pavon and *Chrysophyllum bicolor* Poir. This plant is commonly called star apple or golden leaf tree in English; *caimito, estrella, caimitero* and *cauje* in Spanish; *caimite, caimitier, pomme surette* and *pomme etoile* in French; *pied caimite* and *caimitier a feuilles d'or* in Haitien; *cainito* and *ajara* in Portuguese; *sawoyo, sawoijo, sawohejo* and *sawokadu* in Indonsesian; *sawuduren* and *pepulut* in Malay; *caimito* in the Philippines; and *sataaappoen* in Thailand. It is known as *chicle durian* in Singapore (Morton, 1987). The name "star apple" derives from the fact that when the fruit is cut transversely the seeds are arranged in a star-like shape (Barbeau, 1990).

Origin and distribution

The most probable origin of star apple is the West Indies, from where it was probably spread to Central America long ago. Some authors indicate that it originated in both areas (Barbeau, 1990). This species is found throughout the lowland tropics of Central America, especially the Pacific side of Guatemala, in the Caribbean, especially Haiti, the warmer parts of South America (e.g. Colombia, Ecuador, Venezuela, Bolivia, Brazil, northern Argentina and Peru), Mexico, South-east Asia (including Malaysia, Thailand, Vietnam, Cambodia and the Philippines), Africa, India and Sri Lanka. Star apple is grown on a small scale and/or harvested commercially throughout its distributed range.

Data as to the area of planting and production do not exist. It is normally grown as a dooryard tree or in home gardens. The Philippines in 1987 reported growing 7,600 ha and producing 25,389 t although no formal orchards are established. Star apple is grown on a very small commercial scale in some areas (e.g. South Florida). Star apple fruit are generally harvested from late winter or early spring to early summer (Morton, 1987).

Ecology

Soil

Trees appear well adapted to a wide range of well-drained soil types (Barbeau, 1990) although they perform better in slightly acid soils. They can grow in clayey loams, sand and limestone but require good drainage.

Rainfall

Star apple is a tropical tree that can grow in the warm subtropics. It prefers medium to high rainfall of 1,400 mm year^{-1} or more, but with an interruption by a defined dry period to ensure good flowering that occurs on the current season's growth. Excessive drought can defoliate the tree and reduce fruit size and appearance.

Temperature

The tree prefers the hot humid tropics up to 425 m and a maximum of 1,000 m altitude and does not grow well when temperatures are cool, though it can withstand cool conditions. Trees have a limited tolerance to frost and if exposed to temperatures below 4°C. Mature trees will be seriously injured by temperatures below −2.2°C (Morton, 1987). Large branches will be damaged below −3.3°C, and tree death can occur at temperatures below −3.8°C (Campbell, 1974; Campbell *et al.*, 1977). Young trees can suffer leaf injury at 0°C and may be killed by short exposures to temperatures of −1.1°C to 1.6°C (Crane, 2008).

Light

The tree prefers to grow under full light.

Wind

The tree is fairly sturdy with strong branches so that wind damage is not likely to occur. In commercial plantations, a wind barrier would have to be installed in places where strong winds are a problem, especially to avoid tree deformation.

General characteristics

Tree

This species can reach heights between 8 and 30 m. It has a short trunk that can reach 1 m in diameter and has a dense crown formed by brown-hairy branchlets that will exudate a white-gummy latex if wounded or if a leaf is pulled away. The 5–15 cm long and 3–6 cm wide leaves, with a 1.5 cm petiole, are elliptic or oblong-elliptic in shape (Fig. 4.1). The leaves are slightly leathery, alternate and with a rich green color on the upper surface and a golden brown silky pubescence beneath when they are mature, giving the tree a very ornamental appearance (Morton, 1987).

Flowers

The tree has inflorescences that appear in the leaf axils on the current season's shoots; each can have two to 35 flowers (Verheij and Coronel, 1992), and some flowers are solitary (Vargas *et al.*, 1999). The small, fragrant, inconspicuous flowers have a greenish-yellow, yellow or purplish-white color from petals that form a five-lobed tubular corolla. The calyx can have five to six pubescent sepals and there are five stamens and a stigma with seven to 11 lobes (Morton, 1987). The thin pedicels are about 1–2 cm long. The flowers are normally pollinated by insects and are self-fertile.

Pollination and fruit set

Flowering normally occurs in late summer or at the beginning of autumn. The flowers are hermaphroditic and usually self-fertile, although sometimes

Fig. 4.1. Star apple (*Chrysophyllum cainito L.*) showing leaves, the small flowers in a cluster in the leaf axils and fruit. The fruit in transverse section shows the internal star pattern of the seeds (used with permission from León, J. (2000) *Botánica de los Cultivos Tropicales*. Editorial Agroamérica, Instituto Interamericano de Cooperación para la Agricultura (IICA), San José, Costa Rica).

isolated trees are fruitless (Campbell, 1974). Flowering usually coincides with the main shoot growth period during the rainy season and fruit mature 4–5 months later.

Fruit

The fruit is a round, oblate, ellipsoidal or somewhat pear-shaped berry 5–10 cm in diameter (Fig. 4.1). When ripe, the fruit can be red-purple or dark purple in the purple types or pale green to yellowish-brown in the green types. It has a glossy, smooth, leathery skin that feels rubbery in the hand and adheres to the inner rind. This inner rind can be 6–10 mm thick or half of that in green-colored fruit. In purple fruit, the inner rind is dark purple and adheres to the outer peel, whereas in green fruit it is white.

The inside of the fruit consists of a soft, white, milky sweet pulp surrounding three to 11 gelatinous seed cells (Martin *et al.*, 1987; Morton, 1987). When cut transversely, these seed cells are seen to radiate from the central core, appearing star-like, hence the name. It can take from 120 to about 150 days from fruit set to maturity (Martin *et al.*, 1987; Verheij and Coronel, 1992). Fruit development is slowed by drought or cool temperatures. The seeds are about 2 cm long and 1.25 cm wide, flattened, obovoid or oval and pointed; their color is black when fresh and light brown after drying.

Cultivar development

Caimito ($2n = 52$) has both purple and green color types with significant variation within each color type in yield, fruit size, shape, color and quality and period of maturity. A few named cultivars exist and include: 'Younghans', 'Blanco', 'Grimal', 'New Combe', 'Weeping', 'Lunti' and 'Lila'. Another variety brought from Haiti is 'Haitian' that is vegetatively propagated and has done well in Florida.

Varietal improvement is needed. Prospects for cultivar development are good since such large variation exists in phenotype. In Florida and similar areas, cold resistance or tolerance could be an important aspect for breeding work.

Cultural practices

Propagation

SEXUAL Star apples are normally propagated by seeds that remain viable for several months and will germinate 20–40 days after sowing (Barbeau, 1990). The ideal is to use fresh seeds that do not need any previous treatment and are buried about 1 cm. As soon as the cotyledons have expanded the seedlings can be transplanted to nursery bags. Fruit production from seedlings takes 5–8 years and plants normally become very tall. The variability in production and fruit quality is a problem in orchard development by seed propagation.

ASEXUAL Asexual propagation can be done by using cuttings. Air layers have also been used and can take 4–7 months to root (Vargas *et al.*, 1999). Trees may also be grafted or budded, normally onto star apple seedlings (Crane, 2008); in that case the grafted or budded trees can start producing as soon as a year after transplanting to the field, though normally 3–4 years. Grafting can be done using the cleft method and budding can be done with the patch or the Forkert method using buds without the petiole attached (Alix and Duarte, 1999). Veneer grafting is also effective (Campbell, 1974). An ornamental relative *Chrysophyllum oliviforme* (satin leaf or caimitillo) can also be used as a rootstock for slowing or stunting tree growth (Morton, 1987; Hoyos, 1989).

Orchard establishment

Normal field preparation should be followed that includes sub-soiling, if necessary, plowing and discing or harrowing until a good tilth is attained. After this the location of the planting holes will have to be marked or staked. Recommended orchard plant spacing range for seedling trees is from 10 to 12 m between trees and rows. With asexually propagated plants, distances recommended are significantly shorter, about 7 m between trees and rows.

Irrigation

Newly planted trees should be irrigated soon after planting and frequently during the first 6 months to ensure establishment (Morton, 1987) unless the planting is done at the start of the rainy season. Any period of drought should be avoided especially during the initial years and bucket irrigation can be used if no irrigation system is in place. When trees begin bearing, irrigation is recommended during dry periods to improve fruit size and quality (Verheij and Coronel, 1992).

Pruning

Young trees are not extensively trained other than allowing three to four main branches to form the major framework of the tree; this is achieved by topping the plants when they reach about 1 m height in the field and then leaving the best spatially distributed branches in the vertical and in the horizontal planes. Maintenance pruning includes eliminating damaged or diseased branches, branches growing too close to the ground and any branch growing in the wrong direction or into the canopy. Water sprouts should also be eliminated, especially during the early years when they tend to appear more abundantly. Mature trees may be pruned by hand or with machinery to limit tree size, since normally no more than 4 m in height should be allowed.

Fertilization

Young trees should be fertilized with 150–200 g of ammonium sulfate or similar material applied twice a year. As trees begin to bear, the amount of fertilizer should be increased to 500 g applied twice a year and up to 3 kg year^{-1} for mature, heavy-bearing trees (Crane, 2008).

Pest management

DISEASES *Fusarium solani* may kill young trees and damage limbs of mature trees. The fruit may be attacked by a dry rot (*Lasiodiplodia theobromae*) prior to harvest and also by stem-end decay caused by *Pestalotia* sp. and *Diplodia* sp.

According to Morton (1987) leaves may be attacked by *Phomopsis* sp., *Phyllosticta* sp. and algal leaf spot (*Cephaleuros virescens*).

INSECTS Insect pests include mealy bugs, chewing insects, carpenter moths, scales, twig borers and fruit flies. An unidentified webworm larva has been observed feeding on flowers (Crane, 2008). Larvae of small insects are found in many ripe fruit. Spider mites can also attack the foliage, especially when it is hot and dry. Birds can be very harmful to ripening fruit (Vargas *et al.*, 1999) and bats may also cause damage to them.

Harvesting and postharvest handling

Harvesting

The harvest season is from late winter to early summer. Immature fruit do not ripen properly if harvested green and have a high flesh latex content and are very astringent (Vargas *et al.*, 1999). Fruit do not abscise when ripe and are normally harvested when they become soft to the touch. The ripe fruit can be harvested by being pulled or by clipping the peduncle. The peduncle should be clipped off prior to packing. An adult tree can produce up to 60 kg of fruit per year and even 90–113 kg (Campbell, 1974).

Postharvest treatments

The fruit may be stored at 3–6°C and 90% relative humidity for up to 3 weeks (Morton, 1987; Verheij and Coronel, 1992).

Utilization

The star apple is commonly eaten fresh or the pulp is preserved and processed into beverages, candies, sherbet and ice cream (Martin *et al.*, 1987; Morton, 1987; Verheij and Coronel, 1992). The peel of star apple is inedible and when opening the fruit it is recommended none of the bitter latex from the peel contacts the pulp (Morton, 1987; Subhadrabandhu, 2001). The fruit is 53–63% edible pulp (Verheij and Coronel, 1992) and the pulp moisture ranges from 78 to 86% (Table 4.1). Several medicinal uses have been reported for the pulp of star apple including its use as an anti-inflammatory in laryngitis and pneumonia. The seeds are toxic. The wood is used in the lumber industry.

Abiu

Pouteria caimito (Ruiz & Pav.) Radlk. Synonyms are: *Lucuma caimito* Roem & Schult., *Achras caimito* Ruiz & Pavón, Banth., *Guapeda caimito* Pierre, *Labatia caimito* Mart. and *Pouteria leucophaea* Baehni. The plant is also known as *caimito, caimito amarillo, caimo* or *madura verde* in Colombia; *luma* or *cauje* in Ecuador; *temare* in Venezuela; *caimo* or *caimito* in Peru and *abi, abiu, abio, abieiro* or *caimito* in Brazil. In English, it is also called abiu.

Table 4.1. Composition of 100 g edible portion of different American fruits of the Sapotaceae family.

Proximate		Star apple or caimito[a]	Abiu[b]	Egg fruit or canistel[c]	Green sapote[c]	Lucuma[d]
Energy	cal	68		138.8		99–122
	kJ		586			
Moisture	g	82.0	61.0	60.6	68.10–69.5	64.8–72.3
Carbohydrate	g	14.5	36.3	36.7		
Protein	g	0.8	1.8	1.7	0.15–0.28	1.4–1.5
Lipid	g	1.6	0.4	0.13	0.24–0.28	0.50
Fiber	g		0.9	0.1	1.20–1.6	1.0–1.3
Ash	g		0.9	0.9	0.69–1.38	0.7–0.9
Minerals						
Potassium	mg					470.0
Calcium	mg	21.0	22.0	26.5	18.6–35.70	16.0
Iron	mg	0.8	1.0	0.92	0.57–0.74	26.0
Phosphorus	mg	17.0	41.0	37.3	22.1–23.6	0.4
Sodium	mg					6.0
Vitamins						
Thiamin	mg	0.03	0.02	0.17	0.01–0.01	0.01
Riboflavin	mg	0.04	0.02	0.01	0.027	0.14
Niacin	mg	1.0	34.0	3.72	1.9	1.9–2.2
Ascorbic acid	mg	11.0	49.0	58.1	49.9–62.3	5.4–25.0
Carotene	mg	5.0		0.32	0.031–0.069	
	IU		78.0			

[a]Villachica *et al.*, 1996.
[b]Leung and Flores, 1961.
[c]Morton, 1987.
[d]Lizana, 1990.

Origin and distribution

Abiu is a native of western Amazonia (Clement, 1989). The fruit was known in pre-Columbian times by the natives of that area and was used by them before the conquest (Ferreira, 1974). From its original area it has been dispersed to Colombia, southern Brazil and later to Central America and many other tropical parts of the world by the European colonists (Cavalcante, 1991; Patiño, 2002). It is not grown to any large extent commercially. Small commercial orchards are occasionally found in Ecuador, western Brazil, Colombia, Venezuela and some parts of the Peruvian Amazon, but national statistics do not mention it specifically. It is well adapted to the climate of São Paulo State, Brazil (Donadio, 1983)

Ecology

Soil

Abiu does best on well-drained loamy to clayey soils with a mildly acid to neutral pH and a high content of organic matter. It can produce on less favorable soil conditions (Campbell *et al.*, 1967). The plant tolerates many pH levels but it is sensitive to saline water or soils.

Rainfall

The plant comes from regions with 1,700–3,000 mm of well-distributed yearly rainfall (Villachica *et al.*, 1996). The tree is able to withstand seasonal drought if not excessive. Unlike *Manilkara* (sapodilla or chiku), a short dry spell is not required by abiu to flower. The tree needs supplemental irrigation for good yields and fruit size, if grown in arid areas or areas with low rainfall.

Temperature

Contrary to most tropical species, a mature tree will withstand –2°C with little damage (Morton, 1987). Best growth occurs in the tropical areas of the world and below 800–1,000 m elevation or in warmer areas in the subtropics. In certain areas of Colombia, it is grown in warm valleys that are at 1,900 m altitude (Morton, 1987). It does well with a mean annual temperature greater than 18–20°C. In some areas of the world like the Northern Territory of Australia, scorching of leaves can occur under very hot and dry conditions (Lim and Ramsay, 1992).

Light

Like many tropical trees, it requires some shade during its first year or two in the field while larger trees can grow in full sun (Marshall, 1991).

Photoperiod

No information is available, but apparently adaptation to day lengths in tropical and subtropical areas seems to be ample since abiu flowers and bears fruit anywhere in these regions (Martin and Malo, 1978).

Wind

It is fairly resistant to winds because of its compact canopy.

General characteristics

Tree

Abiu is a small to medium tree with a pyramidal canopy that can grow to 5–10 m. Eventually, the tree can reach 25–30 m and in its adult stage can have a diameter of 10 m. The 10–15 cm long and 3–6 cm wide leaves are alternate, entire, glossy and pointed at both ends (Fig. 4.2), with an acute apex that can be curved. The leaves are dark green above and lighter green below and tend to cluster at the end of shoots that have short internodes (Fig. 4.2), producing a dense foliage. The bark is gray to brown, rough and exudes white latex when wounded.

Flowers

The flowers come in clusters that are almost sessile and appear in the axils of the leaves along the branches. As with other plants of this family

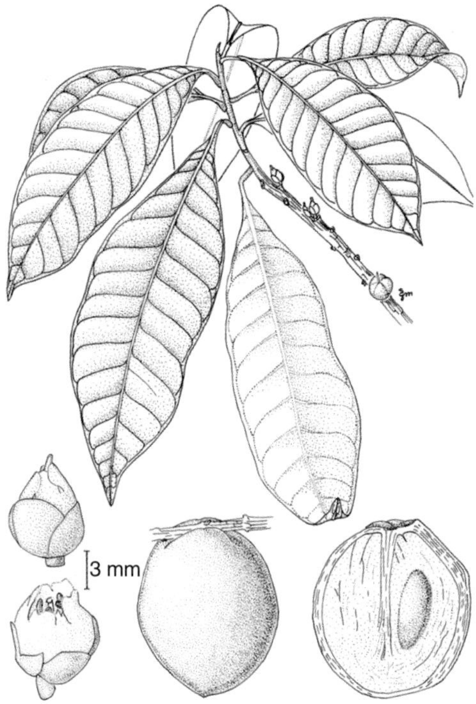

Fig. 4.2. Abiu (*Pouteria caimito*) showing leaves, the small flowers clustered on the branch and fruit (used with permission from Nakasone and Paull (1998) *Tropical Fruits*. CAB International, Wallingford, UK).

the flowers are enclosed by two bracts and four overlapping light green sepals. The cylindrical, white corolla tube is 5 mm long and never opens completely, enclosing four free stamens that alternate with four staminodes (Fig. 4.2). The style is white and arises from a superior ovary and ends in a four-lobed stigma that pushes through the petals at anthesis and extends 2–3 mm above the corolla tube before it becomes receptive (Lim and Ramsay, 1992). The ovary contains four to five locules, with one ovule each, and is 10–12 celled.

Abiu can flower throughout the year and it is common to see fruit of various stages of development on the same tree and even on the same branches. In central Amazonia, three sequential flowering periods occur from February (beginning of the rainy season) to August (middle of the dry season), with fruiting periods going from the middle of the rainy season to the end of the dry season. The main flowering occurs between September and November.

Pollination and fruit set

The abundant flowers are insect pollinated, probably by bees, but fruit set is poor (1.5–3%) according to Falcao and Clement (1999). Fruit set occurs more abundantly during the warmer months.

Fruit

The fruit is round to elliptic (Fig. 4.2), 5–12 cm long and normally weighs from 50 to 500 g, although up to 1,000 g has been reported in a semi-domesticated race (Clement, 1989). The immature fruit has a green skin that changes to light yellow when ripe. This leathery skin, which is about 5 mm thick when the fruit is mature, encloses a sweet, translucent, white, jelly-like flesh and one to five large oblong shiny black seeds. Fruit development takes about 2–3 months (Scholefield, 1984; Falcao and Clement, 1999), depending on rainfall and temperatures. All parts of the fruit contain a sticky white latex and the cut surface rapidly browns when the fruit is immature. As the fruit matures, the latex disappears from the pulp and remains only in the peel. In some populations, the latex is also missing from the peel.

Cultivar development

Abiu occurs in the wild and in cultivation, showing great variation in morphology. No reports exist about breeding programs for this crop. Some abiu selections have been made based upon observations and fruit tasting. Selection criteria used include early and regular bearing, round fruit (> 180 g), few seeds, low levels of skin latex, firm sweet translucent flesh (13–18% total soluble solids) and good shelf life.

Cultivars

There are some named varieties outside of its region of origin. In Queensland, Australia there are some varieties that include 'Inca Gold', and 'Cape Oasis' with other selected seedlings (Parker, 1986), and in Brazil they have selected 'Graudo' (Donadio *et al.*, 2002). In Hawaii there is a selected superior type named 'Gray' (Alix and Duarte, 1999).

Cultural practices

Propagation and nursery management

SEXUAL The seeds are short lived and considered recalcitrant (Villachica *et al.*, 1996). Seedlings will need 3–5 years to fruit and, as with most fruit trees, great variation in form, yield and fruit quality exists. With fresh seeds around 90% germination occurs in about 15–60 days. Plants should be transplanted when the cotyledons have opened (Alix and Duarte, 1999).

ASEXUAL Vegetative propagation produces uniform planting material and reduces the time needed to flowering, bearing in about 2 years. Grafting is difficult because the terminal scion has very short internodes (Alix and Duarte, 1999) and produces abundant latex in the cortex. At present, most trees are propagated by seed in Amazonia and for more advanced domesticated populations, where the landraces come relatively true from seed.

Orchard establishment

No special field preparation is necessary. Trees are planted out when 3–5 months old and 60 cm tall. Spacing of 6 × 8 (Vargas *et al.*, 1999) to 8 × 8 or 10 × 10 m have been recommended in various locations, but at too close spacing trees may need to be thinned as they grow older. Reports that indicate that the trees can thrive well on a 2–3 m inter-row spacing probably refer more to asexually propagated trees. When planted in a windy location the trees have to be staked. In high sunlight areas, shade has to be provided during the first 6–12 months.

Irrigation

The ideal is to plant the trees at the beginning of the rainy season for better establishment. Young trees should be given additional irrigation, especially during the initial dry seasons, so they grow more vigorous and productive. Adult trees give better yields and larger fruit if they receive supplemental irrigation during a long dry season.

Pruning

A tree with a central leader and whorls of laterals requires little pruning. Grafted trees may require some initial trimming to stimulate lateral growth. Pruning of mature trees should be limited to removing dead and thin branches and branches that grow too close to the ground or are curved. The other option is to form the tree with three or four main scaffold branches by cutting off the tip when the plants are 1 m tall; this will induce lateral branching and reduce tree height.

Fertilization

Recommendations for fertilization at planting time include the addition of manure (10 kg), lime (0.5 kg), superphosphate (0.1 kg) and potassium chloride (0.1 kg) put under the root ball (Villachica *et al.*, 1996).

A similar fertilization schedule to that used for chiku has been suggested for abiu (Lim and Ramsay, 1992) and mature chiku trees require at least 1.5 kg N, 0.5 kg P_2O_5 and 0.5 kg K_2O per tree per year, applications being split into two or three per year, with one before the wet season and one just before the end or after each harvest (Marshall, 1991). Manure can also be used and liming and other nutrients may be needed.

Pest management

DISEASES Diseases are generally a minor problem.

INSECTS The fruit is attacked by fruit flies as they ripen thus creating a problem for exporting to some countries. The fruit fly *Anastrepha serpentina* has been reported in eastern Brazilian Amazonia while *A. obliqua*, *A. fraterculus* and *Ceratitis capitata* have been reported for other parts of Brazil (Manica, 2000). Twig borers, leaf-eating larvae, scales, aphids and mealy bugs attack various parts of the tree. Bats can also be a problem for which nets have to be used, especially as the fruit reach the ripening stage. Birds are attracted by the color of the fruit.

Weed control

Weeds are more of a problem with young plants.

Harvesting and postharvest handling

Yields

A productive mature tree can yield up to 200–500 fruit year^{-1}, with fruit weighing from 200 to 250 g (40–125 kg per tree), and a good 6-year-old tree 60–85 kg per tree (Lim and Ramsay, 1992; Vargas *et al.*, 1999). Trees can remain productive for 20 years.

Harvesting

Abiu fruit are harvested when they become bright yellow and they continue to ripen after harvest. Full ripening occurs in 1–5 days, when the fruit pulp does not have the sticky latex. The translucent flesh becomes jelly-like, with a pleasant, somewhat caramel-flavored pulp.

Postharvest treatments

Not much information exists about this species. Fruit can be stored for about a week at 12°C. The tough leathery skin can be easily bruised, but if handled carefully it helps in obtaining a fairly good shelf life.

Utilization

Abiu is used fresh, sometimes in salads with other fruit. The ripe mucilaginous pulp can be used dried or in sherbets, jams, fruit salads and yogurt. The pulp offers fair amounts of vitamins A and C (Table 4.1). The fruit has to be completely ripe otherwise the latex will stick in the mouth and the pulp will be astringent.

The leaves are used in popular medicines and the pulp can be used for treating coughs, bronchitis and pulmonary infections. The latex is used against digestive system parasitic worms and as a purgative (Vargas *et al.*, 1999).

Canistel

Canistel, *Pouteria campechiana* (HBK) Baehni (Sapotaceae), is native to southern Mexico. Common names for this tree include: canistel, egg fruit and yellow sapote (English); *janne d'oeuf* (French); *siguapa, zapotillo, zapote mante, fruta huevo* and *sapote amarillo* (Spanish); *siguapa* and *sapotillo* (Costa Rica); *cakixo, canizte, kanis* and *kantze* or *kantez* (Guatemala); *fruta de huevo* and *zapote amarillo* (Colombia); *zapotillo amarillo* and *guicume* (El Salvador); *borracho, canistel,* and *tiesa* or *toesa* (Philippines); *kaniste, guicomo, caca de niño, costiczpotl, cucumu, mamey de Campeche, huicumo, zapote de niño* and *zapote borracho* (Mexico).

The taxonomy of canistel has been modified numerous times with the tree being assigned to different genera such as *Pouteria campechiana* var. *nervosa* Baehni; *P. campechiana* var. *palmieri* Baehni; *P. campechiana* var. *salicifolia* Baehni; *Lucuma campechiana* HBK.; *L. heyderi* Standl.; *L. laeteviridis* Pittier; *L. multiflora* Millsp.; *L. nervosa* A.DC.; *L. palmieri* Fernald; *L. rivicoa* Gaertn.; *L. rivicoa* var.

angustifolia Miq.; *L. salicifolia* HBK.; *Richardella salicifolia* Pierre; *Sideroxylon campestre* T.S. Brandeg.; *Vitellaria salicifolia* and *Vitellaria campechiana* Engl. (Morton, 1987).

Origin and distribution

The apparent origin of this species seems to be southern Mexico, particularly the Yucatan Peninsula, Belize, Guatemala and El Salvador (Azurdia, 2006). It now is widely distributed in most of Central America, Cuba and the Caribbean (Martin and Malo, 1978). It has also been introduced in some South-east Asian and East African countries. It is grown more as a backyard tree or in small commercial orchards. There are no statistical figures on areas and volumes of production; the Philippines apparently has the highest volume registered officially (Azurdia, 2006).

Ecology

Soil

This plant grows satisfactorily in a wide range of well-drained soil types. It withstands acid soils in Central America as well as alkaline soils on the Peruvian coast. It can also grow in calcareous, lateritic, sandy and heavy clay soils. It grows in the Florida Keys in shallow calcareous soils that lose much of their moisture during the dry season (Martin and Malo, 1978). It is better adapted to poor thin soils than most other fruit trees.

Temperature

The species is best adapted to warm, lowland tropics although it will grow well in higher elevations, for example at 1,000 m in Honduras and 1,400 m in Guatemala, and in the subtropics with no severe frosts. It is moderately cold tolerant to temperatures of −1.7°C and −5.0°C for young and mature trees, respectively (Crane, 2008), and tolerates frosts in South Florida (Martin and Malo, 1978).

Rainfall

Canistel grows in places with moderate to high annual rainfall (Verheij and Coronel, 1992) but trees are also well adapted to monsoon climates with pronounced wet and dry periods (Martin *et al.*, 1987), as well as arid climates where irrigation is available. The specific water requirements for mature trees have not been reported. The tree is moderately tolerant to flooding.

Light and photoperiod

The plant prefers to grow in full sunlight. It flowers in various day lengths of the tropics and subtropics.

Wind

The plant is not very resistant to strong winds (Azurdia, 2006). Pruning the tree to limit its height increases resistance to wind damage.

General characteristics

Tree

This is normally an 8 m tall tree that can reach 12–20 m with 1 m trunk diameter with a slender or spreading crown. The bark is brown and furrowed and contains latex as most of the plants of this family. The tree is evergreen with simple, alternate 10–28 cm long by 3–8 cm wide leaves that come in whorls at the end of the branches and are tapered at both ends (Fig. 4.3). Leaves are rather thin, glossy and oblanceolate to obovate with a bluntly pointed apex (Morton, 1987). The roots of the mature trees spread beyond the drip line of the canopy.

Flowers, pollination and fruit set

The inconspicuous flowers are bisexual, fragrant and appear solitary or in small clusters in the leaf axils or at leafless nodes on slender pedicels. They have five sepals and a tubular corolla with five to six lobes (Fig. 4.3). Flowers are whitish to cream in color and hairy, with five stamens and one ovary and its stigma. Anecdotal evidence suggests canistel is insect pollinated and self-compatible (Crane, 2008).

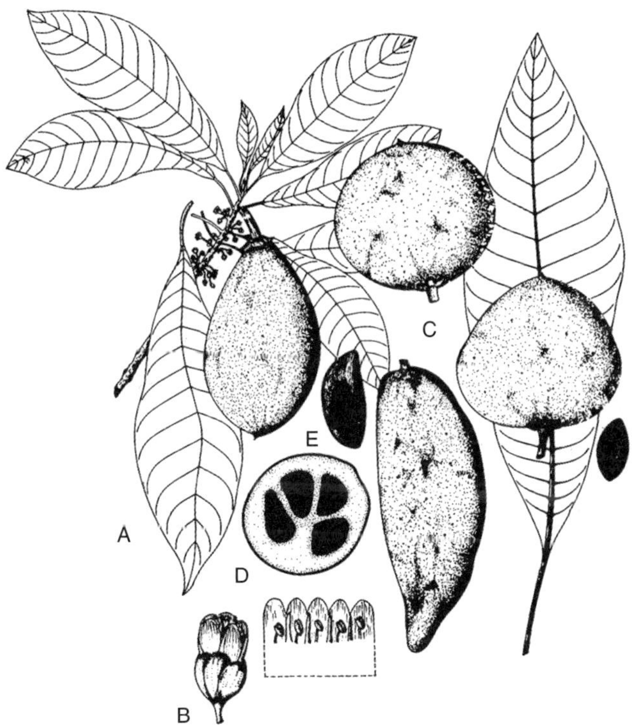

Fig. 4.3. Canistel (*Pouteria campechiana*) showing (A) leaves, (B) flowers, (C) different fruit shapes, (D) transverse section through the fruit and (E) seed (used with permission from León, J. (2000) *Botánica de los Cultivos Tropicales*. Editorial Agroamérica, Instituto Interamericano de Cooperación para la Agricultura (IICA), San José, Costa Rica).

In many places, the tree can flower for several months in the tropics (Morton, 1987). Flowering in subtropical areas tends to be more seasonal. Blooming in Mexico extends from January to June, while in Cuba it takes place in April and May but some trees flower all year round. In Florida, fruit are harvested from December to March.

Fruit

The fruit is very variable in size (8–15 cm long and 5–12 cm wide); some selected varieties can have fruit that weigh up to 700 g, while the normal fruit weighs 50–70 g. The fruit can be oval, ovoid, spindle-shaped or almost round, with or without a pointed apex and often bulged on one side (Fig. 4.3). The young fruit is green and on ripening the skin turns to various yellow tones (pale orange, lemon or golden). The skin is thin and glossy but sometimes it can have some russetting that is brown or reddish. Beneath the thin skin is the yellow-colored pulp that can be mealy or slightly sticky because of latex content. The flavor is sweet or a bit musky. The fruit can have one to four oval to oblong, 2–5 cm long and 1.5–3.5 cm wide seeds (Fig. 4.3) that have a glossy coat except in the ventral part that is the light brown and dull or grayish hilum. Time from flowering to harvest is 5–6 months.

Cultivar development

There has been no real effort to improve this fruit because it is considered of low importance. A few named cultivars exist with round and large fruit that are superior to the spindle-shaped small fruit found normally. The Montgomery Foundation has a small collection of cultivars that were described by Campbell and Wasielewski (1999): 'TREC 9680' has the largest fruit, ca. 400 g, and the highest production with 74 kg in 2 years, 'Bruce' has fruit that are ca. 375 g and produce 58 kg in 2 years, others are 'Fairchild 1', 'Fairchild 2', 'Fitzpatrick', 'Keisau' and 'Ross'. Other varieties selected in Florida are: 'Oro', 'USDA 1', 'Trompo', 'Saludo', 'DuPuis' (Wasielewski and Campbell, 1997; Vargas *et al.*, 1999; Crane *et al.*, 2001). Fruit flesh varies in different varieties from dry, sometimes mealy, with fair eating quality to moist flesh with good eating quality, and with few to many seeds.

Cultural practices

Propagation

SEXUAL Trees may be propagated by seeds that are short lived and should be sown within a few days after removal from the fruit. Germination goes down from almost 100% for fresh seed to 14% after 1 month when seeds are kept at room temperature (Duarte and Villagrán, 2002). With intact seed coats, 100% germination can be obtained after 5 months. Removal of seed coats has a negative effect on survivability. Soaking seeds in gibberellic acid (GA) had no positive effect on germination percentage or seedling size. Seedling growth is rapid and they begin to bear fruit after 3–7 years (Morton, 1983).

ASEXUAL Superior canistel cultivars are generally grafted on seedling rootstock. Grafted trees will begin fruit production in 2–4 years (Martin *et al.*, 1987). Side veneer or cleft grafting are the best methods for seedlings with at least 0.8 cm stem diameter at 25 cm from the ground (Vargas *et al.*, 1999). Scions obtained from terminal parts of the branch with swollen buds should be used and the latex should be wiped off with the grafting knife just before the graft parts come in contact (Alix and Duarte, 1999). Inverted root grafting, a modification of a technique developed for lucuma, consists of taking the germinated seed when it has a root of about 8–10 cm; the root tip is cut off leaving 4–5 cm and cleft grafted like a stem; then the grafted seed is returned to its bag in an inverted position and the graft grows to form a stem as in other grafts (Wasielewski and Campbell, 1999a). Marcottage can also be used successfully (Crane, 2008).

Orchard establishment

No special land preparation is required. Normally trees are planted at spacings of 6 × 6 or 7 × 7 m, although grafted material can be planted at closer distances (Vargas *et al.*, 1999). Mulching is beneficial during the early years.

Pruning

The ideal is to form a vase-shaped tree by taking off the tip of the stem when it reaches about a meter from the ground. After this three or four branches are selected to become the scaffold branches. Mature trees may be pruned by hand or with machinery to limit tree size and eliminate dead, diseased, excessively low branches or those growing in an inappropriate direction. Large branches should also be cut back to keep tree shape. In pruning trials done in Florida canistel has proven to be more manageable than sapodilla and mamey sapote trees, with all three responding positively to most pruning techniques (Wasielewski and Campbell, 1999b).

Fertilization

No specific fertilizer recommendations have been reported for canistel production; however, practices common to similar fruit or citrus appear to suffice. The tree is susceptible to iron deficiency when grown in alkaline and high-pH soils. Mulching has been reported to be beneficial for young trees (Crane, 2008).

Pest management

DISEASES Diseases normally are not a major problem in canistel. Diseases reported include leaf rust (*Acrotelium lucumae*), leaf spot or scab (*Elsinoe lepagei*), root rot (*Pythium* sp.), leaf necrosis (*Gloeosporium* sp.) and fruit spot or anthracnose (*Colletotrichum gloeosporioides*) (Verheij and Coronel, 1992).

INSECTS Various scales and mealy bugs occasionally attack leaves and fruit. Fruit flies (*Anastrepha* sp.) can be a problem, attacking fruit as it ripens.

Harvesting and postharvest handling

No exact yield data has been found but 5–6-year-old trees can produce 30–50 kg per year.

Harvesting

Since mature fruit tends to fall when ripe, it should be harvested with care at maturity (bright orange) when still hard. The peel is very fragile when the fruit is soft and is easily injured. Harvesting should be done with a knife or pruning shears and fruit carefully placed into the picking bag. Sometimes fruit can be pulled from the tree but this normally causes damage. The stalk wound will start bleeding latex for some time and care has to be taken that this latex does not contact other fruit.

Postharvest treatment

Postharvest handling has not been extensively studied. Mature fruit at room temperature generally take 3–10 days to fully ripen. Fully mature fruit may be stored at 13–15°C and 85–90% relative humidity for up to 3 weeks (McGregor, 1987).

Utilization

The fruit is appealing, keeps well and is more nutritious than many other tropical fruits (Morton, 1983). It can be eaten fresh but normally it is consumed as a part of desserts in ice cream, milk shakes and custards. Fruit pulp may also be consumed as a vegetable with salt, pepper, lemon juice or mayonnaise, and as a component of baked goods. The pulp may also be dried to a powder and used as a food additive.

The ripe pulp is highly nutritious and high in niacin and carotene (provitamin A), and fairly high in ascorbic acid and proteins (Table 4.1). The pulp of unripe fruit is dry and mealy in texture contrasting with the pulp of fully ripe fruit, which is very soft and creamy, sometimes sticky. The bright yellow pulp has a pleasant sweet flavor; however, seedlings, selections and cultivars vary in their degree of muskiness (mild to pronounced) and stickiness (due to latex content).

Various parts of the canistel tree are used for medicinal purposes. The bark is processed for use in skin rashes and derivatives from the seed for scalp dermatitis and ulcers. A decoction of the astringent bark is used as a febrifuge in Mexico. The seeds apparently have substances to fight seborrheic dermatitis and ulcers. The timber is also of good quality for rafters. Canistel has also been tested as a possible rootstock for mamey sapote because seeds of this species are difficult to obtain in large numbers and it is well adapted to calcareous soils of Florida (Ogden and Campbell, 1980).

Green Sapote

The taxonomy of green sapote has been modified a number of times. The currently accepted name is *Pouteria viridis* (Pittier) Cronquist (Sapotaceae); synonyms are *Calocarpum viride* Pitt. and *Achradelpha viridis* O.F. Cook. This minor fruit species is known as green sapote in English and as *injerto, injerto verde, raxtul* in Guatemala, *yashtul, mameicito* in Mexico and *zapote injerto* in Costa Rica. In Belize, it is called *white* or *red faisan*. The quality of the fruit can be even better than that of mamey sapote.

Origin and distribution

This tree is native to the highlands of Guatemala, El Salvador and Honduras and southern Mexico (Chiapas). It can be found occasionally in Costa Rica and Nicaragua. The greatest plant diversity is found in Guatemala (Campbell and Ledesma, 2002). No data are available for production area or yields. This plant is generally grown on a very small scale of a few trees in Central America. It is more commonly found in the wild and as a backyard tree in farm or village homes and can occasionally be bought in local markets. This species has never attained export importance and has been less successful than mamey sapote in spite of the fact that the fruit can be superior. It was difficult to grow in South Florida in the past because freezes have killed plants, but now there are some better-established plants (Campbell and Ledesma, 2002).

Ecology

Soil

The trees adapt well to a wide range of soils with good drainage. They are intolerant of flooding or excess water in the soil.

Rainfall

This species prefers to be in places with well-distributed rainfall (Popenoe, 1948).

Temperature

Green sapote is native to the higher altitudes, around 900–2,100 m, of Central America, mainly Guatemala and Honduras and more rarely in Costa Rica and Panama. The tree is best adapted to cool, tropical areas (Popenoe, 1948; Campbell and Ledesma, 2002) that are too cold for canistel or abiu, which prefer the lowland tropics. This does not mean the tree is cold hardy since it has limited cold tolerance, with mature trees suffering twig and leaf damage and young trees being severely damaged or even killed by temperatures lower than −2.2°C to −3.8°C (Campbell *et al.*, 1977).

Light and photoperiod

The tree prefers to grow under full sun and is adapted to the day lengths of the tropics and subtropics, since it flowers and produces fruit there.

Wind

The tree is not especially resistant to strong winds.

General characteristics

Tree

The green sapote is an evergreen erect tree that can reach 12–24 m in height (Martin *et al.*, 1987). The young stems are densely covered with brown hairs and the leaves, as in many of its relatives, are clustered towards the shoot apices of fruiting branches and are irregularly alternate along non-fruiting limbs. Leaves are oblanceolate, 10–25 cm long and 5–7 cm wide, wavy and pointed (Fig. 4.4). Leaves are hairy along the upper midrib and downy-white beneath.

Flowers, pollination and fruit set

The flowers are borne in groups of two to five in leaf axils and along leafless branches (Fig. 4.4). They are tubular, five-lobed, silky-hairy and pinkish to white in color with nine to ten sepals (Morton, 1987; Crane, 2008). Trees may bloom during late winter and early spring (Whitman, 1965). The flowers appear to be self-fertile and insect pollinated (Whitman, 1965; Martin *et al.*, 1987).

Fruit

The fruit varies from nearly round to ovoid to ellipsoid; it is pointed at the apex, 8–20 cm long and 6–8 cm thick (Fig. 4.4), with a thin (almost membranous), olive-green-colored peel that clings tightly to the pulp and presents red-brown spots. At maturity, the outer peel becomes translucent with a brownish-orange to amber color underneath. The ripe pulp is smooth, sweet and orange to red colored. The fruit can contain one or two dark brown, shiny, 4–6 cm long, elliptic or ovate seeds (Fig. 4.4B). Green sapote fruit take about 12 months from fruit set to harvest. Fruit are generally harvested in Florida during the fall and winter from February to June in Guatemala, from March to August in Nicaragua and in Costa Rica from January to June (Azurdia, 2006).

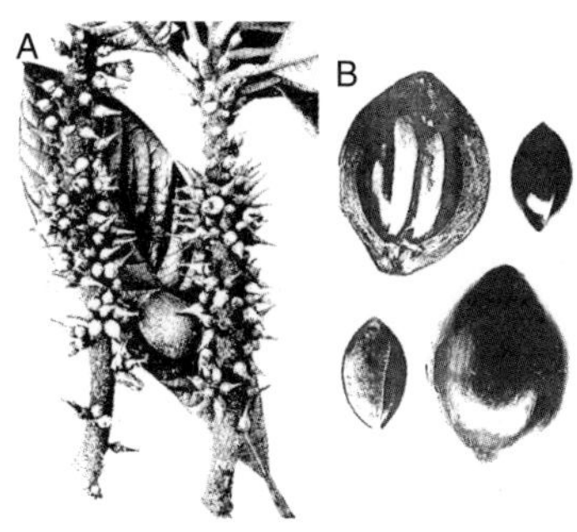

Fig. 4.4. Green sapote (*Pouteria viridis*): (A) leaves, flowers and fruit; (B) transverse section of the fruit and seed. (From Pittier, H. (1914) New and noteworthy plants from Colombia and Central America. *Contributions from the United States National Herbarium* 18 (2), 69–86.)

Cultivar development

Very little selection or breeding has been done with this species. There are no named cultivars though there appears to be considerable variation in productivity, fruit size and peel texture within the species. Trees with superior characteristics have been described (Granados-Friely, 1994; Whitman, 1965).

Cultural practices

Propagation

SEXUAL Green sapote is normally propagated by seed (Martin *et al.*, 1987). Removing or cracking the hard seed coat prior to planting speeds up germination (Whitman, 1965). The seedlings take 8–10 years to begin fruiting (Popenoe, 1948; Morton, 1987).

ASEXUAL The tree can also be propagated by grafting (Martin *et al.*, 1987). Grafting success can be improved by preparing scion wood by girdling 15 cm below the shoot apex and removing all but the terminal three to four leaves 20–30 days prior to grafting. Green sapote may be grafted on mamey sapote (*Pouteria sapota*) and in South Florida, trees grafted onto this rootstock appear to grow more vigorously than seedling trees of green sapote (Whitman, 2001). Grafting onto canistel (*Pouteria campechiana*) has not been successful (Ogden and Campbell, 1980). Inverted root graft seems to work well as for other fruits of the Sapotaceae, resulting in precocity and size reduction (Wasielewski and Campbell, 2000; Campbell and Ledesma, 2002).

Orchard establishment

The normal field preparation should be used. No specific plant spacing has been recommended in the absence of commercial orchards. Probably 6–7 m will be adequate for grafted plants.

Irrigation

Irrigation will be necessary during prolonged dry seasons or in arid areas.

Pruning

Young trees are generally not trained but the ideal would probably be to let the plant grow up to about a meter after transplanting and eliminate the tip so that it will produce side branches. From these branches three are to be selected and trained. Mature trees may be pruned by hand or with machinery to limit tree size. Pruning should include eliminating all dead, diseased or too low hanging branches and any branch that is growing into the canopy or deforming its contour or inserted at a too narrow angle.

Fertilization

No specific fertilizer recommendations have been reported for green sapote production; however, practices common to similar fruit trees such as mamey sapote or citrus appear to suffice (Crane, 2008). The use of animal manure is encouraged.

Pest management

DISEASES Little information is available about tree or fruit diseases, suggesting in backyard plantings these do not pose a problem. An unidentified fungus has been observed to attack the graft union of green sapote grafted onto mamey sapote that was controlled by fungicide application to the graft union (Crane, 2008).

INSECTS No major insect problems appear to plague the tree. Occasionally, Cuban May beetles (*Phyllophaga bruneri*) feed on leaves, but this is more of a problem for young trees with a limited number of leaves than mature trees. Fruit flies (*Anastrepha* sp.) can attack ripening fruit and could become a major problem in larger established orchards.

Weed control

Care has to be taken to keep the area around the young plants free of weeds, to minimize competition. This can be achieved by using a hoe or mulching with either a black plastic sheet or organic mulch, taking care not to put it directly in contact with the stem base, to avoid possible rotting problems. The rest of the area can be kept under control with brush cutters or machete.

Harvesting and postharvest handling

No exact data is available on tree yields in the absence of commercial orchards (Azurdia, 2006). According to Balerdi and Shaw (1998) individual trees have been estimated to produce from 34 to 45 kg.

Harvesting

The fruit of green sapote take about 5–9 months from fruit set to harvest depending upon temperature (Martin *et al.*, 1987). Fruit should be picked while still hard and the peel has becomes translucent with a brownish-orange color showing through the outer peel or when the fruit starts coloring yellow or orange in the types with reddish-colored fruit (Azurdia, 2006). An orange color under the persistent calyx also indicates maturity. Fruit should then be allowed to soften before it can be consumed or used for processing. Fruit on a tree do not ripen uniformly, therefore pickings should be programmed every 5–6 days.

Harvesting should be made using knife or pruning shears. Sometimes the fruit can be pulled from the tree but this normally damages the fruit. The stalk wound will start bleeding latex for some time and care has to be taken that this latex does not contact other fruit.

Postharvest treatment

Hard mature fruit take 4–14 days to fully ripen at room temperature (Crane *et al.*, 2001). There is no information on postharvest storage, but harvested fruit should be put in a shaded cool place. Once fruit turns soft its storage life is very short (~2 days). Sometimes the fruit can be opened, the seeds removed and the pulp frozen in plastic bags; this way pulp can be available for 2–3 months (Azurdia, 2006).

Utilization

The pulp of green sapote can be eaten fresh, preserved or utilized in desserts. The orange to red flesh has 68–70% moisture content (Balerdi and Shaw, 1998). The major pulp sugars include sucrose followed by glucose and fructose (Chan and Heu, 1975). Green sapote pulp has recently been shown to be high in antioxidant activity (Ma *et al.*, 2004) and a reasonable source of vitamin C (Table 4.1). The latex has been collected to be used like that from sapodilla for the chewing gum industry.

Lucuma

Lucuma (*Pouteria lucuma* (Ruiz & Pav.) Kuntze) was formerly placed in the genus *Lucuma* as *L. obovata*. Other synonyms are *P. insignis* Baehni, and *L. bifera* Mol., *Richardella lucuma* Aubr., and *Achras lucuma* Ruiz & Pavón. The common names are *lucuma* in Chile and Peru; *lucma* in Ecuador; *lucuma* or *rucma* in Colombia; and *mamón* in Costa Rica (Morton, 1987). This fruit is similar to that of canistel or egg fruit but the pulp contains less moisture and latex.

Origin and distribution

This is a subtropical fruit tree that originated in the slopes of the Ecuadorian and Peruvian Andes mountains (Popenoe, 1934). The fruit was well known to the Incas and in its native range but little known outside its homeland. The fruit is represented in ancient pottery from the pre-Inca time indicating it was already being used. In some parts of Peru, it plays a significant role in the basic diet of the poor. It is mainly being cultivated in Peru and Chile where the population is familiar with its use but it has not expanded to other parts of the world, except as plants in collections or backyard orchards. In Peru, the bulk of the production is in the Central Andean department of Ayacucho and in the central coast in the vicinity of Lima. It is estimated that there are around 650 ha technically cultivated, plus the non-technical production in the Andes that amounts to about another 200 ha that is not included in the statistics, being scattered trees and backyard trees in many houses. In Chile, it is grown mainly in the central area of Quillota and in the Elqui River Valley near La Serena with around 200 ha under cultivation and many backyard trees.

Ecology

Soil

The plant adapts itself to many types of soils but sandy or sandy loams with a high content of organic matter and good drainage are the most suited. It has moderate tolerance to salinity and alkalinity. The ideal pH is around 6.5–7.5.

Rainfall

The tree is adapted to fairly dry conditions and it can withstand fairly long periods of drought without showing symptoms. In the Peruvian Andes, it grows under moderate rainfall that lasts 6–7 months, but for proper production it needs supplementary irrigation. On the Peruvian coast, where it never rains, trees are irrigated and yields are very adequate. It will grow well with a yearly rainfall of about 800–1,000 mm or its equivalent in irrigation. Excess of soil moisture should be avoided.

Temperature

It grows in the tropics/subtropics at high elevation up to 3,000 m. It also grows and produces good quality fruit in the coastal areas of Peru and Chile that have a subtropical climate. The tree is killed by temperatures below −5°C. The ideal temperature for fruit development seems to be between 18 and 24°C (Franciosi, 1992). It does not adapt well to dry or humid tropics (Calzada-Benza, 1993).

Light and photoperiod

It grows in full sunlight and flowers and sets fruit all year round in the absence of drought. It also flowers in central Chile and near the equator, suggesting no photoperiodic requirement.

Wind

Moderate to strong winds tend to deform the plants and cause flower damage or low fruit set. Wind barriers should be used under these circumstances.

General characteristics

Tree

This evergreen tree has a straight trunk reaching up to 8–15 m, even 20 m if not grafted, and a canopy diameter of 6–8 m. The dense and spherical crown has the leaves concentrated at the apex of the young branches with their younger tips having a light or dark brown-colored pubescence. Each alternate leaf is lanceolate, oblong or elliptical, with a flattened base, 12.5–25 cm long and 5–10 cm wide (Fig. 4.5). The leaves have a leathery texture and are dark green on the underside with a very noticeable mid-vein, in some cases with undulated margins. As with most fruit trees in this group the plant exudes latex when a leaf is broken off or is wounded, and the same with immature fruit where latex is present until it completely ripens.

Flowers

The small profuse flowers are solitary or in clusters of two or three in the leaf axils (Fig. 4.5). The flowers are tubular, yellow or green, and invariably hermaphrodite and normally self-fertile. Five to seven hairy sepals are found in two series, the external with three to four sepals and the internal with two to three sepals, covered by a rust-colored pubescence. The sepals persist until the fruit is ripe and normally stay attached to it after harvest. There are five light green petals that fall after anthesis and five stamens that arise from the base of the petals and five staminodes,

Fig. 4.5. Lucuma (*Pouteria lucuma*) showing leaves, flower, fruit and seed (from Ruiz Lopez, H. and Pavon, J. (1799) *Flora Peruviana et Chilensis*. Vol. II. Typis Gabrielis de Sancha, Madrid).

longer than the stamens that arise between the petals. The pentalocular ovary ends in a cylindric style (León, 2000). The petals are longer than the calyx and the pistil elongates still more than the petals to be able to receive pollen (Fig. 4.5).

Flower buds form in the leaf axils of new leaves shortly after the new shoots have formed, and some flower buds form in branchlets that arise in the axils of old leaves. This flowering pattern has to be taken into account at pruning time. Flowering is not simultaneous and in many places the plants flower almost continuously.

Pollination and fruit set

Pollination is done by insects, especially honey bees. To ensure pollination and fruit set one to three beehives per ha are needed.

Fruit

The fruit is oblate, ovate or elliptic, pointed or depressed at the apex, 7.5–10 cm long (Fig. 4.5), with thin, delicate skin, and brownish-green or yellow-green more or less overlaid with russet. The pulp is bright yellow, firm, dry, mealy, sweet and permeated with latex until almost overripe. There may be one to five, usually two, rounded or broad-oval, dark brown, glossy seeds with a broad whitish hilum on one flattish side (Morton, 1987). The fruit follows a sigmoid growth curve and takes almost nine months from anthesis to maturity (Morton, 1987; Franciosi, 1992).

Cultivar development

Little selection or breeding has been done with this species. There are few named cultivars. Considerable genetic diversity exists in the size and shape of the fruit, skin color (green to yellow-green) and flesh color (light yellow to bright yellow), flesh texture and fruit aroma. In Peru, there is a clear distinction between what is called "*lúcuma de seda*" (silk lucuma) meaning a soft pulp and "*lúcuma de palo*" (woody lucuma) meaning a hard compact pulp with a smaller fruit. The latter is used for rootstock and industrialized as meal or frozen. In addition, in the 1970s, selection work was started at the Universidad Nacional Agraria (UNA) La Molina in Lima with the result of three superior clones named: 'Lucuna B-1', 'Lucuna B-2' and 'Lucuna R-3' (Calzada-Benza *at al.*, 1972). Later selections include 'Beltran 4' called sometimes 'Calzada', 'Yema de Huevo', 'San Hilarión', 'La Molina', 'Amarilla 1' and 'Tercer Mundo' that are used by some growers; all of them belong to the soft-pulp type, but there is not a dominant variety. In Chile, which has many backyard trees and few specialized production areas, there are several selections that bear the name of the person that selected it or the place where the tree was found, the most important being 'Montero', an early producer, and 'Rosalía', a late producer; other selections include 'Dispersa', 'Globosa', 'San Patricio', 'Vegara', 'Merced' and 'Bozzolo'.

Cultural practices

Propagation

SEXUAL Seed propagation is used to produce the rootstocks. Seeds are rather fleshy and should not be dried too much so they do not lose their viability. The ideal is to get freshly extracted seeds of good size and to remove the seed coats resulting in fastest and higher germination; cracking the seed coats produces intermediate results in relation to leaving them intact, while soaking the peeled seeds in 100 ppm gibberellic acid for 24 h significantly increases germination rate and seedling growth (Duarte *et al.*, 1976). Germination should start 25–30 days after sowing with some unpeeled seeds taking up to 3 months. The ideal is to put the seeds directly in the nursery bag to avoid transplanting a very small plant with a poor root system that will suffer and often die in the process. Once the seedlings reach a 0.8–1.0 cm diameter at 30 cm from the soil they can be grafted. This takes place normally 10–12 months after sowing.

ASEXUAL Grafting is one of the most used methods for clonal propagation of lucuma (Calzada-Benza, 1993; Franciosi, 1992). Splice or whip-and-tongue grafting are among the most used types of graft. There is a tendency for the cuts to bleed latex that has to be wiped with the grafting knife just before the two cuts are joined. One technique to improve grafting is to let the plants go without watering for 2–3 weeks prior to the process and irrigate them heavily the day before grafting. Root grafting or the inverted seed grafting technique can also be used, where the seedling with a 2-week-old main root is lifted, this root cut to 4–5 cm and cleft grafted with a thin scion, then the grafted seed is planted in an inverted position and roots will form again with the main root becoming the base of the grafted plant; this way precocity is obtained.

Leafy cuttings have also been rooted successfully using terminals with mature leaves and treating them with naphthalene acetic acid (NAA) at 4,000 ppm; this material was planted in a disinfected substrate of equal volumes of sand, peat moss and Styrofoam shavings and put under mist or hermetic polyethylene chamber with 50% shade in both cases. After 50 days, 78% and 80% rooting was obtained for one clone under mist and polychamber respectively, while the other clone under mist rooted 69% (Duarte, 1992).

Orchard establishment

The normal field preparation should be used with plowing and harrowing. In fields with a hardpan or plow-pan problem sub-soiling is necessary. Holes 50 × 50 cm are prepared at distances of 6 × 7 m to 6 × 5 m with shorter distances in poorer or sandy soils (Franciosi, 1992). The 50–80 cm grafted plants normally come in plastic nursery bags, which should be carefully removed and the plants installed with the substrate level slightly above the soil level to avoid water accumulation at the stem base. On the Peruvian coast with sandy or very sandy new irrigation soils, raised beds or humps are prepared in the plant rows that are 30–40 cm high and about 1.50 m wide at their base and 1.0 m at the top. The plants are planted in the middle of the bed. These beds are normally prepared with the incorporation of manure or compost.

Irrigation

Irrigation will be necessary during prolonged dry seasons or in arid areas. Although lucuma is a very drought tolerant plant, it does not set fruit under these conditions. The traditional irrigation system has been by furrows that can go around both sides of the plant or a circular or square furrow is made to wet the area under the canopy. At present, most of the modern plantings use drip irrigation with normally two lines of drippers along the hump and with drippers every 25–30 cm.

Pruning

Lucuma tends to be a symmetrical tree but needs some pruning. After planting trees should be topped at about a meter from the ground to induce the development of several primary branches, and three or four of them that are well spatially oriented, arising as separated as possible from each other vertically on the stem, are retained. In the second year, these side branches are cut back to 30–40 cm to induce profuse secondary branching. All shoots going into the canopy as well as water sprouts and rootstock sprouts below the graft union are eliminated. Maintenance pruning is then required to keep the tree in production.

Fertilization

During the last two to three decades some recommendations have been established that vary according to the soil analysis, the plant stage of growth and yields. No standards for leaf analysis have been developed. Present recommendations are to apply around 75 g of N, 45 g of P_2O_5, 40 g of K_2O and 10 g of MgO per plant for every year of age of the plant up to the tenth year and if possible to add manure at about 2 kg per plant for every year of age of the plant up to 12–15 years. The addition of 200 g triple superphosphate to the planting hole is also recommended. For dry applications normally 33% of the yearly amount of chemical fertilizer and manure should be applied three times a year, except for phosphorus that is applied as a single dosage once a year. The times of application are before the heavy flowering season, with the other two following at 3-month intervals. For this a small furrow should be made close to the drip line in order to bury the fertilizer and the manure, and after that a heavy irrigation should follow. The use of drip irrigation means fertigation is possible with many mini applications during the year.

Pest management

DISEASES Powdery mildew is one of the main problems in the humid Peruvian coast, though not in the Andes where humidity is low. Algae can grow on the foliage in very humid areas and is controlled with copper fungicides.

INSECTS The main insect in lucuma is the fruit fly (*Anastrepha serpentina*). Other insects include the hemispherical scale (*Saisettia coffeae*) and whiteflies (*Aleurothrixus floccosus*). In some cases, caterpillars like *Clutomulus*, normally controlled by parasites, can become a problem.

Weed control

Care has to be taken to keep the area around the young plants free of weeds, to minimize competition. Mechanical or chemical means are used.

Harvesting and postharvest handling

Yields

A tree can produce about 500 fruit per year with yields of up to 15–18 t ha^{-1} $year^{-1}$ in the tenth year. Fruit can weigh 250–300 g or up to 1 kg in exceptional cases.

Harvesting

Fruit is ready for harvest about 8–9 months from anthesis. On the Peruvian coast, the main harvest starts in October–November and can end in February–March. In the Andes, harvest starts in December and ends in May, similar to northern Chile, while in central Chile it starts in April and ends in November, depending on the specific selections.

Harvest starts when the peel of the fruit starts turning yellowish; this stage is too advanced for long shipments where a less advanced maturity (green-mature) is needed. Some types will show the yellowing initially only in the peel under the persistent calyx. Pulp color also turns from clear yellow to orange or orange-yellow as fruit ripens. Fruit harvested too early shrivel, do not ripen properly and latex will freely flow at the point the peduncle is cut. Many times ripe fruit tends to have a cracked peel. Fruit should be hand-picked and treated with the utmost care when put into a harvest bag before being transferred to a cushioned box and taken to the packing house.

Lizana *et al.* (1986) recommended that fruit be harvested when the penetration force is between 36 and 54 N or the total soluble solids content by refractometry is 10–12%. Different standards would need to be established for each variety and final fruit usage, fresh or processed.

Postharvest treatment

Lucuma is a climacteric fruit (Lizana *et al.*, 1986) and must be picked when mature, as immature fruit do not ripen properly. The pulp becomes soft and the soluble solids content increases from about 9–10% to 16–17% (Lizana, 1990). The soft pulp and its tendency to lose water rather quickly does not lend this fruit for long transit or storage times, thus fruit for export is normally frozen or dried. Only a small percentage of fruit reaches the fresh fruit market.

Fruit held at 13–18°C can be kept for 2 weeks in good condition. The quality of fruit stored at 7°C for up to a week is not affected by storage temperature while longer periods will cause non-uniform ripening. Dehydration is a major reason for the loss in fruit appearance postharvest and the storage conditions (packing and humidity) need to minimize water loss (Yahia and Gutierrez-Orozco, 2011).

Utilization

This fruit with its smooth, bronze-yellow skin somewhat resembles a persimmon. The bright yellow or orange flesh is usually blended into other foods and tastes and smells like maple syrup. It normally is not eaten out of hand because of its mealy and rather dry texture (similar to a hard-boiled egg yolk). It is used in drinks, puddings, pies, cookies, and cakes where it adds its special flavor. When mixed with milk or ice cream, it adds both color and flavor. The lucuma is high in solids and is a good source of carbohydrate and calories (Table 4.1). It is low in acid, and a good source of fiber, iron, carotene (provitamin A), niacin (vitamin B3), and phenolics. Ripe fruit are dried and milled into flour that is used by the ice cream and pastry industries and for export. Frozen pulp and puree are also produced and exported from Peru. The tree is valued for its dense, durable timber.

References and Further Reading

Caimito or star apple

Alix, C. and Duarte, O. (1999) *Propagación de Especies Frutales Tropicales*. CURLA; PDBL; AFE / COHDEFOR; DICTA; SETCO / PROFORFITH. La Ceiba, Honduras, pp. 87–88.

Barbeau, G. (1990) *Frutas Tropicales en Nicaragua*. Dirección General de Técnicas Agropecuarias, MIDINRA, Editorial Ciencias Sociales, Managua, Nicaragua, pp. 341–342.
Campbell, C.W. (1974) Research on the caimito (*Chrysophyllum cainito* L.) in Florida. *Proceedings of the Tropical Region of the American Society for Horticultural Science* 18, 123–127.
Campbell, C.W., Knight, R.J. Jr. and Zareski, N.L. (1977) Freeze damage to tropical fruits in southern Florida in 1977. *Proceedings of the Florida State Horticultural Society* 90, 254–257.
Crane, J.H. (2008) *Chrysophyllum cainito* Star apple. In: Janick, J. and Paull, R.E. (eds) *Encyclopedia of Fruit and Nuts*. CAB International, Wallingford, UK, pp. 825–827.
Hoyos, F.J. (1989) *Frutales en Venezuela, Nativos y Exóticos*. Monografía No. 36. Sociedad de Ciencias Naturales La Salle, Caracas, Venezuela, pp. 288–290.
Martin, F.W., Campbell, C.W. and Ruberté, R.M. (1987) *Perennial Edible Fruits of the Tropics: An Inventory*. Agriculture Handbook No. 642. US Department of Agriculture (USDA), Agriculture Research Service (ARS), US Government Printing Office, Washington, DC, pp. 59–60.
Morton, J.E. (1987) *Fruits of Warm Climates*. Creative Resource Systems, Winterville, North Carolina, pp. 408–410.
Subhadrabandhu, S. (2001) *Under-utilized Tropical Fruits of Thailand*. RAP (Regional Office for Asia and the Pacific) Publication 2001/26. Food and Agriculture Organization (FAO), Rome, pp. 51–53.
Vargas, O., Alix, C., Lobo, A.D. (authors), Duarte, O. and Sanchez, J. (technical reviewers) (1999) *Frutales y Condimentarias del Trópico Húmedo*. CURLA; PDBL; AFE/COHDEFOR; DICTA; SETCO; PROFORFITH, La Ceiba, Honduras, pp. 298–299.
Verheij, E.W.M. and Coronel, R.E. (eds) (1992) *Edible Fruits and Nuts. Plant Resources of South East Asia No. 2.* PROSEA Foundation, Bogor, Indonesia, pp. 115–117.
Villachica, H., Carvalho, J.E.U., Muller, C.H., Diaz, S.C. and Almanza, M. (1996) Caimito. In: *Frutales y Hortalizas Promisorios de la Amazonía*. Tratado de Cooperación Amazónica, Lima, Perú, pp. 69–74.

Abiu

Alix, C. and Duarte, O. (1999) *Propagación de Especies Frutales Tropicales*. CURLA; PDBL; AFE / COHDEFOR; DICTA; SETCO / PROFORFITH. La Ceiba, Honduras, pp. 88–89.
Campbell, C.W., Malo, S.E. and Goldweber, S. (1967) *The sapodilla*. University of Florida Cooperative Extension, Fruit Crop Fact Sheet No. 1.
Cavalcante, P.B. (1991) *Frutas Comestiveis da Amazonia (Edible Fruits of Amazonia)*. Museu Paraense Emilio Goeldi, Belém, Pará., Brazil.
Clement, C.R. (1989) A center of crop genetic diversity in Western Amazonia. *BioScience* 39, 624–631.
Clement, C.R. and Paull, R.E. (2008) *Pouteria caimito*, abiu. In: Janick, J. and Paull, R.E. (eds) *Encyclopedia of Fruit and Nuts*. CAB International, Wallingford, UK, pp. 832–835.
Donadio, L.C. (1983) Review of South American Fruits. *Proceedings of the Tropical Region of the American Society for Horticultural Science* 27, 21–41.
Donadio, L.C., Moro, F.V. and Servidone, A.A. (2002) *Frutas Brasileiras (Brazilian Fruits)*. Editora Novos Talentos, Jaboticabal, São Paulo, Brazil, pp. 43–46.
Falcao, M.A. and Clement, C.R. (1999) Fenologia e produtividade do abiu (*Pouteria caimito*) na Amazonia Central (Phenology and yield of abiu in Central Amazonia). *Acta Amazonica* 29 (1), 3–12.

Ferreira, A.R. (1974) *Viagem filosofica pelas capitanias do Grao-Pará, Río Negro, Mato Grosso e Cuiabá (Philosophical voyage in the provinces of Grao-Pará, Río Negro, Mato Grosso and Cuiaba)*. Conselho Federal de Cultura, Brasilia.

Leung, W.T.W. and Flores, M. (1961) *Food Composition Table for Use in Latin America*. The Institute of Nutrition of Central America and Panama, Guatemala City and US National Institute of Health, Bethesda, Maryland.

Lim, T.K. and Ramsay, G. (1992) Abiu – a new fruit with potential for the Northern Territory. *Acta Horticulturae* 321, 99–105.

Manica, I. (2000) *Frutas Nativas, Silvestres e Exoticas. I* (Native, Wild and Exotic Fruits. I). (Abiu, amora-preta, araca, bacuri, biriba, carambola, cereja-do-Rio-Grande, jabuticaba). Cinco Continentes Editora, Porto Alegre, Rio Grande do Sul, pp. 13–43.

Marshall, J.R. (1991) *Sapodilla (Chico)*. Rare Fruit Council of Australia. Cairns, Queensland. Fact Sheet No. 4.

Martin, F.W. and Malo, S.E. (1978) *Cultivation of Neglected Tropical Fruits with Promise, Part 5. The canistel and its relatives*. US Department of Agriculture, Science and Education Administration, 12 pp.

Morton, J.F. (1987) *Fruits for Warm Climates*. Creative Resource Systems Inc., Winterville, North Carolina, pp. 406–408.

Nakasone, H.Y. and Paull, R.E. (1998) *Tropical Fruits*. CAB International, Wallingford, UK, pp. 389–400.

Parker, G.H. (1986) *Abiu*. Rare Fruit Council of Australia Fact Sheet No. 13. Rare Fruit Council of Australia, Cairns, Queensland.

Patiño, V.M. (2002) *Historia y Dispersión de los Frutales Nativos del Neotrópico (History and Dispersal of Native Neotropical Fruits)*. Centro Internacional de Agricultura Tropical (CIAT), Cali, Colombia.

Scholefield, P.B. (1984) Abiu. In: Page, P.E. (compiler) *Tropical Tree Fruits for Australia*. Queensland Department of Primary Industry Information Series QI 83013. Queensland Department of Primary Industry, Brisbane, Australia, pp. 216–217.

Vargas, O., Alix, C., Lobo, A.D. (authors), Duarte, O. and Sanchez, J. (technical reviewers) (1999) *Frutales y Condimentarias del Trópico Húmedo*. CURLA; PDBL; AFE/COHDEFOR; DICTA; SETCO; PROFORFITH, La Ceiba, Honduras, pp. 304–305.

Villachica, H., Carvalho, J.E.U., Muller, C.H., Diaz, S.C. and Almanza, M. (1996) Caimito. In: *Frutales y Hortalizas Promisorios de la Amazonía*. Tratado de Cooperación Amazónica, Lima, Perú, pp. 69–74.

Canistel

Alix, C. and Duarte, O. (1999) *Propagación de Especies Frutales Tropicales*. CURLA; PDBL; AFE / COHDEFOR; DICTA; SETCO / PROFORFITH. La Ceiba, Honduras, p. 88.

Azurdia, C. (2006) *Tres especies de zapote en América tropical*. Southampton Centre for Underutilised Crops, University of Southampton, Southampton, UK.

Campbell, C.W., Knight, R.J. Jr. and Zareski, N.L. (1977) Freeze damage to tropical fruits in southern Florida in 1977. *Proceedings of the Florida State Horticultural Society* 90, 254–257.

Campbell, R.J. and Wasielewski, J. (1999) Commercial potential of canistel for lowland tropical America. *Proceedings of the Interamerican Society for Tropical Horticulture* 43, 143–146.

Crane, J.H. (2008) *Pouteria campechiana*, Canistel. In: Janick, J. and Paull, R.E. (eds) *Encyclopedia of Fruit and Nuts*. CAB International, Wallingford, UK, pp. 835–837.

Crane, J.H., Balerdi, C.F., Campbell, C.W. and Regalado, R. (2001) Evaluation of 'Oro' and 'Trompo' canistel (*Pouteria campechiana* Baehni) at the University of Florida Tropical Research and Education Center, Homestead. *Proceedings of the Florida State Horticultural Society* 114, 3–4.

Duarte, O. and Villagrán, L. (2002) Efecto de escarificación, remojo en acido giberélico, posición de siembra y edad de la semilla en la germinación y conformación de plántulas de canistel (*Pouteria campechiana* Baehni). *Proceedings of the Interamerican Society for Tropical Horticulture* 46, 14–16.
McGregor, B.M. (1987) *Tropical Products Transport Handbook*. Agriculture Handbook No. 668. US Department of Agriculture (USDA) Agriculture Marketing Service (AMS), Washington, DC, p. 86.
Martin, F.W. and Malo, S.E. (1978) *Cultivation of Neglected Tropical Fruits with Promise, Part 5. The canistel and its relatives*. US Department of Agriculture, Science and Education Administration, 12 pp.
Martin, F.W., Campbell, C.W. and Ruberté, R.M. (1987) *Perennial Edible Fruits of the Tropics: An Inventory*. Agriculture Handbook No. 642. US Department of Agriculture (USDA), Agriculture Research Service (ARS), US Government Printing Office, Washington, DC, pp. 59–61.
Morton, J.F. (1983) Why not select and grow superior types of canistels? *Proceedings of the Tropical Region of the American Society for Horticultural Science* 27, 43–52.
Morton, J.F. (1987) *Fruits of Warm Climates*. Creative Resource Systems, Winterville, North Carolina, pp. 402–405.
Ogden, M.A.H. and Campbell, C.W. (1980) Canistel as a rootstock for mamey sapote. *Proceedings of the Florida State Horticultural Society* 93, 133–135.
Sauls, J. and Campbell, C.W. (1994) *The Canistel*. Florida Cooperative Extension Service, University of Florida. Fact Sheet HS-61, 2 pp.
Vargas, O., Alix, C., Lobo, A.D. (authors), Duarte, O. and Sanchez, J. (technical reviewers) (1999) *Frutales y Condimentarias del Trópico Húmedo*. CURLA; PDBL; AFE/COHDEFOR; DICTA; SETCO; PROFORFITH, La Ceiba, Honduras, pp. 306–307.
Verheij, E.W.M. and Coronel, R.E. (eds) (1992) *Edible Fruits and Nuts. Plant Resources of South East Asia No. 2*. PROSEA Foundation, Bogor, Indonesia, pp. 258–259.
Wasielewski, J. and Campbell, R.J. (1997) Canistel cultivars in South Florida. *Proceedings of the Florida State Horticultural Society* 110, 141–143.
Wasielewski, J. and Campbell, R.J. (1999a) Comparison of side veneer, cleft and inverted seed grafts of canistel. *Proceedings of the Interamerican Society for Tropical Horticulture* 43, 147–150.
Wasielewski, J. and Campbell, R.J. (1999b) Pruning and shaping sapodilla, mamey sapote and canistel for proper growth and production. *Proceedings of the Interamerican Society for Tropical Horticulture* 43, 143–146.

Green sapote

Azurdia, C. (2006) *Tres especies de zapote en América tropical*. Southampton Centre for Underutilised Crops, University of Southampton, Southampton, UK, 231 pp.
Balerdi, C.E. and Shaw, P.E. (1998) Sapodilla, sapote and related fruits. In: Shaw, P.E., Chan, H.T. Jr. and Nagy, S. (eds) *Tropical and Subtropical Fruits*. Ag. Science Inc., Auburndale, Florida, pp. 78–136.
Campbell, C.W., Knight, R.J. Jr. and Zareski, N.L. (1977) Freeze damage to tropical fruits in southern Florida in 1977. *Proceedings of the Florida State Horticultural Society* 90, 254–257.
Campbell, R.J. and Ledesma, N. (2002) The sapodilla and green sapote potential in Tropical America. *Proceedings of the Interamerican Society for Tropical Horticulture* 46, 55–56.
Chan, H.T. Jr. and Heu, R.A. (1975) Identification and determination of sugars in starfruit, sweetsop, green sapote, jack fruit, and wi apple. *Journal of Food Science* 40, 1329–1330.
Crane, J.H. (2008) *Pouteria viridis*, Green sapote. In: Janick, J. and Paull, R.E. (eds) *Encyclopedia of Fruit and Nuts*. CAB International, Wallingford, UK, pp. 843–844.

Crane, J.H., Balerdi, C.E., Campbell, C.W. and Regalado, R. (2001) Evaluation of fruit from a green sapote (*Calocarpum viride* Pitt.) selection at the University of Florida Tropical Research and Education Center, Homestead. *Proceedings of the Florida State Horticultural Society* 114, 5–6.

Granados-Friely, J.C. (1994) El cultivo del zapote en Guatemala: investigación, desarrollo e industria. *Proceedings of the Interamerican Society for Tropical Horticulture* 38, 142–149.

Ma, J., Yang, H., Basile, M.J. and Kennelly, E.J. (2004) Analysis of polyphenolic antioxidants from the fruits of three *Pouteria* species by selected ion monitoring liquid chromatography-mass spectrometry. *Journal of Agricultural Food Chemistry* 52, 5873–5878.

Martin, F.W, Campbell, C.W. and Ruberté, R.M. (1987) *Perennial Edible Fruits of the Tropics: An Inventory*. Agriculture Handbook No. 642. US Department of Agriculture (USDA), Agriculture Research Service (ARS), US Government Printing Office, Washington, DC, p. 59.

Morton, J. (1987) *Fruits of Warm Climates*. Creative Resources Inc., Winterville, North Carolina, pp. 401–402.

Ogden, M.H. and Campbell, C.W. (1980) Canistel as a rootstock for mamey sapote. *Proceedings of the Florida State Horticultural Society* 93, 133–136.

Popenoe, W. (1948) *Manual of Tropical and Subtropical Fruits*. Macmillan, New York, pp. 340–343.

Wasielewski, J. and Campbell, R.J. (2000) The potential of inverted seed grafts for dwarfing and precocity in members of the Sapotaceae grown in South Florida. *Proceedings of the Florida State Horticultural Society* 113, 12–13.

Whitman, W.F. (1965) The green sapote, a new fruit for south Florida. *Proceedings of the Florida State Horticultural Society* 78, 330–336.

Whitman, W.F. (2001) *Five Decades with Tropical Fruit: a Personal Journey*. Quisqualis Books, Englewood, Florida, pp. 101–105.

Lucuma

Calzada-Benza, J. (1993) *143 Frutales Nativos*. Ediciones Universidad Nacional Agraria La Molina, Lima, Perú, 366 pp.

Calzada-Benza, J., Bautista, V., Bermúdez, J. and Morán, M. (1972) *Cultivo del Lúcumo*. Boletín Técnico No. 2, Programa de Frutales Nativos, Universidad Nacional Agraria – La Molina, Lima, Perú, 45 pp.

Duarte, O. (1992) Propagación del lúcumo (*Lucuma obovata* H.B.K.) por estacas terminales con hojas. *Acta Horticulturae* 310, 235–238.

Duarte, O., Santos, D. and Franciosi, R. (1976) Efecto de diversos tratamientos sobre la germinación y crecimiento de plántulas de lúcumo (*Lucuma obovata* H.B.K.). *Proceedings of the Tropical Region of the American Society for Horticultural Science* 20, 242–249.

Franciosi, R. (1992) *Manual del Cultivo del Lúcumo en el Perú*. Ediciones Fundeagro, Lima, Perú, 84 pp.

León, J. (2000) *Botánica de los Cultivos Tropicales*. Editorial Agroamérica, Instituto Interamericano de Cooperación para la Agricultura (IICA). San José, Costa Rica.

Lizana, L.A. (1990) Lucuma. In: Nagy, S., Shaw, P.E. and Wardowski, W.F. (eds) *Fruits of Tropical and Subtropical Origin, Composition, Properties and Uses*. Florida Science Source Inc., Lake Alfred, Florida, pp. 373–380.

Lizana, L.A., Reginato, G. and Lavanderos, J.C. (1986) Caracterización de la maduración de Lúcuma (*Lucuma bifera* Mol., *L. obovata* HBK). *Proceedings of the Interamerican Society for Tropical Horticulture* 30, 105–120.

Morton, J.E (1987) *Fruits of Warm Climates*. Creative Resource Systems, Winterville, North Carolina, pp. 408–410.

Popenoe, W. (1948) *Manual of Tropical and Subtropical Fruits*. Macmillan, New York, pp. 351–352.

Yahia, E.M. and Gutierrez-Orozco, F. (2011) Lucuma (*Pouteria lucuma* (Ruiz and Pav.) Kuntze). In: Yahia, E.M. (ed.) *Postharvest Biology and Technology of Tropical and Subtropical Fruits, Volume 3. Cocona to mango*. Woodhead Publishing Ltd., Cambridge, UK, pp. 443–449.

5 Solanaceae

The Solanaceae or nightshade family has a worldwide distribution and includes herbs, shrubs and trees that are of important economic value. The most important are in the genus *Solanum* and include potato (*S. tuberosum* L.), tomato (*S. lycopersicon* L.) and eggplant (*S. melongena* L.). Related species in the *Solanum* genus with edible fruit include the naranjilla (*S. quitoense* Lam.), a native of the Andean highlands, the pepino or melon shrub *S. muricatum* Alt. and the tree tomato *S. betaceum*, formerly *Cyphomandra betacea* Sendt., named tamarillo in New Zealand. Other species of importance are in the genus *Capsicum* that includes chillis and bell peppers, and *Physalis* with its edible fruit. The edible fruit of Cape gooseberry *Physalis peruviana* has become popular and is being exported from some countries. Solanaceae also has a number of ornamental species such as *Petunia* and *Lycianthes*, as well as plant species with psychoactive alkaloid drugs, *Datura* (*Brugmansia*), *Mandragora* and *Atropa belladonna*. Tobacco *Nicotiana tabacum* L. is also in this family.

Cocona

Cocona, *Solanum sessiliflorum*, formerly *S. topiro* Humb. & Bonpl., was mistakenly called *S. hyporrhodium* (Martin *et al.*, 1987). In Spanish, it is called *cocona, topiro, lulo grande, cubiyu, cocanilla, pepino, tamarillo, cubiu* or *tupiro*, in Portuguese it is *cubiú*, and cocona or peach tomato in English, while in the Tupi language it is *kubi´u* (Donadio *et al.*, 2002).

Origin and distribution

Cocona comes from the upper Amazon region, specifically the eastern foothills of the Andes at altitudes of 200–1,000 m. This area extends from the higher

Orinoco River region in Venezuela to the upper reaches of the Amazon in Ecuador, Colombia, Peru and Brazil.

This plant was cultivated widely in the upper Orinoco and Amazon basins before the arrival of the Europeans. Presently, it is still cultivated on a small scale in the same region and used both as a fruit and as a vegetable. It is little known outside of this area. It has been taken to several places including Florida, South Africa, Costa Rica and Puerto Rico but has not gained importance. Trials have been done in Honduras under low seasonal rainfall conditions and heavy soils with success.

Ecology

Soil

Cocona prefers light well-drained soils, rich in organic matter and with a pH close to 6.0. It does not tolerate flooded or wet soils that favor fungal root diseases. The water table should be at least 50–60 cm below the surface. It does produce in heavy and light, even sandy soils but they are not the most suited. As with many other Solanaceae, cultivation is not recommended on land that has been used for other members of this family because of the potential for root rot and nematode problems to be present. It is preferable to rotate the land with a non-susceptible crop for a couple of years to eliminate or reduce the inoculum level. The plant does not tolerate root knot nematodes (Martin *et al.*, 1987).

Rainfall

Cocona comes from hot and wet tropical lowland areas with an annual precipitation of 2,000–3,000 mm, up to 4,500 mm (Donadio *et al.*, 2002). It can be grown with irrigation in areas with much less rainfall as well as in arid areas with adequate temperatures using irrigation.

Temperature

Cocona is adapted to hot humid tropics, although not the low Amazonia tropics subject to flooding. It also grows well in the dry tropics from 500–1,000 m altitude with irrigation. The higher it is planted within its natural limits, the greater the plant longevity. Average temperatures for good growth range from 24 to 26°C. This species requires warmer areas than naranjilla, although the native ranges of the two species overlap.

Light and photoperiod

It can be grown under some shade but it does best under full sunlight. No information is available on photoperiodic responses. The plant flowers almost continuously throughout the year, indicating that day length does not influence flowering.

Wind

It has soft large leaves that are very susceptible to strong wind damage.

General characteristics

Plant

Cocona, like naranjilla, is a herbaceous shrub that becomes semi-hardwood with age. The main difference is that the latter is normally a shorter plant, much-branched, with no spines and its leaf veins are green instead of purple. The cocona plant can reach 2 m in height and the canopy almost 2 m in diameter. It tends to form several primary branches from its base that are densely white-hairy on the underside when young. The primary root of sexual origin is pivotal and is part of a fairly superficial root system not deeper than 50 cm. The alternate, ovate leaves with scalloped margins are downy with hard hairs on the upper surface, with prominent veins on the underside and softer hairs. The leaves have a very large leaf blade (70–90 by 35–40 cm) which is oblique to the base (Fig. 5.1) and attached to the stem by a 2–5 cm petiole (Duarte, 1997). The plant can live from 3 to 5 years. In the dry tropics of Central America with heavy soils, high disease and insect pressure it lives for 12–14 months and gets no higher than 1.0–1.2 m and can be managed as an annual (Duarte, 1997).

Flowers, pollination and fruit set

Flowers appear in the leaf axils in clusters of five to eight. Each flower is about 2.5–4.0 cm wide and they open sequentially on the stem. The flowers have five dark green sepals and five pale greenish-yellow petals that form a star with five

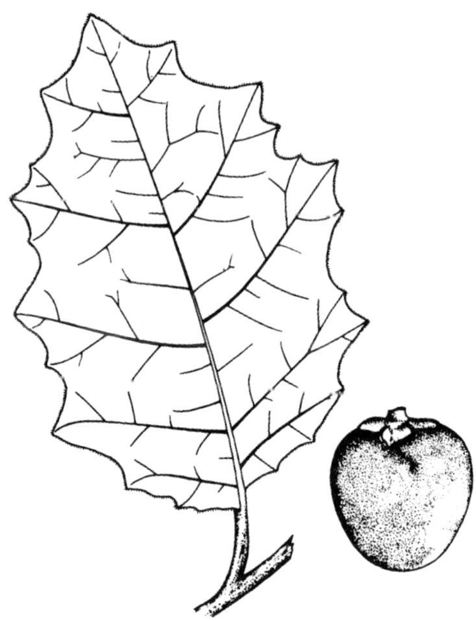

Fig. 5.1. Cocona (*Solanum sessiliflorum*) showing the large leaves and fruit (used with permission from León, J. (2000) *Botánica de los Cultivos Tropicales*. Editorial Agroamérica, Instituto Interamericano de Cooperación para la Agricultura (IICA), San José, Costa Rica).

yellow stamens and a 2–5 mm peduncle. Once the plant starts flowering, it continues to flower for the rest of its life with variable intensity and fruit production follows the same pattern. A plant can produce around 1,000 flowers during its lifetime but only 5–10% will set fruit. Cocona is self-fertile although bees visit the flowers and natural crosses are common.

Fruit

The fruit is a berry with several forms from ovoid (Martin *et al.*, 1987) to spheric, oblate, oblong or conical-oval. Each fruit is 6–12 cm long and 4–8 cm wide, weighing from 24 to 240 g, with a bluntly rounded apex (Fig. 5.1). The thin, tough skin is coated with a peach-like fuzz formed by trichomes that stay strongly attached until the fruit matures; once the skin becomes smooth acquiring a golden to orange-yellow, burnt-orange, red, red-brown or deep purple red color these hairs are easily removed by rubbing with the hands or a piece of cloth.

When cut transversely, a cream-colored thick (4–8 mm) hard flesh layer is found under the skin ("hoof") surrounding four compartments containing the fairly liquid, jelly-like, creamy pulp with numerous seeds. This hoof and the compartment walls are fairly hard and can be used either alone or mixed with the liquid pulp to prepare other products. The hard flesh has a faint tomato-like aroma and the liquid pulp has a pleasant acidity. The small (2–4 mm), cream-colored seeds are flat and are not noticed during eating. There are about 600–1,200 seeds in a fruit. The time from anthesis to maturation is about 100 days. Fruit at different stages of development are normally present on the plant year round.

Cultivar development

Cytogenetics, genetics and breeding

Cocona ($2n = 24$) has a high genetic diversity shown by the form, size, color, pubescence, flavor and aroma of the fruit (Villachica *et al.*, 1996). The plant is apparently not found in the wild having been observed only in cultivation. A wild relative, *S. topiro* var. *georgicum* Heiser, grows in the lowlands of eastern Ecuador and Colombia and has spines on the stem, branches and leaves and spontaneously hybridizes with the cultivated form. This species is believed to be the ancestor of cocona. Cocona has a strong dominance by the female parent in fruit characteristics so that crossing plants with large fruit used as the female parent results in large-fruited offspring, independent of the characteristic of the pollen parent plant, and this influence continues in the second generation (Villachica *et al.*, 1996).

Crosses of a large fruit cocona have been made with naranjilla and have resulted in the 'Puyo' and 'Palora' hybrids. The hybrids have a significantly higher yield (18 t ha^{-1}) than naranjilla (4–5 t ha^{-1}) in Ecuador and southern Colombia because of nematode and fungal infections that attack the traditional naranjilla. 'Puyo' has a greenish juice with no seeds and a thinner skin than the 'Palora' hybrid which has orange juice (Rodríguez *et al.*, 1994).

Selection and evaluation

The Instituto Nacional de Pesquisas da Amazônia (INPA) at Manaus in Brazil has a collection of cocona strains from Peru and Brazil. Studies from Peru have shown the existence of more than 25 biotypes and 11 of those have been selected for higher yield, larger fruit, better fruit taste and resistance to nematodes.

Cultivars

There are not many defined cultivars. There are agro-types according the shape and size of the fruit. Peru distinguishes four types: (i) medium yellow fruit; (ii) round-yellow fruit; (iii) small purple-red fruit; and (iv) pear-shaped fruit (Morton, 1987). The producer normally selects the best fruit from the higher yielding plants to obtain the seeds. According to Donadio *et al.* (2002) INPA released a superior cultivar named 'Alejo' that can produce 60 t ha^{-1}.

Cultural practices

Propagation

SEXUAL Coconas are normally seed propagated (Martin *et al.*, 1987). This method is fairly safe for maintaining the genetic uniformity since seeds are obtained from the best fruit, coming from healthy and productive plants that represent the desired type. Special care has to be taken that the mother plant is free of bacterial diseases that can be seed-transmitted. Seed propagation is not difficult since there is no dormancy. The pulp of mature fruit is placed in a blender with water and blended for a short period. This mixture is passed through a sieve and the seeds are separated by flotation eliminating all pulp residues that float. After seed cleaning, the seeds are dried in the shade between thin layers on newspaper; once dried, they are ready to be sown. Fermenting the pulp containing the seeds for 2 days or washing the seeds does not improve germination (Duarte *et al.*, 2000a). Production starts about 10–12 months after sowing (Martin *et al.*, 1987) with about 2–3 months in the propagation trays or bags, and 6–9 months in the field.

ASEXUAL Terminal or sub-terminal leafy cuttings with only the top leaf left and leaf-bud cuttings with half of the leaf removed root well during the warmest season. Treating cuttings with 1,000–3,000 ppm indole butyric acid (IBA) enhances rooting of terminal, sub-terminal and leaf-bud cuttings; basal wounding also stimulated rooting (Duarte *et al.*, 2000a). However, the vigor of plants from cuttings is not as good as that of seedlings. Grafting onto nematode or root rot resistant stocks is possible.

Orchard establishment

This is done the same way as for planting vegetable crops, normally plowing and harrowing. When cocona plants are established as associate crops or in land where no machinery is available, it is necessary to remove weeds and prepare the planting holes at the chosen distance.

Transplanting can be done directly from trays containing small (10–15 cm) seedlings; this can be used when irrigation is available or rains are continuous. Larger plants (20–30 cm) in bags can also be used when planted in poorer soils. Under appropriate growing conditions, plants should be set at 1.0–2.5 m apart (Villachica *et al.*, 1996). The plant will normally live for 3–4 years; however, if soil fungal inoculum and/or nematode pressure is high it will normally last only 12–15 months as plants progressively die as they begin to bear fruit.

In Honduras, where plants normally lived 12–15 months and were managed as annuals, several distances were tested in a fairly heavy soil with non-optimal irrigation. The highest number of fruit per plant was obtained with distances of 1.50 m by 1.25 m but total yield was higher at 1.5 m between rows and 0.5 m between plants (13,333 plants per ha) resulting in an equivalent of 24 t ha^{-1}. No major differences were found in fruit weight (~100 g) at the six densities tested (Duarte and Sandoval, 2002).

Irrigation

Most cocona plants are grown without irrigation. In places with a long dry season or with little or no rain, furrow or drip irrigation is needed. The plant is very succulent, has a very large leaf area and therefore does not tolerate long periods of drought. Ideally, plants should be irrigated no less than every 7–10 days although with drip irrigation this would be almost daily. Water consumption is very high.

Pruning

Pruning practices are normally limited. Basal shoots are left to develop to form a low canopy. Maintenance pruning consists in eliminating the lower and old yellowish leaves, to improve ventilation, as well as any damaged or diseased parts.

Fertilization

Cocona prefers rich soils with high organic matter content. A soil analysis should be done to determine the proper amounts to be applied. In sandy soils, the addition of large amounts of manure is recommended. Recommendations include the application of 150 kg ha^{-1} of N, 60–120 kg of P_2O_5 and 120 kg of K_2O split into three to four applications for N and K, and only one yearly application of phosphorus. In Honduras, with plants managed as annuals, yields increased proportionally to N applied; 375 kg ha^{-1} N produced more fruit though the marginal yield increase cannot be justified by the extra fertilizer cost (Duarte *et al.*, 2000b).

Pest management

DISEASES Cocona, as with many Solanaceae species, is very susceptible to bacterial attack that can start in the seedbed. Clean seed and disinfected media should be used and watering managed carefully. Cocona is very susceptible to fungal attacks by fungus such as *Fusarium*, *Phytophthora*, *Rhizoctonia* and *Pythium*, which are very frequent in seedbeds or when the plant is produced under stress. Risk is reduced by rotating the crop so that it is not planted in soils where a species of this family was sown or planted the year before. Crop rotation is essential, together with good drainage and irrigation management. Anthracnose caused by *Colletotrichum gloeosporioides* can damage foliage and fruit under very humid and rainy conditions as well as leaf spots caused either by *Septoria solanicola* or *Cercospora*. Fusarium wilt could also become a problem as well as *Phytophthora infestans* (Gallozzi and Duarte, 2007). Bacterial rot caused by *Pseudomonas solanacearum* can be very serious and infected plants should be uprooted and burned far away from the field.

INSECTS The main insects are *Diabrotica* spp. that attacks young plants and if not controlled can eat most of the foliage. Mealy bugs (*Planococcus pacificus*) attack the fruit and the pedicel. *Corythaica cyathicollis* Costa larvae form groups in the lower side of the leaf and produce a yellowing of this area that later dries out and falls leaving holes in the leaf blade; *Phrydenus muriceus* Germar is a curculionid that bites the young fruit stunting its growth and its larvae burrow into branch terminals. Additionally, the root knot nematode *Meloidogyne incognita* can severely reduce production. Brazil is trying to select for nematode resistance and their negative effects can be minimized by the use of abundant organic matter and crop rotation. Other insects can include scales and aphids (Gallozzi and Duarte, 2007).

Weed control

Under rainfall irrigation there should be partial weed control during the dry season to reduce competition. However, it is important that some weeds should be left to maintain the microclimate, as long as the weeds are not hosts for disease and insects. The area around each plant should be clean to reduce competition. Cut weeds can be used as mulch.

Orchard protection

Normally in the areas where this crop is grown there are no strong winds and protection from the neighboring tree is frequently found.

Harvesting and postharvest handling

Plants can be harvested after 6–8 months in the field for 1–3 years. According to Villachica *et al.* (1996) plants with small fruit (25–40 g) produce 87–119 fruit, those with medium-sized fruit (40–60 g) produce 83–95 fruit, and those with large fruit (141–215 g) between 24 and 39 fruit year^{-1}. In the Amazon of Peru and Brazil, plants at 1.5 m by 0.75 m produce 22 t ha^{-1}. In Honduras, Duarte and Sandoval (2002) estimated 40 t ha^{-1} at 13,333 plants ha^{-1}, each tree yielding 3 kg. Soil-borne diseases and nematodes are the main reason for low yields. Results from Brazil indicate that in virgin Amazon forest with distances of 1 m × 1 m, yields of 40 t ha^{-1} year^{-1} can be obtained (Gallozzi and Duarte, 2007).

Harvesting

Cocona plants will start production 6–8 months after transplanting. For the local market, fruit are collected when more than half colored, while for distant markets fruit will be harvested when the color starts to appear; this normally results in some loss of internal quality. Normally a sharp knife or harvest shears are used but some people twist and then pull the fruit to break its very short peduncle, but this sometimes results in the base of the fruit being torn and the opening serves as an entrance for fungal infection. Harvested fruit should be put into bags or boxes under shaded conditions. Harvest is normally done once a week. Fruit pubescence should not affect harvesters although harvesting during warm hours in a dry climate can result in the hairs getting into the respiratory tract creating some discomfort, hence it is better to harvest during the early morning hours.

Postharvest treatment

Cocona is a climacteric fruit that is rather hard and can withstand quite rough handling after harvest mainly due to its thick pulp and hard peel. Fruit can be stored at room temperature (27–30°C) with good ventilation for 5–7 days. In the home refrigerator, fruit can be held for 1 month without loss of aroma and flavor. Frozen pulp can be held for up to 6 months without appreciable changes in color, flavor or aroma. Under storage at 15°C and 50% RH, the fruit can be held for about 3 weeks, after which dehydration becomes significant and senescence starts that leads to loss of firmness. Fruit coatings improve postharvest life and impart gloss (Duarte, 2011).

Utilization

Ten kg of fruit can yield 2.5 l of fleshy pulp or 1.5 l of jelly or 7.5 l of juice (Duarte, 2011). The ratio of "hoof" to pulp with seeds varies from 2.27 to 2.96, with the "hoof" being around 70–75% and pulp between 25 and 30% of the total weight, with smaller fruit tending to have a lower ratio. Seed weight is about 17% of the total weight of the liquid flesh and 5.3% of the total fruit weight (Duarte and Sandoval, 2002).

Fruit are rarely eaten out of hand or uncooked since they are acid, astringent and have little sugar but have a fair content of niacin and ascorbic acid (Table 5.1). The major use is in the preparation of refreshments using the inner flesh containing the seeds and mixing it in a blender with water and sugar. This fleshy pulp can also be used to make a hot meat sauce mixing it with salt, vinegar, pepper, and hot pepper. The "hoofs" can be boiled in syrup to eat as a sweet or can be canned as peach halves. The cooked "hoofs" are used to make jellies and jams. They can also be used to make refreshments by boiling and mixing them in a blender with the fleshy pulp, sugar and water, but the resulting drink is not as good as that of the pure fleshy pulp juice (Duarte, 2004). The fleshy pulp or the boiled and mashed "hoofs" can be frozen for export. The juice and pulp do not oxidize as readily as naranjilla and this is an advantage.

Tree Tomato

Although the tree tomato *Solanum betaceum* Cav. is now placed in the genus *Solanum* (Bohs, 1995), it is not a close relative of the tomato. Tree tomato was

Table 5.1. Composition of 100 g edible portion of different American solanaceous fruits.

Proximate	Cocona[a]	Tree tomato[b]	Cape gooseberry[b]	Tomatillo[b]	Pepino[c]
Energy (kcal)	41.0		73		32
Moisture (g)	88.5	83–88	78.9	90.4–91.7	92.0
Carbohydrate (g)		10.3	19.6	5.8	6.3
Protein (g)	0.9	1.5	0.054	0.2–0.7	0.4
Lipid (g)		0.06–1.28	0.16	0.6	1.0
Fiber (g)	9.2	1.4–4.2	4.9	0.6–1.7	0.4
Ash (g)	0.7	0.61–0.84	1.01	0.6–0.69	0.3
Minerals					
Calcium (mg)	16.0	3.9–11.3	8.0	6.3–10.9	18.0
Potassium (mg)		0.28–0.38			
Iron (mg)	1.5	0.66–0.94	1.23	0.57–1.4	0.8
Phosphorus (mg)	30.0	22–65	55.3	21.9–40	14.0
Vitamins					
Thiamin (mg)	0.06	0.038–0.137	0.101	0.054–0.106	0.08
Riboflavin (mg)	0.1	0.035–0.048	0.032	0.023–0.057	0.04
Niacin (mg)	2.25	1.1–1.38	1.73	2.1–2.7	0.5
Ascorbic acid (mg)	4.5	23.3–33.9	43	2–4.8	32.0
Carotene (mg)	0.18	0.371–0.653	1.613	0.061–0.074	

[a]Gallozzi and Duarte, 2007.
[b]Morton, 1987.
[c]Leung and Flores, 1961.

long known as *Cyphomandra betacea* Sendt. (synonym *C. hartwegi* Sendt.). Like the tomato, the tree tomato was returned to the genus *Solanum* on the basis of molecular studies (Bohs and Olmstead, 1999). The genus *Cyphomandra* of small shrubs and trees used to contain 35 species; this former genus and its species are now all under *Solanum* as a section. Other members of this section are grown as ornamentals and for their edible fruit similar to tree tomato (*S. cajanumense*, *S. sibundoyense*).

Regional names for tree tomato include: *tomate*, *tomate extranjero*, *tomate de árbol*, *tomate granadilla*, *granadilla*, *pix*, and *caxlanpix* (Guatemala); *tomate de palo* (Honduras); *arvore do tomate*, *tomate de arvore* (Brazil); *lima tomate*, *tomate de monte*, *sima* (Bolivia); *pepino de árbol*, *tomate de árbol* (Colombia); *tomate dulce*, *tomate de árbol* (Ecuador); *tomate cimarrón* (Costa Rica); *berenjena*, *sachatomate*, *tomate de árbol* (Peru) and *tomate francés* (Venezuela, Brazil) (Morton, 1987). In 1967, New Zealand started calling the fruit "tamarillo" and this has now become the accepted commercial name.

Origin and distribution

The species is native to the Andes of Peru, Chile, Ecuador, Colombia and Bolivia where wild populations of the tree tomato occur along with the closely related edible-fruited species cited above and others. The tree tomato is cultivated in home gardens and small orchards in this region. The tree was probably spread by the Spanish throughout Latin America then to the tropics and subtropics of Africa, South Asia, South-east Asia, China, Australia, New Zealand and the USA. It was brought to New Zealand in 1891, which now exports the fruit to temperate-zone countries using the same channels developed for kiwifruit and using the name tamarillo.

Ecology

Soil

The tree prefers light, deep, fertile well-drained soils with a pH of 5–8.5 (Prohens and Nuez, 2000). In Haiti, it grows well on deep lateritic soil. A high organic matter content is an advantage, although if in excess it results in a too succulent plant that is prone to wind breakage (Federación Nacional de Cafeteros de Colombia, undated). In New Zealand's shallow soils, it is planted on raised beds (Sale, 1984b).

Rainfall

The tree is found in areas with a rainfall between 600 and 4,000 mm. Its large soft leaves and shallow root system limit its drought tolerance. It is also very intolerant of waterlogged conditions. In many areas of Ecuador and Colombia, production depends upon rainfall only unless long dry periods occur.

Temperature

The tree tomato is subtropical and grows well between 1,500 and 3,000 m in Ecuador at the equator and at lower elevations as you move to cooler areas north or south. In Puerto Rico, it occurs between 300 and 1,000 m, and 300–2,300 m in India. It grows well in the central coastal area of Peru where lowest temperatures reach 5°C while in the Peruvian Andes it grows in frost-free locations above 2,500–3,000 m where night temperatures can reach close to the frost point for a short time in the winter. Temperatures should remain above 10°C for growth and ideally with a mean between 15°C and 20°C. Frost (–2°C) kills the small branches and foliage of mature trees with the tree recovering. Young trees and seedlings in their first year are killed by frost.

Light and photoperiod

It is grown in direct sunlight. Trees can flower and set fruit throughout the year suggesting no photoperiodic response.

Wind

The plant being shallow-rooted is damaged by wind and is easily blown over. The branches are brittle and readily broken by wind gusts, especially when laden with fruit. The leaves are fairly large and thin and easily torn. Windbreaks are needed.

General characteristics

Plant

The tree tomato is a small, fast-growing, shallow-rooted plant that reaches 3–6 m. The wood is very brittle and easily broken. The evergreen musky, odorous, ovate, simple alternate leaves are large initially becoming smaller at later stages (10–35 cm long by 4–12 cm wide); they have a long petiole and are more or less heart-shaped at the base and pointed at the apex (Fig. 5.2). The leaves are thin, softly hairy, with conspicuous coarse veins. The plants continue to be productive for 5–6 years, sometimes up to 10–12 years.

Flowers

The flowers are borne in small, loose clusters near the branch tips. Each fragrant flower is 1.25–2 cm wide with five pale pink or lavender, pointed petals, five prominent yellow stamens, and a green-purple calyx. Similar to the tomato, the flowers occur in successive trusses. In New Zealand flowering begins in the start of summer (November) and goes through to the end of autumn (May).

Pollination and fruit set

The tree is normally self-pollinating with cross-pollination being carried out by bees with improved fruit set (Albornoz and Morales, 1989). When planted in

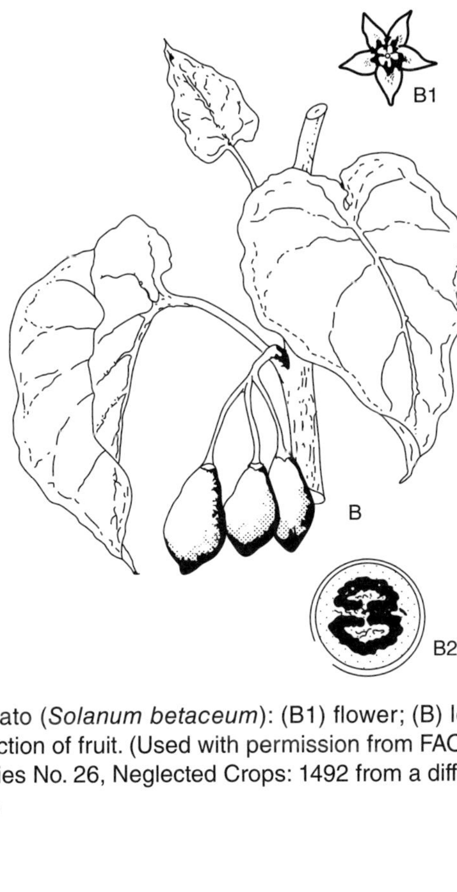

Fig. 5.2. Tree tomato (*Solanum betaceum*): (B1) flower; (B) leaves and fruit; (B2) transverse section of fruit. (Used with permission from FAO Plant Production and Protection Series No. 26, Neglected Crops: 1492 from a different perspective, 1994 (A001/2013).)

greenhouses with no wind to shake the flowers, and no bees or other insects to transport the pollen, yields tend to be low (Sale, 1984b). Unpollinated flowers drop prematurely. On average about 12% of the flowers on a truss set fruit but only about 3.3% develop to maturity (Lewis and Considine, 1999).

Fruit

The long-stalked, pendent egg-shaped fruit are borne singly, or in clusters of 3–12, and are pointed at both ends and capped with the persistent conical calyx (Fig. 5.2). The fruit are dull red, purple, yellow or orange, 5–10 cm long by 4–5 cm wide. Occasionally the fruit may have faint dark, longitudinal stripes. Flesh color varies accordingly from orange-red or orange to yellow or cream-yellow. The tough skin has an unpleasant flavor while the outer layer of flesh is slightly firm, succulent and bland. The pulp surrounding the seeds is soft, juicy, and subacid to sweet. The pulp is black in dark purple and red fruit and yellow in yellow and orange fruit (Morton, 1987). It takes 21–24 weeks from anthesis to fruit maturity, depending upon temperature (Pratt and Reid, 1976).

The seeds are thin, nearly flat, circular, larger and harder than those of the true tomato and distinctly bitter; they can have yellow or purple arils (Manzano, 2005). The fruit has a slightly resinous aroma and the flavor suggests a mild or under-ripe tomato with a faintly resinous aftertaste.

Cultivar development

Cytogenetics, genetics and breeding

Spontaneous polyploidy occurs with a frequency of 0.24% (Pringle and Murray, 1992) in seedling populations of a normal diploid ($2n = 2x = 24$). These polyploids have small, almost seedless fruit.

Selection and evaluation

Considerable variation is found in fruit and inflorescence traits, and seed number, which also have the greatest heritability values. This variation is found within five cultivar groups, orange, orange pointed, purple, red, and red conical, for numerous traits. An odd, red conical accession had small fruit containing very few seeds (Acosta-Quezada *et al.*, 2011). Some improvement of the plant has occurred in New Zealand, but more is needed in the flavor of the fruit before it can become a rival to traditional fruits.

Cultivars

The red and purple types are preferred in Europe because of their color, even though they are more acidic. In New Zealand, there are several named varieties grouped into three categories according their type: red, golden or yellow. The golden types are actually amber in color. In the red types the varieties include 'Red Beau', an oval medium-sized fruit with good flavor on a tree with less vigor than most other reds. 'Oratia' is a large-fruited cultivar, tending to round-heart shape; 'Red Delight' is another large-fruited red very similar to 'Oratia'; 'Ted's Red' is an attractive scarlet red, with good size and flavor; and 'Kerikeri Red' is a small fruit known for its sweetness and ability to set heavily in multi-fruited trusses. Golden varieties have a milder flavor than the reds, and tend to appear sweeter in taste. Golden varieties include 'Goldmine' with a golden skin and a red blush that turns red late in the season or under cold growing conditions; the fruit is medium sized and oval in shape. 'Amberlea Gold' has an amber-colored skin and no dark pigmentation internally. It has a pleasant flavor and a satisfactory size. The main yellow variety is 'Bold Gold', a mutation of 'Oratia', with large fruit compared to all other golden types and which does not show virus mottling, though the plant is susceptible. The flavor is poor compared to other tamarillo cultivars.

In Ecuador and Colombia, the main producers in Latin America, there are types based on fruit color rather than formally described and named varieties. Orange-yellow, red and dark red or purple-colored types are common. The purple types are called Mora because of the similarity in color with blackberries. In Colombia, 'Rojo Común' (Common Red) is the most popular type with a red-orange peel and an 80 g fruit; 'Amarillo Común' (Common Yellow) has yellow peel, oval form and 70 g fruit; 'Morado' or 'Tomoro' (Purple) has purple peel, pulp and seed arils, 90 g fruit, and is the most susceptible to diseases but preferred by local consumers (Osorio, 1992).

Cultural practices

Propagation

SEXUAL Seeds are normally used for propagation though the tree does not always come true from seed. For best germination, freshly extracted seeds should be washed and left to dry in the shade for 1–7 days then sown, resulting in around 80% germination. Fermenting the pulp and seeds before sowing resulted in a lower germination percentage (Duarte and Alvarado, 1997). Seeds produce a high-branched and erect tree, while cuttings develop into a shorter, bushy plant with low-lying branches (Morton, 1987). Bearing begins in 1.5–2 years.

ASEXUAL Cutting propagation is possible with 3,000 ppm indole butyric acid (IBA) improving rooting significantly (Duarte and Alvarado, 1997). Cuttings should be taken from 1–2-year-old wood, 10–25 mm thick and 45–75 cm long. The leaves are removed and the base cut square below a node. The treated cuttings can be planted directly in the field. The tree should not be permitted to fruit in the first year. In Ecuador, some commercial plantings use grafted plants with *Nicotiana glauca* being the rootstock for nematode and soil fungus resistance. Tissue culture methods have also been developed.

Orchard establishment

No special field preparation is necessary and the normal orchard preparation is used. If a hardpan layer is found, sub-soiling is necessary. If soil is too shallow, raised beds are prepared. Seedlings or cuttings not directly planted in the field are transplanted when at least 30 cm tall. In Colombia and Ecuador, plants are normally prepared from seed that is sown in "Speedling" or similar cell trays and when 10–15 cm tall they are transplanted into bags until they reach 30–40 cm.

If good irrigation is available, they can be transplanted directly from the trays when plants reach around 15–25 cm.

New Zealand uses mechanized production and plants are spaced 1–1.5 m between plants and 4.5–5 m between rows. In areas without mechanization, the rows are spaced more closely together. In Colombia and Ecuador, they normally use 2.0 m by 2.0 m, 3.0 m by 3.0 m, or 1.5 m by 3.0 m, and even 4.0 m by 4.0 m depending on soil type, topography and if hand labor or machinery are used. In New Zealand, they transplant during the autumn in the warmest areas while they prefer to do it in spring in frost-prone areas (Sale, 1984b). In Latin America, it is planted year round.

Irrigation

Irrigation is necessary during long dry periods. Mulch helps to limit water loss from the soil.

Pruning

Seedling trees are trained by pruning back in the first year to about 0.5–1.0 m to encourage side branching. Further pinching back of new shoots may be necessary to obtain a sturdy framework. If trees are left to grow naturally, they tend to branch too high (Sale, 1984b). In Colombia, they prune back to 30–40 cm and try to leave four branches. Pruning is normally done in the early spring after the previous crop has been harvested and danger of frost has passed. This pruning is completed before the end of spring as the time of pruning influences when the next flowering will occur. Early pruning or not pruning give early flowering that occurs on new shoot growth.

A light pruning that involves removing up to half of the old canopy leads to minimal regrowth and rapid branching on which flowers occur. Pruning intensity needs to balance the amount of new growth and the number of flowers. Excessive flowering can lead to a heavy early maturing crop of small- to medium-sized fruit. Pruning branches back to near the original branch forks leads to vigorous regrowth and a smaller crop of larger fruit that matures later. Maintenance pruning includes eliminating branches that are damaged, those that have already produced or are directed into the center of the canopy and all basal water sprouts. In some cases, the lower third of the plant is defoliated to create better ventilation and fruit ripening.

Branches are tied in the first year to the main stem or other branches to provide a more rigid umbrella-type structure. Frequently, branches with a heavy fruit load break and tying them up reduces stem breakage and subsequent fruit loss (Osorio, 1992).

Fertilization

The recommendation is for 0.25–1.0 kg per tree of NPK (5:6:6), half in early spring and half in midsummer. When the tree is 5–6 years old the application rate is increased to 1–1.5 kg per plant. In New Zealand, 110–170 kg N; 35–55 kg P and 50–100 kg K ha^{-1} year^{-1} are applied (Sale, 1984b). Lime is used in Colombia to increase the pH of many soils and often 500 g of triple superphosphate is added to the bottom of the planting hole. The trees are then fertilized every 3–4 months with N and K plus foliar applied microelements. A typical program would be to apply every 3 months 150–200 g of a 10-30-10 fertilizer plus 50 g urea during the first year, and increase the application to about 300–400 g of 10-30-10 and 75 g urea every 3 months during the second and third years. Some growers just apply 8–10 kg manure per plant every year. Fertilizers should be applied, if there is no fertigation, by placing them into a narrow and shallow ditch made initially 30 cm away from the stem and then 10–15 cm farther away with every subsequent application in order to place the fertilizer closer to the active roots and reduce fertilizer loss (Federación Nacional de Cafeteros de Colombia, undated).

Pest management

DISEASES The major diseases are tamarillo mosaic virus that causes pale mottling on leaves and sometimes on the fruit, powdery mildew (*Erysiphe* sp. and *Oidium* sp.) that causes defoliation, and sooty mold. Diseases in Colombia and Ecuador include anthracnosis (*Colletotrichum gloeosporioides*) that attacks the fruit, and root rots caused by *Sclerotium* and *Pythium*. There is also bacterial wilt caused by *Pseudomonas*. In Ecuador, they report *Phytophthora* blight as well as *Alternaria* attacking the aerial parts with the former being much more damaging while the latter affects mostly the foliage (Albornoz and Morales, 1989).

Five viral diseases occur that include tamarillo mosaic potyvirus (TaMV), which causes mosaic symptoms on leaves, reduces fruit yield and causes blotches on the fruit skin making it difficult to market. Potato Leaf Roll Virus (PLRV) causes yellowing, wilting and rapid death (Federación Nacional de Cafeteros de Colombia, undated). Another virus is Cucumber Mosaic Virus (CMV) that is frequently isolated with the TaMV potyvirus. The viruses produce several symptoms and affect production to different degrees according to the age of the plant, health, time of the year and number of viruses present (Sale, 1984c).

INSECTS The major pests in New Zealand are whitefly, aphids and green vegetable bug. In 2009, the New Zealand industry was hit with a new pest, a wind-borne tomato/potato psyllid insect and the liberibacter it carries, causing rapid tree dieback. In Colombia and Ecuador, the fruit is attacked by fruit flies (*Anastrepha* spp.), a fruit borer (*Neoleucinodes elegantalis*), a stem borer (*Faustinus* sp.), the chinch bugs *Corythucha* that cause leaf yellowing and *Leptoglosus zonatus* that feeds on the fruits and causes them to drop. A root knot forming insect (*Margarodes*) and a small fruit fly (*Lonchea*) occasionally attack the plant. Additionally red spider mites (*Tetranychus*) and nematodes (*Meloidogyne*) can be a problem (Federación Nacional de Cafeteros de Colombia, undated).

Weed control

The shallow root system limits deep cultivation, though light cultivation is desirable to eliminate weeds until there is sufficient tree growth to shade them. Both herbicides and mechanical control with hoes or machetes are used, making sure not to damage the stem or the roots.

Orchard protection

Windbreaks are highly recommended for this shallow-rooted, large-leafed tree with brittle branches.

Harvesting and postharvest handling

Yields

In Brazil, each tree is expected to yield 20–30 kg of fruit annually (Morton, 1987). In Colombia and Ecuador yield can be around 8–10 t ha^{-1} $year^{-1}$ in the second year and can be as high as 16–20 t in the following years (Federación Nacional de Cafeteros de Colombia, undated). In New Zealand, it is estimated that in the second year yields will be about 8 t ha^{-1}, while in the third and fourth years they should be 16 t ha^{-1} (Sale, 1984b).

Harvest

In the Andean countries, the plants can produce continuously for 2–3 years. Normally harvest is done once a week. Plants from seed will start producing 8–10 months after transplanting. The fruit are ready for harvest in about 4 months from flowering. Fruit are harvested based upon peel and pulp color. Other harvest indices that relate to skin color include firmness, juice content and soluble solids content. The amount of red skin pigment increases with age from 15 weeks, but harvesting of immature fruit appears to stop red pigment development. Fruit harvested immature at 12–19 weeks or younger shriveled in storage (Pratt and Reid, 1976). If harvested at the mature green stage, the final flavor is poor with a lower juice and soluble solids content; they are inferior to those harvested at the dark purple stage. The fruit do not ripen simultaneously and several pickings are necessary. The fruit are clipped, leaving about 2.5 cm of stem attached.

Postharvest treatment

The fruit is non-climacteric (Schotsmans *et al.*, 2011) and storage recommendations are at 3–4.5°C with 90–95% RH for 4–8 weeks with an additional week for marketing. The fruit can be held at 7°C for up to 32 days (Espina and Lizana, 1991). After 13 days at 18°C, there is a rise in respiration, apparently because of the onset of senescence. Chilling injury occurs in fruit stored below 3°C with fungal decay on the stem and calyx occurring when stored above 4.5°C (Cantwell, 2004). Symptoms of chilling injury include pitting and a scald-like browning of the skin, calyx and stem. The most common storage decays are caused by the latent infections *Colletotrichum acutatum* and *C. gloeosporioides*.

Utilization

The mucilaginous, juicy and seedy pulp has a sweet-acid taste, somewhat like tomato. The fruit are occasionally eaten raw but are more commonly baked or grilled as a vegetable and included in cooked dishes and made into relishes and chutneys. In Latin America, the fruit is eaten in several forms: as a hot sauce when ground together with hot chili pepper and salt; as a juice by mixing the pulp in a blender with water or milk and sugar; as a dessert boiling the entire or sliced peeled fruit in heavy syrup. Mixing the pulp with frozen evaporated milk, sugar and tasteless gelatin obtains a very good tasting foamy jelly. The fruit contains good quantities of several vitamins and is rich in iron and potassium (Table 5.1). An average tamarillo contains less than 40 calories.

Cape Gooseberry

The Cape gooseberry or goldenberry (*Physalis peruviana* L.) is a minor Andean fruit found in markets from Venezuela to Chile. It was one of the Inca fruits (Veitmeyer, 1991). The genus *Physalis* includes annual and perennial herbs bearing globular fruit. Each fruit is enclosed in a large papery calyx-derived husk at maturity. The temperate/subtropical *Physalis* genus contains around 100 species; many are called ground-cherries, though only a very few species are of economic value. Some species are grown as ornamentals because of the bright-colored husks and some have poisonous fruit.

Origin and distribution

The fruit is neither a gooseberry nor from the Cape of Good Hope. Plants were grown by settlers in the Cape in 1807, then carried to Australia and New Zealand in the 19th century and given that name. Other common English names include golden husk, ground cherry, Peruvian cherry, Peruvian ground cherry, strawberry tomato, winter cherry and in Hawaii poha. In Latin America, it has numerous names: in Peru it is *capuli, aguaymanto, tomate siilvestre, tomatito de perro* or *uchuba*; in Boliva *capulí* or *motojobobo embolsado*; in Ecuador *uvilla*; in Colombia *uvilla, uchuva, vejigón* or *guchavo*; in Venezuela *topotopo* or *chuchuva*; in Chile *capuli, amor en bolsa*, or *bolsa de amor*; and in Mexico, *cereza del Peru* (Morton, 1987). In South Africa, it is known as *pompelmoes, apelliefie* or *alkekengi*; in Gabon *coqueret*; in the Philippines *lobolobohan*; and in India *taparee, tiparee*, or *makowi*. Normally production is limited to backyard plants or in small orchards. It is also grown in Hawaii, California, China, South Africa, East Africa, India, Malaysia, New Zealand, Australia, and Great Britain. Fruit is exported to Europe from Colombia, the main producer in the world with 534 ha in 2003 (Sanabria, 2005). Jam was exported in the past from South Africa to England.

Ecology

Soil

The plant grows well in any well-drained soil (pH 4.5–8.2) but does best on sandy to gravelly loams (Morton, 1987). It does not like heavy nor excessively wet soils. On highly fertile alluvial soil, the plant becomes very vegetative and the fruit fail to color properly.

Rainfall

A minimum of 800 mm is necessary during the growing season. Higher rainfall, up to 4,300 mm, increases growth and yield if the soil is well drained. The plants become dormant during drought. For proper production 1,000–2,000 mm of well-distributed rainfall is needed, otherwise irrigation is required.

Temperature

The plant grows well with an annual average temperature from 13 to 18°C. Day temperatures of 27–30°C apparently do not affect fruit set. Between Chile and Colombia, it grows wild at altitudes of 1,500–3,000 m, and in Venezuela it grows in the Andes and the coastal range between 800 and 3,000 m. Further north from the equator in Hawaii, it is found growing wild between 300 and 2,400 m with temperatures between 27 and 30°C; it also grows well in Mediterranean climates. Frost can kill the plant and growth is slow at temperatures below 10°C.

Light and photoperiod

The plants grow in full sun but can grow under partial shade including that under greenhouse benches. Day length does not play a significant role in flowering as it yields well both near the equator and at high latitudes.

Wind

Wind can cause significant damage to the plant.

General characteristics

Plant

The plant is a herbaceous or soft-wooded perennial and usually reaches 1.0–1.6 m, occasionally 1.8–2.0 m if pruned and guided. It has a sympodial growth habit. The spreading branches are ribbed, often purplish. The nearly opposite, velvety, heart-shaped, pointed, randomly toothed leaves are 6–15 cm long and 4–10 cm wide (Fig. 5.3). After fruit ripening leaves turn yellow and fall. Most of the fibrous root system is found between 10 and 15 cm depth while some main roots can go down to 50–80 cm.

Flowers, pollination and fruit set

The yellow hermaphrodite flowers appear solitary in the leaf axils and are bell-shaped and nodding, 2 cm wide, with five dark purple-brown spots in the throat. The cupped calyx is purplish-green, hairy and five-pointed. After the flower falls, the calyx expands ultimately forming a straw-colored husk much larger than the fruit it encloses (Morton, 1987). Flowering occurs year round in frost-free warmer areas starting 70–80 days after sowing, while the time between flower primordia initiation and anthesis is about 3 weeks. Flowers are readily pollinated by insects and winds.

Fruit

The fruit is a berry 1.25–2 cm wide, with smooth, glossy, orange-yellow skin and juicy pulp containing 100–300 very small yellowish seeds (Fig. 5.3). When fully ripe, the fruit is sweet but with a pleasing grape-like tang. The husk is bitter and inedible. The fruit take 85–100 days to develop from anthesis.

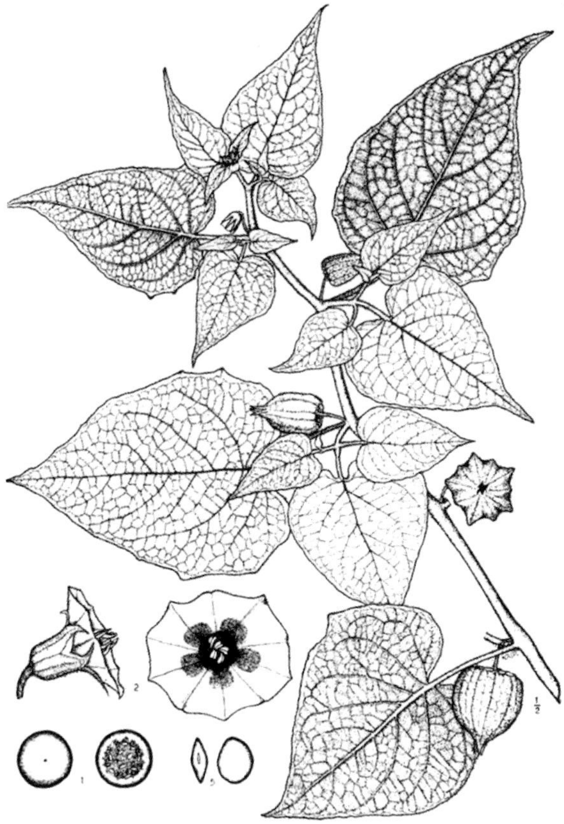

Fig. 5.3. Cape gooseberry (*Physalis peruviana*) showing leaves, flower and fruit (used with kind permission from Isa Degener, appeared in Otto Degener's book *Flora Hawaiiensis* in 1938).

Cultivar development

Cytogenetics, genetics and breeding

The chromosome number of this plant is $2n = 48$. However, apparently there is variability with wild types having $2n = 24$, the "Colombia" type $2n = 32$ and the "Kenya" ecotype $2n = 48$. The main collections of ecotypes are in Colombia with more than 200 and Peru with about 12 (Ligaretto *et al.*, 2005).

Selection and evaluation

There is considerable variation in size, shape, and flavor of the fruit, time of maturity, and plant form, providing useful germplasm for breeding improvement.

Cultivars

Selected strains for commercial use are not widely available. In Australia, they have named cultivars such as 'Golden Nugget' and 'New Sugar Giant'. US cultivars include 'Peace', 'Giant Groundberry', 'Goldenberry', 'Giant Poha Berry', 'Giallo

Grosso' and 'Giant' (Fischer, 2000). Farmers often propagate selected plants with desired characteristics. The recommended practice is to use the best plants of the previous crop; in this way superior selections have been obtained but not many named varieties exist.

In the late 1980s, Colombia introduced two ecotypes from Africa, one from Kenya and the other from South Africa. The researchers found that these introduced ecotypes had an average fruit weight of 6–10 g, and a short plant with larger leaves, while the fruit of the Colombia ecotype weighed 4–5 g but with a much better color and sugar content on a taller plant with smaller leaves (Fischer and Almanza, 1993). The local ecotype is the one cultivated in Colombia.

Cultural practices

Propagation

SEXUAL The plant is mainly grown from seed. Germination is reported to be low, although Colombia reports 85–90% germination from seeds left standing for 2 weeks before sowing (Almanza, 2000). Normally seeds are sown in seedbeds or more recently in trays with 200 cells and from there transplanted to poly-bags that contain about 0.5 kg of media and where the plant grows for another 30–60 days until it reaches 20–25 cm for transplanting. Germination starts after 10–15 days. Flowering occurs 65–75 days after planting.

ASEXUAL This method is not favored. The plant can be propagated from 1-year-old stem cuttings (15–20 cm long) treated with rooting hormones with a 37.7% success rate. The plants from cuttings flower earlier and yield well but are less vigorous than seedling plants. The difference may be due to the absence of a taproot and their shallow adventitious root system, resulting also in a weaker anchorage and shorter plant life. Fruit from plants derived from cuttings tend to be larger but more prone to cracking. Air layering has also proven to be successful.

Orchard establishment

The plant can live several years but normally it is grown as an annual as the yield declines significantly after the first year. In some situations, farmers will grow them up to 3–4 years. Field preparation is similar to that used for tomatoes with plowing, harrowing and furrowing where the plants are to be established. Transplanting is done when the plants in the bag are about 30 cm high. Plant spacing will depend on the plant training and supporting system.

There are two planting systems used: one is with free-standing plants and the other uses trellises. In the free-standing system, plants are spaced 1.5–2 m apart in rows with rows 2–3 meters apart. Several types of trellis systems are used that normally include formation pruning. The two main trellis types used in Colombia are: (i) the "V" type; and (ii) the "Hanging plant" type. In the "V" type, two 3 m posts are stuck into the ground at an angle every 7.5 m, crossing their bases at ground level, and the slanted posts form a "V" above ground. Each side of the "V" is ~2.5 m and on each "V" branch two or three rows of No. 14 wire are fixed, separated by 40 cm to support the "V" branches. As the plant grows, the lower wire line can be moved to a higher position. The "Hanging plant" type uses two rows of posts separated by 1.0–1.2 m with the effective height of the posts being 1.60 m installed every 6–7 m along the line. A No. 14–16 wire is run along the top of both post lines. Sometimes only one row of posts is used with a horizontal piece of wood affixed to the top of each post forming a "T" that is about 1.0–1.2 m wide. At both ends of the horizontal "T" bar, a No. 14–16 wire is attached and strings tied to these wires are used to hang the branches. Planting distances in the trellis systems are usually 2.5–3.0 m between rows and 2 m between plants (Angulo, 2000; Miranda, 2005). Greenhouse production can also be performed, with higher costs but also higher yields.

Irrigation

In dry seasons, irrigation is necessary to keep the Cape gooseberry plant in production. Initial recommendations were to furrow irrigate during the dry periods, as young plants need about 2-5 l per plant. The use of tensiometers has been recommended to establish irrigation times and needs. A more precise method has been suggested consisting of calculating the difference between rainfall and evapotranspiration. Evapotranspiration is determined with evaporimeters or a combination of atmometer and evaporimeter, and the calculated need is 19.44 l per adult plant. In Colombia, the main world producer, very few people use drip irrigation and even fewer fertigation.

Pruning

Four pruning types are described when a trellis is used (Miranda, 2005). The first pruning is formation pruning carried out to promote the growth of several branches that will produce the productive or tertiary branches. It is usually done when the plant is young after 1.0–1.5 months in the field by pinching the apex when it is at 15–20 cm, to encourage the appearance of four–six branches; all basal sprouts are eliminated. This is performed especially with the "V" trellis system. In the "Hanging plant" system normally no pinching is done so that not too many branches have to be tied to the strings. In plants from cuttings, normally no formation pruning is done as the plant tends to branch laterally. The second type is maintenance pruning, which is done five–six times a year to eliminate all non-productive or excessively long branches. Sanitary pruning is the third type carried out to eliminate all diseased or damaged material. The final pruning type involves renovation pruning to start another cycle and rejuvenate the plants. Many growers prune the plants back after the first harvest to reduce pest infestations and to allow fruit to form on new growth.

Fertilization

The plant requires initially high amounts of N. In Colombia, they recommend adding 1–2 kg dried poultry manure into the planting hole. One month after transplanting, 100–150 g of a 10-30-10 or triple 15 are applied per plant and this is repeated every 3–4 months coinciding with production peaks. At each fertilizer application 1–2 kg poultry manure should be added. Just before flowering, potassium in the form of nitrate or sulphate should be applied or burnt rice hulls incorporated into the soil. The ever-increasing incidence of *Fusarium* seems to be enhanced by the ammonia form of nitrogen, thus using urea should be avoided. Urea is applied to the foliage with boron to help boron transport in the plant (Angulo, 2000). In South Africa, they recommend 25 t of cattle manure plus 500 kg of a 3-1-5 fertilizer per ha, plus some additional top dressings during the season (Davis and Whitehead, 1974).

Pest management

DISEASES The most troublesome disease is powdery mildew. In Colombia, *Sclerotinia sclerotiorum*, *Phoma* and *Cercospora* attack the foliage and the calyx rendering the fruit useless. *Alternaria* can be a foliage problem. Fruit can be damaged by *Botrytis*. Bacterial leaf spot (*Xanthomonas* spp.) occurs in Queensland and in Colombia where also *Ralstonia solanacearum* can be a problem. The plants are prone to root rots and viruses on poorly drained soil or if carried over to a second year. A strain of tobacco mosaic may affect plants in India, while in Colombia Potato Virus X causes mosaic symptoms. A phytoplasmic disease has been identified in Colombia (Zapata *et al.*, 2005).

INSECTS Reported insect pests include: cutworms (*Spodoptera*, *Agrotis*, *Feltia*), potato tuber moth, flea beetles and whitefly. Leaf miners (*Liriomyza* sp.), aphids and thrips can damage the foliage, while *Heliothis* larvae and the rust mite (*Aculops* sp.) can damage the fruit. Red spider mite attacks are frequent. Root knot nematode (*Meloidogyne* spp.) can be an important problem in certain places.

Weed control

Weed control is important, especially at the initial growth stages. Care has to be taken to keep the area around the plant as clean as possible to avoid competition. Herbicides are not much used except before planting. Weeding by hand, pulling or with a hoe are used taking care not to damage the roots or stems.

Orchard protection

The large shaggy bush requires some shelter from strong wind.

Harvesting and postharvest handling

Yields

A single plant may yield 300 fruit. At a planting density of 228–900 plants ha^{-1} a yield of approximately 3,000 kg ha^{-1} can be expected. Yields are about 5–7 t ha^{-1} in the first calendar year, rising to 15 t ha^{-1} in the second and dropping to 10 t ha^{-1} in the third and fourth years if the crop is maintained for that long. Normally the crop productive life will end after 18–20 months with about 20 t ha^{-1}. Harvest is estimated to require 400 labor days ha^{-1}.

Harvesting

Harvest starts 4–7 months after transplanting. Fruit maturity is judged on the change in the fruit skin color from green to yellow or orange. The ideal is to harvest between the green-yellow stage and the dark yellow or orange stage. Changes in the color of the calyx are not indicative of fruit maturity. Fruit are allowed to dry in the field before harvesting in rainy weather or if wet with dew, otherwise they are dried immediately after harvest to reduce postharvest diseases. The fruit are usually picked by hand every 1–3 weeks by twisting the fruit. For fresh fruit export or in warmer climates harvest is done twice a week during peak seasons to insure uniform quality. Mature fruit abscise and can be picked up from the ground though food safety is an issue due to soil contamination. Shaking the plants and gathering the fallen fruit on plastic sheets helps yield more uniformly mature berries.

Postharvest treatment

The fruit is climacteric (Fischer *et al.*, 2011) and the berries continue to ripen after picking with climacteric peak occurring 12–14 days after harvest. The fruit have a long postharvest life of 3–4 months with the calyx attached at 2°C. At 4–10°C the fruit can be held up to 5 weeks. Fruit stored at 0°C and 7°C during 15 and 33 days and removed to 18°C for 1 week did not differ in soluble solids content but had a lower acidity and a higher pH that was directly related to storage time and temperature. Harvesting fruit with a green pedicel is preferred to those with a brown pedicel (Lizana and Espina, 1991). Removal of the calyx significantly reduces postharvest life. Over 2–3 weeks the fruit achieve a uniform, bright golden yellow color and the soluble solids increase from 11% to as much as 16% (Fischer *et al.*, 1997; Trinchero *et al.*, 1999).

Utilization

The fruit is high in vitamin A, B, and C content (Table 5.1), is a rich source of carotene, phosphorous, and iron. It may be eaten fresh, in salads or in cocktails. Removed from the paper-like husks, the attractive yellow marble-sized fruit makes an extremely tasty jam. It is canned whole and preserved as jam, made into sauce, and used in pies, puddings, chutneys and ice cream. In Colombia, the fruit are stewed with honey and eaten as dessert, and a raisin-like dehydrated fruit is produced and this is being followed in other countries; the dried fruit can be covered with chocolate, making a very tasty product. The use of osmo-dehydration followed by hot air results in a better quality dried product or it can be followed by pasteurization for canning the fruit in syrup.

A leaf decoction is taken as a diuretic and anti-asthmatic in Colombia. In South Africa, the heated leaves are applied as poultices on inflammations and as an enema to relieve abdominal ailments in children.

Tomatillo

The tomatillo or husk tomato (*Physalis philadelphica* Lam.) originates from Mexico and is now cultivated in Mexico and the highlands of Guatemala where it is a staple seen in local markets. Its use in Mexico dates back to pre-Columbian times and was often as a sauce. Synonyms include *P. ixocarpa* Brot., ex. Hornem., and *P. aequata* Jacq. It has been introduced into the USA, Europe, South Africa, India, Kenya and Australia. In many places, it is now regarded as a weed.

Common names for this tropical species include: in English, tomatillo, husk tomato, jamberry, and ground cherry; in Spanish, *tomate de cascara, tomate de fresadilla, tomate milpero, tomate verde, tomatillo* (Mexico), and *miltomate* (Mexico, Guatemala).

Ecology

It grows well in soil suitable for tomatoes but prefers good drainage. This tropical species is grown in summer in more temperate areas. The plant needs full sun and some varieties require long days and others short days to flower.

General characteristics

Plant

This semi-woody annual may grow to 1.2–1.5 m, but is often prostrate and spreading. The branches and leaves are smooth. The leaves are ovate, pointed at the apex, wedge-shaped at the base, sometimes with wavy margins, and 6.25 cm long by 3 cm wide (Fig. 5.4).

Flowers, pollination and fruit set

The flowers are borne singly in the leaf axils and have a fused five-toothed green calyx. The yellow flower petals are 1.25–2 cm long and wide (Fig. 5.4), with dark brown spots in the throat. Tomatillo plants are highly self-incompatible.

Fruit

As the fruit develops, the calyx enlarges to more or less tightly enclose the fruit and at maturity it becomes straw colored and papery. The berry is slightly oblate and 2.5–6.25 cm wide (Fig. 5.4). When ripe, the fruit's thin skin may be yellow, purple, or more rarely reddish, or still green. The flesh is pale yellow, crisp or soft, and acid, subacid, sweet, or insipid, and contains many tiny seeds.

Cultivar development

The $2n$ value is 24. Traits of interest in breeding programs are: fruit yield; fruit weight; fruits per plant; equatorial and polar fruit diameter; total soluble fruit solids; fruit firmness; pH; and vitamin C content. There are at least two improved Mexican varieties, 'Rendidora' and 'Rendidora mejorada'; both have smaller and more uniform growth habit with few or no empty fruit that are firmer and lime-green.

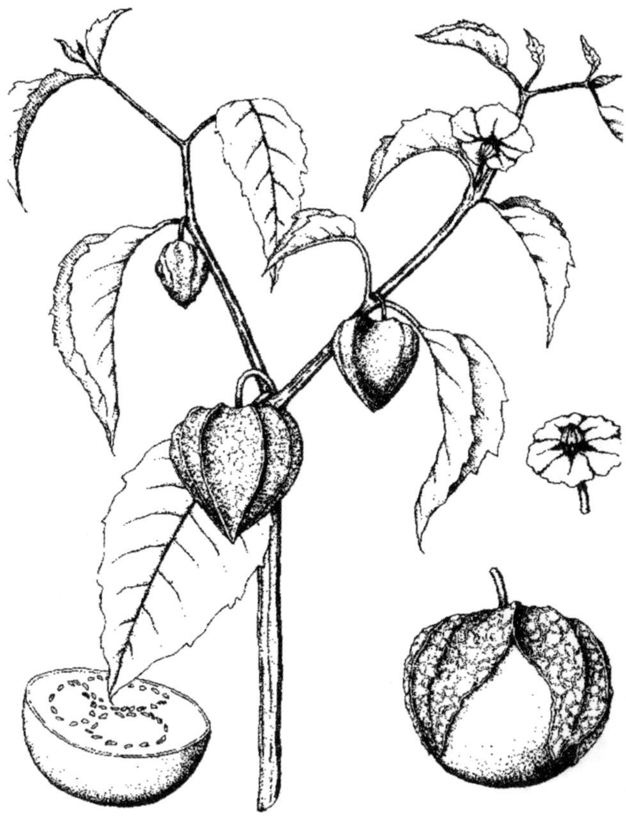

Fig. 5.4. Tomatillo (*Physalis philadelphica*) showing flower, leaves and fruit (used with permission from FAO Plant Production and Protection Series No. 26, Neglected Crops: 1492 from a different perspective, 1994 (A001/2013)).

Cultural practices

The plant is normally propagated from seed, though some seed dormancy may occur. Transplanting is common, especially in areas subject to frost. Cultural practices are very similar to those of tomatoes with similar spacing, fertilization and most commonly irrigation.

Harvesting and postharvest handling

Considerable production occurs in Mexico and Guatemala for local use, export and processing.

Harvesting

Individual plants may produce 64–200 fruit in a season. The freshness and greenness of the husk are major quality criteria. Fruit should be firm and bright green.

Postharvest treatment
Fruit can be kept for several months.

Utilization

Tomatillos are the key ingredient in fresh and cooked Latin American green sauces and used more as a vegetable than a fruit. The green color and tart flavor are the main culinary contributions of the fruit. Purple- and red-ripening cultivars often have a slight sweetness, unlike the green- and yellow-ripening cultivars, and are therefore somewhat more suitable for fruit-like uses in jams and preserves, and are also canned whole.

Pepino

The pepino or pepino dulce (*Solanum muricatum* Aiton) is a common fruit in the markets of Colombia, Ecuador, Peru, Bolivia and Chile. The synonyms include *Solanum variegatum* Ruíz and Pavón and *Solanum guatemalense* Hort., amongst others; this last name was given because in colonial times it was introduced into Central America and Mexico. The species is known only in cultivation or as escapes and has been cultivated in the Peruvian Andes since pre-Columbian times. It is represented in numerous ceramic pieces from the pre-Inca time that can be seen in the museums of Peru. During the colonial period and in the beginnings of the republican times the authorities banned its cultivation because it was a "crop of the Indian aboriginals". In some parts of Peru there is still the belief that eating the fruit after having had some drinks can cause death; therefore its pejorative name is *mataserrano* or Andean aboriginal killer. This is one of the reasons why in the Andean countries this plant has not been researched and its cultivation improved as with other native fruit crops. More research has been done on this crop in New Zealand and California. Chile has recently paid more attention to this adaptable crop and the fruit is now being marketed in Europe, Japan and the USA. In Peru, it is being grown and marketed to a certain extent, especially from the central coast.

The fruit comes in a variety of shapes, sizes, colors and qualities. The common names include in English pepino, Peruvian pepino, pear melon, melon pear, melon shrub, tree melon, sweet cucumber and mellow fruit. In Spanish, it is known as *pepino, pepino dulce, pepino blanco, pepino morado, pepino redondo, pepino de fruta, pepino de agua, mataserrano* and *pera melón*.

Ecology

Soil
Soils with good drainage and a pH above 6.0 are desirable. When grown in very fertile soils, fruit set and fruit quality are adversely affected. It is tolerant of moderate salinity.

Rainfall

A rainfall of at least 1,000 mm is regarded as the minimum if well distributed over several months. The plant is not drought tolerant. In the Peruvian central coast with no rainfall, it is furrow irrigated. The plant is very sensitive to drought so that care has to be taken in this aspect.

Temperature

This tropical species originated at higher elevations in the Andes where the climate is subtropical. In Peru, Colombia, Bolivia and Ecuador, it grows in the Andean valleys at altitudes between 900 and 2,800 m. In Chile, New Zealand, Peru and California it is grown near sea level preferably in frost-free areas, otherwise shelter is needed. In these cool coastal areas or where autumn and winter temperatures are between 17 and 21°C and there is high humidity, commercial production occurs. It can tolerate temperatures greater than 30°C for short periods as well as temperatures slightly below 0°C; the latter can cause loss of most of the leaves. If both day and night temperatures are high, fruit set and size are significantly reduced. If the nights are below 10–12°C no fruit set occurs (INIA, 1999).

Light and photoperiod

It can grow in semi-shaded places but it prefers sunny locations. No specific information exists about photoperiodic behavior but it grows well near the equatorial line as well as in high latitudes such as Chile and New Zealand; apparently it does not respond to photoperiod (Nuez Viñals and Ruiz Martínez, 1996).

Wind

It does not tolerate strong winds that can cause scarring of the fruit. The sprawling canopy grows in close contact with the ground so that moderate winds will not cause problems. Otherwise, wind protection will be needed.

General characteristics

Plant

This highly variable species is a sprawling, perennial herb that reaches about 0.60–1.0 m in height with a fibrous and shallow root system that normally concentrates in the first 60 cm of the soil. Several stems may arise from the woody base that can develop roots when in contact with the soil. The leaves may be either simple or compound with the number of leaflets varying from three to seven in the compound form (Fig. 5.5). Leaf color is dark green to purple-green and in most cases they are pubescent. In Peru, most of the varieties are glabrous.

Flowers

The hermaphrodite flowers have five persistent sepals and five petals that vary from white to pale purple to bright blue with whitish margins. Flowers occur in groups of 5–20, sometimes 40, in pseudo-terminal clusters. The stamens and yellow anthers are shorter than the corolla and dehiscent through apical pores with the pistil protruding in the middle of the anthers (Sanchez-Vega and Tapia, 1997).

Fig. 5.5. Pepino (*Solanum muricatum*) showing (A) stem with leaves, (A1) flower and (A2) fruit (used with permission from FAO Plant Production and Protection Series No. 26, Neglected Crops: 1492 from a different perspective, 1994 (A001/2013)).

Pollination and fruit set

Pollen is not usually very abundant. Since the stigma is longer than the anthers and the anthers are below the corolla, pollination requires transfer by an insect or human hand that leads to outcrossing. Under greenhouse conditions there could be a pollination problem due to the little movement of the branches (Nuez Viñals and Ruiz Martínez, 1996). Fruit can be produced parthenocarpically with fruit set being greater when self- or cross-pollination occurs. Parthenocarpic fruit are normally the result of pollen sterility. Poor fruit set is often a problem due to high temperatures at anthesis.

Fruit

The fruit is a berry that varies from globose to pointed oval (Fig. 5.5) with a skin background color varying from creamy to yellow-orange. Purple, gray, or green striping or blush colorations also occur on the skin; sometimes the purple color covers most of the fruit surface with the background color appearing in strips and in some varieties fruit can be completely purple or cream. The flesh may be greenish, yellow, salmon, or nearly clear. The fruit weigh between 100 and 600 g.

In many cases, the fruit are parthenocarpic, and in those with seeds these appear in the central part of the fruit, being kidney-shaped, yellow to brownish in color and 2–3 mm in size. Seedless fruit are lighter and smaller than seeded fruit.

Cultivar development

Cytogenetics, genetics and breeding

The genetic make-up of pepino is $2n = 24$. It has a very wide intra-specific variability, as well as many morphological variations with different leaf shapes and fruit colors, flavors, forms and consistencies (INIA, 1999). New Zealand has had an active breeding program since 1999. Selections have been made by the growers in Peru and Chile. The varieties grown outside of Latin America often are selections from material introduced from Chile.

Cultivars

Chile has a number of named varieties all producing similar purple-striped, egg-shaped fruit that are not sweet with soluble solids around 8%. The same varieties are also grown in California and New Zealand. In New Zealand, the most common cultivated varieties are 'El Camino' and 'Suma'. 'El Camino' was a selection in 1982 from material introduced from Chile. It has medium to large egg-shaped fruit with a green-brownish color and regular purple stripes. 'Suma' is a vigorous cultivar producing heavy crops of medium to large globose fruit, with regular purple stripes and an attractive appearance. Their flavor is mild and sweet. In California, 'New Yorker' is the most widely grown cultivar originating from material brought from Chile. Since 1984, however, 'Miski Prolific' derived from 'Miski' of New Zealand has become equally popular. Its flesh is a deep salmon color, sweet and aromatic, without the soapy flavor of some varieties, and its skin is creamy-white with light purple stripes; it produces well without pollination and the plant is very vigorous in yielding early fruit. There are a few seeds in either variety. Other varieties in California include 'Rio Bamba' from Ecuador and 'Vista', a cross of 'Rio Bamba' with another South American parent. Australia introduced 'Temptation', 'Nara Gold', 'Pepino Gold', 'Wayfarer Special', 'Golden Splendour' and 'Colossal'. Chile has some other varieties such as 'Toma'. In Ecuador, they grow the varieties 'Cacho', 'Castilla' and 'Morado'. Colombia has 'Morado' and 'Amarillo' while in Peru the most popular are 'Melón', 'Corazón de Toro' and 'Corazón de Paloma'; there is also 'Oreja de Burro' and 'Morado listado' (Nuez Viñals and Ruiz Martínez, 1996; INIA, 1999).

Propagation

SEXUAL Though easily propagated by seed, this is not done commercially. Many varieties have parthenocarpic fruit and those that have seeds will result in a highly variable progeny. Plants propagated by seed also take longer to start fruiting.

ASEXUAL Normally the plant is vegetatively propagated by easily rooted cuttings. The cuttings are 20–35 cm long with four to five leaves at the tip and allowed to dry in the shade for 2–3 days to allow the cut wounds to dry. In areas with cold winters such as Chile, the cuttings are taken before the first frost and they are propagated under shelter of plastic or woven canes. Moist sand is used for rapid rooting and 5–6 cm of the cutting is buried into the moist sand. This depth of buried cutting avoids damaging the roots at transplanting time in the spring. In warmer climates, as in northern Chile or in the growing areas of Peru, the cuttings are planted directly into the field, sometimes in groups of three to four to make sure that at least one is successful. The use of auxins apparently improves rooting (INIA, 1999). The cuttings root in about 3 weeks and have 3–4 months of vegetative growth before flowering. Tissue culture is also possible.

Orchard establishment

Pepino is grown normally like tomato and eggplant as a free-standing sprawling plant or as a pruned crop on trellises, especially for greenhouse production. Field preparation follows the normal procedure for any vegetable crop; this includes plowing, harrowing and preparing the furrows if they will be used.

In the open field planting distances usually are 1.0–1.2 m between rows and 0.6–1.0 m for plants in the row with about 16,666 and 10,000 plants per ha. In a trial performed in Chile, it was found that planting up to 40,000 plants per ha resulted in almost double the total yields in kg and in number of marketable fruit with less than 5% reduction in fruit size (Corporación de Fomento de la Producción, 1982). Though a perennial, it is usually cultivated as an annual. Sometimes a ratoon crop is done but yields are much lower.

For greenhouse production, planting is normally done with 0.5 m between plants, and trellises and pruning are used. The trellises used under greenhouse conditions normally consist of a wire line on top of the plants with strings attached to tie the three shoots that are normally left in a vertical position. This system requires more labor but results in higher yields and fruit quality. Another type of trellis that can be used under field or greenhouse conditions is to use wires at each side of the plant row. The plants are prepared by eliminating all lower shoots that cannot be tied to the wires. In this case, plants are at 0.8–1.0 m in the row and rows are at 1.5–2.0 m apart and the wires at each side are at 0.25, 0.50 and 0.80 m heights. In this system, the use of labor is much less and fruit has no contact with the ground but is less exposed to light (Nuez Viñals and Ruiz Martínez, 1996).

A common practice in Peru is to mound the soil around the plants once the plants have attained a certain size, to increase root formation and to get better anchored plants.

Irrigation

As with tomatoes, irrigation is essential in dry areas. Since roots are seldom deeper than 60 cm, it is necessary to provide light but frequent irrigation especially in loamy or sandy soils. Drip irrigation is very appropriate for this crop though in many places furrow irrigation is still used. The fruit and foliage should be kept dry whenever possible and irrigation should be reduced once the fruit are close to harvest time in order to obtain better flavor associated with higher total soluble solids and lower water content.

Pruning

On trellises, pruning is carried out to generate new growth and flowering. In a greenhouse using trellis, normally three vertical branches are left eliminating lateral shoots and leaving three to five inflorescences. The difficulty is that the basal fruit inhibit or delay the fruiting of the more terminal flowers, although some varieties do not show this behavior. Traditional cultivation does not include pruning.

Fertilization

Since pepino is a very hardy plant, in many parts of Chile, Ecuador and Peru farmers barely fertilize their crops. It is believed that excess of N reduces yields due to excessive vegetative growth, though no scientific data supports this conclusion. In Peru, side dressing is usually used with furrow irrigation and mounding the soil to cover the fertilizer is a normal practice. Some of the procedures include the addition of 20 t ha^{-1} compost or cattle manure about 2 months before planting. At planting time, 40–60 kg N, 75–80 kg phosphoric acid and 150 kg of K per ha are added. This is usually followed by an additional 50 kg of N per ha at the start of flowering. In Australia, they apply around 160 kg N, 130 kg of P_2O_5 and 130 kg of K_2O per ha. In poor soils 50% more can be applied. In New Zealand, the same fertilizer dosages as for tomato are used with about 150 kg of N, 100–140 of P_2O_5 and 120–160 of K_2O per ha (Nuez Viñals and Ruiz Martínez, 1996).

Pest management

DISEASES In Peru the most common diseases are caused by *Alternaria* and *Phytophthora infestans* that affect the foliage. The control measures include avoiding excess of water, nitrogen fertilizer and other stresses. Weed and aphid control also help. Virus diseases are common and perpetuated by asexual propagation. About seven different viruses have been identified, but their economic significance has not been established and they do not seem to be as important as in tomato production. Axillary bud *in vitro* cultivation and chemotherapy have been tried successfully to create virus-free material (Nuez Viñals and Ruiz Martínez, 1996).

INSECTS Aphids, spider mites and whitefly are problems in California and New Zealand. In Chile, *Agrotis* damages the roots. Other insects that can cause problems are *Pseudococcus obscurus* that damages the plants while the larvae of *Symmetrischema plaesiosema* can burrow in leaves and fruit. Spider mites are also important. In Peru, whiteflies, aphids, leaf eaters, leaf miners and spider mites are common problems.

Weed control

Weed control is essential during early growth. Care has to be taken if done mechanically not to damage the superficial roots or the foliage and fruit.

Orchard protection

Pepino fruit can be scarred by neighboring branches due to winds, so it is important to protect the plants with wind barriers or planting in less windy locations.

Harvesting and postharvest handling

Yields

Yields of 30–60 t ha^{-1} and even 80 t ha^{-1} under trellis can be obtained, although in some places normally yields are around 15–20 t ha^{-1}.

Harvesting

Fruit development will take from 2 to 3 months. Fruit are ready for harvest when they develop a pale yellow or cream background color; at this time the fruit total soluble solids is around 7–12%. Penetrometer tests can be made (0.2 N m^{-1}) as well as checking for minimum juice content (Nuez Viñals and Ruiz Martínez, 1996). Overripe fruit have a poor flavor. The fruit is very susceptible to mechanical injury and bruising, and thus needs to be harvested and handled carefully. Since fruit mature sequentially, several harvests are necessary during the season.

Postharvest treatment

The fruit are considered climacteric and can continue to ripen after harvest if temperatures are above 15°C. The fruit also responds to ethylene treatments. Some growers put the harvested fruit on a bed of straw or sand and cover them with a canvas, thus elevating temperature and ethylene concentration and promoting a more complete ripening; after that they take them to the market. *Botrytis*, *Fusarium*, *Alternaria* and *Penicillium* can attack the harvested fruit in Chile. The fruit are susceptible to chilling injury and can be stored at 7.5–10°C for 4 weeks (Ahumada and Cantwell, 1996). At temperatures less than 5°C chilling injury occurs especially on less ripe fruit. Fruit can be held for 75 days at 5°C and 85–95% relative humidity (Corporación de Fomento de la Producción, 1982). For ship transportation, it is recommended to keep the fruit at 5–7°C and 95% relative humidity (INIA, 1999).

Utilization

The fruit is consumed fresh and in various dishes. In Latin America, it is normally eaten out of hand after peeling and in many cases after cooling in the refrigerator. It has a pleasant aroma and it is slightly sweet which makes eating it cold very refreshing. Sometimes it is used in fruit salads. New Zealanders have served it with soups, seafood, sauces, prosciutto, meats, fish, fruit salads, and desserts. The fruit can also be frozen, jellied, dried, canned, or bottled. Often the fruit is peeled as in some varieties it has a disagreeable bitter flavor. Overripe fruit has a "soapy" flavor. The seeds are small, soft and edible and the number of seeds varies with variety.

The pepino is as good a source as many citrus fruits for vitamin C and a fair source of vitamin A (Table 5.1). It is recognized for its diuretic properties and iodine content that makes it useful to treat goitre. The fruit are normally subacid to 10–12% soluble solids content.

References and Further Reading

Cocona

Andrade, Jr. M.C. de and Andrade, J.S. (2012) Physicochemical changes in cubiu fruits (*Solanum sessiliflorum* Dunal) at different ripening stages. *Ciência e Tecnologia de Alimentos* 32, 250–254.

Donadio, L.C., Moro, F.V. and Servidone, A.A. (2002) *Frutas Brasileiras*. Editora Novos Talentos, Jaboticabal, São Paulo, Brazil, pp. 146–148.

Duarte, O. (1997) *Cocona*. Escuela Agrícola Panamericana – El Zamorano, Honduras.

Duarte, O. (2004) *Usos de la Cocona*. CCPA, Escuela Agrícola Panamericana – El Zamorano, Honduras.

Duarte, O. (2011) Cocona (*Solanum sessiliflorum* Dunal) In: Yahia, E.M. (ed.) *Postharvest Biology and Technology of Tropical and Subtropical Fruits, Volume 3. Cocona to mango*. Woodhead Publishing Ltd., Cambridge, UK, pp. 1–7.

Duarte, O. and Sandoval, X. (2002) Rendimiento y rentabilidad de 6 densidades de plantación en cocona (*Solanum sessiliflorum* Dunal) en Honduras. *Proceedings of the Interamerican Society for Tropical Horticulture* 46, 21–22.

Duarte, O., Leiva, R. and Huete, M. (2000a) Estudios de la propagación sexual y asexual de la cocona (*Solanum sessiliflorum* Dunal). *Proceedings of the Interamerican Society for Tropical Horticulture* 44, 98–101.

Duarte, O., Troncoso, J. and Huete, M. (2000b) Efecto de 6 dosis de nitrógeno en la producción de la cocona (*Solanum sessiliflorum* Dunal). *Proceedings of the Interamerican Society for Tropical Horticulture* 44, 95–97.

Gallozzi, R. and Duarte, O. (2007) *Naranjilla (lulo) y Cocona. Guía práctica de manejo agronómico, cosecha, poscosecha y procesamiento*. Como parte de: Cultivos de Diversificación para pequeños productores de frijol y maíz en América Central, producidos por el I.I.C.A. y COSUDE, 49 pp.

Heiser, C. (1985) *Of Plants and People*. University of Oklahoma Press, Norman, Oklahoma.

Heiser, C. (2001) Interspecific hybridization and the improvement of the naranjilla, *Solanum quitoense*. In: van den Berg, R., Barendse, G., Van der Weerden, G. and Mariani, C. (eds) *Solanaceae V; Advances in Taxonomy and Utilization*. Nijmegen University Press, Nijmegen, the Netherlands, pp. 307–310.

Heiser, C. (2008) Solanaceae, Cocona and other relatives of the naranjilla. In: Janick, J. and Paull, R.E. (eds) *Encyclopedia of Fruit and Nuts*. CAB International, Wallingford, UK, p. 858.

Martin, F.W., Campbell, C.W. and Ruberté, R.M. (1987) *Perennial Edible Fruits of the Tropics: An Inventory*. Agriculture Handbook No. 642, US Department of Agriculture, pp. 62–63.

Morton, J.F. (1987) *Fruits of Warm Climates*. Creative Resource Systems Inc., Winterville, North Carolina, pp. 428–430.

National Research Council (1989) *Lost Crops of the Incas: Little-known Plants of the Andes with Promise of Worldwide Cultivation*. National Academy Press, Washington, DC.

Pahlen, A.V.D. (1977) Cubiu (*Solanum topiro* Humbl. & Bonpl.), uma fruteira da Amazônia. *Acta Amazonica* 7, 301–107.

Rodríguez, V., Soria, N. and Heiser, C. (1994) Híbrido 'Palora', nueva opción para el cultivo de Naranjilla (*Solanum quitoense*) en Ecuador. *Proceedings of the Interamerican Society for Tropical Horticulture* 38, 194–196.

Silva Filho, D.F. (1998) *Cocona (Solanum sessiliflorum Dunal): Cultivo y utilizacion*. Caracas, Venezuela: Secretaria Pro-Tempore. Tratado de Cooperacion Amazonica. 114pp.

Silva Filho, D.F. (2002) Discriminação de etnovariedades de cubiu (*Solanum sessiliflorum* Dunal Solanaceae) da Amazônia, com base em suas características morfológicas e químicas. INPA/UFAM. Manaus, AM. Tese de Doutorado. 117 pp.

Silva Filho, D.F., Noda, H. and Clement, C.R. (1993) Genetic variability of economic characters in 30 accessions of cubiu (*Solanum sessiliflorum* Dunal) evaluated in Central Amazonia. *Revista Brasileira de Genética* 16(2), 409–417.

Silva Filho, D.F., Anunciação Filho, C.J., Noda, H. and Reis, O.V. (1996) Variabilidade genética em populações naturais de cubiu da Amazônia. *Horticultura Brasileira* 14(1), 9–15.

Silva Filho, D.F., Yuyama, L.K.O., Aguiar, J.P.L., Oliveira, M.C. and Martins, L.H.P. (2005) Caracterização e avaliação do potencial agronômico e nutricional de etnovariedades de cubiu (*Solanum sessiliflorum* Dunal) da Amazônia. *Acta Amazonica* 35(4), 399–405.

Villachica, H., Carvalho, J.E.U., Muller, C.H., Diaz, S.C. and Almanza, M. (1996) Cocona. In: *Frutales y Hortalizas Promisorios de la Amazonía*. Tratado de Cooperación Amazónica, Lima, Perú, pp. 97–102.

Weber, H., Sarruge, J.R., Haag, H.P and Dechen, A.R. (1981) Nutrição mineral de hortaliças. XXXVII. Deficiências de macronutrientes em cubiu (*Solanum topiro* Humb. & Bonpl.) *Anais da Escola Superior de Agricultura Luiz de Queiroz (Brazil)* 38, 481–506.

Yuyama, L.K.O., Aguiar, J.P., Macedo, S.H.M., Gioia, T. and Silva Filho, D.F. (1997) *Composição Centesimal de Diversas Populações de Cubiu (*Solanum sessiliflorum* Dunal) da Estação Experimental do Instituto Nacional de Pesquisas da Amazônia, INPA*. In: Anais do II Simpósio Latino Americano de Ciências de Alimentos. Campinas, São Paulo, Brazil.

Tree tomato

Acosta-Quezada, P., Martínez-Laborde, J. and Prohens, J. (2011) Variation among tree tomato (*Solanum betaceum* Cav.) accessions from different cultivar groups: implications for conservation of genetic resources and breeding. *Genetic Resources and Crop Evolution* 58(6), 943–960.

Albornoz, G. and Morales, R. (1989) *Normas para el Cultivo del Tomate de Árbol (Cyphomandra betacea Cav. Sendt.) en el Ecuador*. Boletín Informativo No. 1. Universidad Central del Ecuador, Facultad de Ciencias Agrícolas, Quito, Ecuador. 20 pp.

Bohs, L. (1989) Ethnobotany of the genus *Cyphomandra* (Solanaceae). *Economic Botany* 43,143–163.

Bohs, L. (1995) Transfer of *Cyphomandra* (Solanaceae) and its species to *Solanum*. *Taxon* 44, 583–587.

Bohs, L. and Olmstead, R. (1999) *Solanum* phylogeny inferred from chloroplast DNA sequence data. In: Nee, M., Symon, D., Lester, R. and Jessup, J. (eds) *Proceedings of the Fourth International Solanum Conference*, Royal Botanic Gardens, Kew, UK, pp. 97–110.

Cantwell, M. (2004) *Tamarillo*. In: US Department of Agriculture, Agriculture Handbook Number 66. Available at: http://www.ba.ars.usda.gov/hb66/tamarillo.pdf (accessed 14 July 2013).

Duarte, O. and Alvarado, E. (1997) Tratamientos para mejorar la propagación de Tomate de Árbol (*Cyphomandra betacea* Sendt.) por semillas y estacas. *Proceedings of the Interamerican Society for Tropical Horticulture* 41, 248–251.

Espina, S. and Lizana, L.A. (1991) Comportamiento del tamarillo (*Cyphomandra betacea*). *Proceedings of the Interamerican Society for Tropical Horticulture* 35, 285–290.

Federación Nacional de Cafeteros de Colombia (undated) *El Cultivo del Tomate de Árbol*. Federación Nacional de Cafeteros de Colombia, Bogotá, Colombia, 22 pp.

Lewis, D.H. and Considine, J.A. (1999) Pollination and fruit set in the tamarillo (*Cyphomandra betecea* (Cav.) Sendt.) 1. Floral biology. *New Zealand Journal of Crop and Horticultural Science* 27, 101–112.

Lewis, D.H. and Considine, J.A. (1999) Pollination and fruit set in the tamarillo (*Cyphomandra betecea* (Cav.) Sendt.) 2. Patterns of flowering and fruit set. *New Zealand Journal of Crop and Horticultural Science* 27, 113–123.

Manzano, J. (2005) Caracterización de frutos de tomate de árbol (*Cyphomandra betacea* Cav. Sendt.) y sus relativos en zonas montañosas de Venezuela. *Proceedings of the Interamerican Society for Tropical Horticulture* 48, 149–151.

Morton, J.F. (1987) *Fruits of Warm Climates*. Creative Resource Systems Inc., Winterville, North Carolina, pp. 437–440.

Osorio, G. (1992) Avances en el cultivo del Tomate de Árbol (*Cyphomandra betacea*). *Acta Horticulturae* 310, 199–203.

Pratt, H.K. and Reid, M.S. (1976) The tamarillo: fruit growth and maturation, ripening, respiration, and the role of ethylene. *Journal of the Science of Food and Agriculture* 27, 399–404.

Pringle, G.J. and Murray, B.G. (1992) Polyploidy and aneuploidy in the tamarillo, *Cyphomandra betacea* (Cav.) Sendt. (Solanaceae) I. Spontaneous polyploidy and features of the euploids. *Plant Breeding* 108, 132–138.

Prohens, J. and Nuez, F. (2000) The tamarillo (*Cyphomandra betacea*). *Small Fruits Review* 1, 43–68.

Prohens, J., Ruiz, J.J. and Nuez, F. (1996) Advancing the Tamarillo harvest by induced postharvest ripening. *HortScience* 31, 109–111.

Sale, P.R. (1984a) *Tamarillos: General Requirements, Varieties, and Propagation for Commercial Production*. Aglink Horticultural Produce and Practice No. 296. Ministry of Agriculture and Fisheries Media Services, Private Bag, Wellington, New Zealand.

Sale, P.R. (1984b) *Tamarillos: Orchard Management, Planting, Training, Pruning, Nutrition. Harvesting*. Aglink Horticultutral Produce & Practice No. 297. Ministry of Agriculture and Fisheries Media Services, Private Bag, Wellington, New Zealand.

Sale, P.R. (1984c) *Tamarillos: Pests and Diseases*. Aglink Horticultural Produce and Practice No. 298. Ministry of Agriculture and Fisheries Media Services, Private Bag, Wellington, New Zealand.

Schotsmans, W.C., East, A. and Woolf, A. (2011) Tamarillo (*Solanum betaceum* (Cav.)). In: Yahia, E.M. (ed.) *Postharvest Biology and Technology of Tropical and Subtropical Fruits, Volume 4. Mangosteen to white sapote*. Woodhead Publishing Ltd., Cambridge, UK, pp. 427–441.

Cape gooseberry

Almanza, J. (2000) Propagación. In: Flórez, V.J., Fischer, G. and Sora, A.D. (eds) *Producción, Poscosecha y Exportación de Uchuva (*Physalis peruviana *L.)*. Universidad Nacional de Colombia, Facultad de Agronomía, Santa Fe de Bogotá, Colombia pp. 27–40.

Angulo, R. (2000) Siembra, soporte, poda y fertilización. In: Flórez, V.J., Fischer, G. and Sora, A.D. (eds) *Producción, Poscosecha y Exportación de Uchuva (*Physalis peruviana *L.)*. Universidad Nacional de Colombia, Facultad de Agronomía, Santa Fe de Bogotá, Colombia, pp. 41–49.

Benavides, M.A. and Mora, H.R. (2005) Los insectos-plaga limitantes en el cultivo de la uchuva y su manejo. In: Fischer, G., Miranda, D., Piedrahita, W. and Romero, J. (eds) *Avances en el Cultivo, Poscosecha y Exportación de la Uchuva (*Physalis peruviana *L.) en Colombia*. Universidad Nacional de Colombia, Facultad de Agronomía, Bogotá, pp. 83–96.

Cailes, R.L. (1952) The cultivation of cape gooseberry. *Journal of Agriculture of Western Australia* 1, 363–365.

Chia, C.L., Nishina, M.S. and Evans, D.O. (1987) *Poha*. Commodity Fact Sheet Poha-3(A), Hawaii Cooperative Extension Service, Hawaii Institute of Tropical Agriculture and Human Resources, University of Hawaii at Manoa, Honolulu, Hawaii, 2 pp.

Council of Scientific and Industrial Research (CSIR) (1969) Physalis. In: *The Wealth of India: Raw Materials*, Vol. 8. Publications & Information Directorate, CSIR, New Delhi, pp. 38–40.
Davis, B.R. and Whitehead, V.B. (1974) *Cape Gooseberry Cultivation*. Information Bulletin Number 210. Fruit and Fruit Technology Research Institute, Stellenbosch, South Africa. Mimeo, 4 pp.
Fischer, G. (2000) Crecimiento y desarrollo In: Florez, V.J., Fischer, G. and Sora, A.D. (eds) *Producción, Poscosecha y Exportación de Uchuva (*Physalis peruviana *L.)*. Universidad Nacional de Colombia, Facultad de Agronomía, Santa Fe de Bogotá, Colombia, pp. 9–26.
Fischer, G. and Almanza, P.J. (1993) La uchuva (*Physalis peruviana* L.) una alternativa promisoria para las zonas altas de Colombia. *Agricultura Tropical* 30(1), 79–87.
Fischer, G., Lüdders, P. and Gallo, F. (1997) Quality changes of the cape gooseberry fruit during its ripening. *Erwerbsobstbau* 39, 153–156.
Fischer, G., Herrera, A. and Almanza, P.J. (2011) Cape gooseberry (*Physalis peruviana* L.). In: Yahia, E.M (ed.) *Postharvest Biology and Technology of Tropical and Subtropical Fruits, Volume 2. Acai to citrus*. Woodhead Publishing Ltd., Cambridge, UK, pp. 374–396.
Klinac, D.J. (1986) Cape gooseberry (*Physalis peruviana*) production systems. *New Zealand Journal of Experimental Agriculture* 14, 425–30.
Klinac, D.J. and Wood, F.H. (1986) Cape gooseberry (*Physalis peruviana*). *The Orchardist of New Zealand* 60, 103.
Legge, A.P. (1974) Notes on the history, cultivation, and uses of *Physalis peruviana*. *Journal of the Royal Horticultural Society* 99, 310–314.
Ligarreto, G.A., Lobo, M. and Correa, A. (2005) Recursos genéticos del género *Physalis* en Colombia. In: Fischer, G., Miranda, D., Piedrahita, W. and Romero, J. (eds) *Avances en el cultivo, poscosecha y exportación de la uchuva (Physalis peruviana* L.*) en Colombia*. Universidad Nacional de Colombia, Facultad de Agronomía, Bogotá, pp. 9–27.
Lizana, L.A. and Espina, S. (1991) Efecto de la temperatura de almacenaje sobre el comportamiento en poscosecha de frutos de fisalis (*Physalis peruviana*). *Proceedings of the Interamerican Society for Tropical Horticulture* 35, 278–284.
Mazumdar, B.C. (1979) Cape-gooseberry, the jam fruit of India. *World Crops* 31 (January/February), 19, 23.
Miranda, D. (2005) Criterios para el establecimiento, los sistemas de cultivo, el tutorado y la poda de la uchuva In: Fischer, G., Miranda, D., Piedrahita, W. and Romero, J. (eds) *Avances en el Cultivo, Poscosecha y Exportación de la Uchuva (*Physalis peruviana *L.) en Colombia*. Universidad Nacional de Colombia, Facultad de Agronomía, Bogotá, pp. 29–53.
Morton, J.F. (1987) *Fruits of Warm Climates*. Creative Resource Systems Inc., Winterville, North Carolina, pp. 434–437.
Morton, J.F. and Russell, O.S. (1954) The cape gooseberry and the Mexican husk tomato. *Proceedings of the Florida State Horticultural Society* 67, 261–266.
Sanabria, S. (2005) Situación actual de la uchuva en Colombia. In: Fischer, G., Miranda, D., Piedrahita, W. and Romero, J. (eds) *Avances en el Cultivo, Poscosecha y Exportación de la Uchuva (*Physalis peruviana *L.) en Colombia*. Universidad Nacional de Colombia, Facultad de Agronomía, Bogotá, pp. 1–8.
Trinchero, G.D., Sozzi, G.O., Cerri, A.M., Vilela, F. and Fraschina, A.A. (1999) Ripening-related changes in ethylene production, respiration rate and cell-wall enzyme activity in goldenberry (*Physalis peruviana* L.), a solanaceous species. *Postharvest Biology and Technology* 16, 139–145.
Vietmeyer, N. (1991) Lost crops of the Incas. *New Zealand Geographic* 10, 49–67.
Wang, J.K. (1966) *Equipment for Husking Poha Berries*. Technical Bulletin No. 60. Hawaii Agricultural Experiment Station, Honolulu, 18 pp.
Zapata, J.L., Saladarriaga, A., Londoño, M. and Diaz, C. (2005) Las enfermedades limitantes en el cultivo y poscosecha de la uchuva y su control. In: Fischer, G., Miranda, D.,

Piedrahita, W. and Romero, J. (eds) *Avances en el Cultivo, Poscosecha y Exportación de la Uchuva (*Physalis peruviana *L.) en Colombia*. Universidad Nacional de Colombia, Facultad de Agronomía, Bogotá, pp. 97–110.

Tomatillo

Cantwell, M., Flores-Minutti, J. and Trejo-González, A. (1992) Developmental changes and postharvest physiology of tomatillo fruits (*Physalis ixocarpa* Brot.). *Scientia Horticulturae* 50, 59–70.

Cartujano-Escobar, F., Jankiewicz, L., Fernandez-Orduna, V.M. and Mulato-Brito, J. (1985a) The development of the husk tomato plant (*Physalis ixocarpa* Brot) I. Aerial vegetative parts. *Acta Societatis Botanicorum Poloniae* 54, 327–338.

Cartujano-Escobar, F., Jankiewicz, L., Fernandez-Orduna, V.M. and Mulato-Brito, J. (1985b) The development of the husk tomato plant (*Physalis ixocarpa* Brot) 11. Reproductive parts. *Acta Societatis Botanicorum Poloniae* 54, 339–349.

Del Monte, D. de G.J.P. (1988) Presencia, distribución y origen de *Physalis philadelphica* Lam. en la zona centro de la Península Ibérica. *Candollea* 43, 93–100.

Díaz-Pérez, J.C., Phatak, S.C., Giddings, D., Bertrand, D. and Mills, H.A. (2005) Root zone temperature, plant growth, and fruit yield of tomatillo as affected by plastic film mulch. *HortScience* 40, 1312–1319.

Govindasamy, R., Puduri, V.S. and Simon, J.E. (2011) Willingness to buy new ethnic produce items: A study of latino consumers from Mexico and Puerto Rico in the Eastern United States. *HortTechnology* 21(2), 202–207.

Hernandez, S.M. (1994) Tomatillo. In: Hernando Bermejo, J.E. and León, J. (eds) *Neglected Crops: 1492 from a Different Perspective*. Plant Production and Protection Series No. 26. FAO, Rome, pp. 117–122.

Hudson, W.D. Jr. (1986) Relationships of domesticated and wild *Physalis philadelphica*. In: D'Arcy, W.G. (ed.) *Solanaceae: biology and systematics*. Columbia University Press, New York.

Menzel, Y.M. (1951) The cytotaxonomy and genetics of *Physalis*. *Proc. Am. Phil. Soc.* 95, 132–183.

Morton, J. (1987) Mexican Husk Tomato. In: *Fruits of Warm Climates*. Julia F. Morton, Miami, Florida, distributed by Creative Resources Systems Inc., Winterville, North Carolina, pp. 434–437.

Mulato-Brito, J., Jankiewicz, L., Fernandez-Orduna, V.M., Cartujano-Escobar, F. and Serrano-Covarrubias, L.M. (1985) Growth, fructification and plastochron index of different branches in the crown of the husk tomato (*Physalis ixocarpa* Brot.). *Acta Societatis Botanicorum Poloniae* 54, 195–206.

Pandey, K.K. (1957) Genetics of self-incompatibility in *Physalis ixocarpa* Brot. – A new system. *American Journal of Botany* 44, 879–887.

Quiros, C.F. (1984) Overview of the genetics and breeding of husk tomato. *HortScience* 19, 872–874.

Vargas-Ponce, O., Pérez-Álvarez, L., Zamora-Tavares, P. and Rodríguez, A. (2011) Assessing genetic diversity in Mexican husk tomato species. *Plant Molecular Biology Reporter* 29, 733–738.

Pepino

Ahumada, M., and Cantwell, M. (1996) Postharvest studies on pepino dulce (*Solanum muricatum* Ait.): maturity at harvest and storage behavior. *Postharvest Biology and Technology*, 7(1), 129–136.

Burge, G.K. (1989) Fruit set in the pepino (*Solanum muricatum* Ait.). *Scientia Horticulturae* 41, 63–68.

Corporación de Fomento de la Producción (1982) *Cultivo del Pepino Dulce*: Antecedentes, Agronómicos, Económicos. AA 82/34. Corporación de Fomento de la Producción, Gerencia de Desarrollo, Facultad de Agronomía de la Pontificia Universidad Católica de Chile, Santiago.

Dawes, S.N. (1984) Pepino breeding and selection. *The Orchardist of New Zealand* 58, 172.

Dennis, D.J., Burge, G.K. and Lill, R. (1985) *Pepino Cultural Techniques: An Introduction*. Aglink Horticultural Produce and Practice No. 208. Ministry of Agriculture and Fisheries Information Services, Private Bag, Wellington, New Zealand, 2 pp.

Heiser, C.B. (1964) Origin and variability of the pepino (*Solanum muricatum*): a preliminary report. *Baileya* 12, 151–158.

Heiser, C.B. (1985a) Green "tomatoes" and purple "cucumbers". In: Heiser, C.B. (ed.) *Of Plants and People*. University of Oklahoma Press, Norman, Oklahoma, pp. 128–141.

INIA (1999) *El Cultivo del Pepino Dulce*. Boletín Técnico 21. Instituto Nacional de Investigación Agraria-Comisión Nacional de Fruticultura, Lima, Perú, 20 pp.

Leung, W.W. and Flores, M. (1961) *Tabla de Composición de Alimentos para Uso en América Latina*. Instituto de Nutrición de Centro América y Panamá (INCAP), Editorial Interamericana, S.A. 132 pp.

Lizana, L.A. and Levano, B. (1977) Caracterización y comportamiento del pepino dulce, *Solanum muricatum*. *Proceedings of the Tropical Region of the American Society for Horticultural Science* 21, 11–15.

Morley-Bunker, M.J.S. (1983) A new commercial crop, the pepino (*Solanum muricatum* Ait) and suggestions for further development. *Royal New Zealand Institute of Horticulture Annual Journal* 11, 8–19.

National Research Council (1989) *Lost Crops of the Incas: Little-known Plants of the Andes with Promise for Worldwide Cultivation*. National Academy Press, Washington, DC.

Nuez Viñals, F. and Ruiz Martínez, J.J. (1996) *El Pepino Dulce y su Cultivo*. Estudio FAO de Producción y Protección Vegetal 136. FAO, Rome, 142 pp.

Prohens, J., Ruiz, J.J. and Nuez, F. (1998) The inheritance of parthenocarpy and associated traits in Pepino. *Journal of the American Society for Horticultural Science* 123(3), 376–380.

Redgwell, R.J. and Turner, N.A. (1986) Pepino (*Solanum muricatum*): Chemical composition of ripe fruit. *Journal of the Science of Food and Agriculture* 37, 1217–1222.

Ruiz, J.J. and Nuez, F. (1997) The pepino (*Solanum muricatum* Ait.): An alternative crop for areas affected by moderate salinity. *HortScience* 32, 649–652.

Sanchez-Vega, I. (1994) Pepino. In: Hernándo Bermejo, J.E. and León, J. (eds) *Neglected Crops: 1492 from a Different Perspective.* Plant Production and Protection Series No. 26. FAO, Rome, pp. 181–191. Available at: http://www.hort.purdue.edu/newcrop/1492/fruits.html#Pepino (accessed 15 July 2013).

Sanchez-Vega, I. and Tapia, M. (1997) Frutales Andinos. In: Tapia, M.E. (ed.) *Cultivos Andinos Subexplotados y su Aporte a la Alimentación*. Oficina Regional de FAO para América Latina y el Caribe. Santiago, Chile, pp. 118–124.

Seidel, H. (1974) Erfahrungen mit dem Anbau von *Solanum muricatum* in Südspanien (Experiences with cropping of *Solanum muricatum* in Southern Spain). *Tropenlandwirt* 75(5), 24–30.

6 Sapindaceae

The Sapindaceae or soapberry family includes temperate, warm subtropical and tropical species of mainly shrubs and trees, rarely herbs. The family has 150 genera and about 2,000 species with the majority native to Asia, with a few in South America, Africa and Australia. The largely temperate families of Aceraceae and Hippocastanaceae containing the maples and the horse chestnuts have only recently been included in the Sapindaceae. The Sapindaceae contains a number of valuable tropical fruit including litchi, longan, rambutan, pulasan and ackee. Other members are important ornamentals, timber trees, nuts, or sources of oils or drugs. In this chapter, two tropical American fruits will be described: Spanish lime and guarana.

Spanish Lime or Kenep

Spanish lime *Melicoccus bijugatus* Jacq. (*Melicoccus bijuga* L.) is a minor fruit native to tropical America used as a fresh fruit and in beverages. The Latin names come from the Greek *mel* (honey) and *kokkos* (seed). Another member of this genus, *Melicoccus lepidopetalus* Radlk., is similarly used as a fresh fruit in Bolivia, Paraguay and northern Argentina.

In Spanish speaking countries it is called *mamoncillo* (the most common name), *grosella de miel, macao, maco, mauco, muco* (Colombia, Venezuela), *mamón* (Central America, Colombia, Venezuela, Argentina), *mamón de Cartagena* (Costa Rica), *quenepa, mamoncillo, limoncillo* (Dominican Republic, Puerto Rico, Colombia), *grosella de miel, guayo* (México) and *tapaljocote* (El Salvador). In English, names include genip, ginep, ginepe, guenepa, guinep (Barbados, Jamaica, Bahamas, Puerto Rico, Trinidad and Tobago), genip lime, genip tree, honeyberry (Guyana), Jamaica bullace plum, kanappy (Puerto Rico), and knippelboom and Spanish lime (Florida). The names in French are *kenépier, kenettier, knépier* and *quenettier*, with *quenette* (French West Indies) and *quenepe* (Haiti). In German, it is *dotter knippe. Limoncillo* and Spanish lime evoke acidity, whereas honeyberry and *grosella de miel* refer to sweetness. Popenoe (1974) explained this apparent contradiction by the existence of very sour types of mamoncillo in Cuba.

Origin and distribution

The species is native to northern South America, from Colombia to the Guianas, and Central America and the West Indies (Martin *et al.*, 1987). It is grown and found growing wild in most of these countries. Spanish lime is widely cultivated in home gardens, and even in urban habitats including streets, in the Antilles, Central and northern South America. It is grown in South Florida and Hawaii mainly as an ornamental or street or backyard tree. The fruit is generally consumed locally although some is exported for ethnic markets, with the main exporter being the Dominican Republic and smaller exports made by Haiti and Jamaica mostly to the USA (Coppens d'Eeckenbrugge and Paull, 2008).

Ecology

Soil

The tree grows on very poor soils including alkaline and rocky soils (Martin *et al.*, 1987). In some countries, it is used in soil reclamation efforts. The best soils are deep, rich, and of calcareous origin. It does very well in the oolitic limestone of Florida and grows spontaneously in dry coastal districts (Morton, 1987).

Rainfall

It is well adapted to low rainfall areas (750–1,500 mm), sometimes up to 2,600 mm, and tolerates 3–5 dry months, apparently needing a pronounced dry period to flower and produce satisfactorily.

Temperature

It can be considered either a tropical or subtropical, as it grows well in both hot tropical lowlands and altitudes of up to 1,200 m in Central America. The mature plant withstands only light frost without injury, allowing it to grow in southern Florida (Martin *et al.*, 1987).

Light and photoperiod

The tree prefers full sun exposure. It flowers at all latitudes indicating it is not photoperiodic.

General characteristics

Tree

The tree is slow-growing and erect, growing to 6–15 m, sometimes 25–30 m, and the trunk can reach 1.7 m. The bark is thick, smooth and gray. The briefly deciduous alternate leaves have four opposite or sub-opposite, elliptical, sharp-pointed leaflets measuring 5–12.5 cm by 3.25–6.25 cm with a frequently conspicuously winged rachis (Fig. 6.1).

Flowers

The fragrant flowers are small, greenish-white to white, and appear on slender racemes (8–10 cm long) that are clustered terminally on short pedicels (4–8 mm). The panicles are highly ramified for the male and generally simple for the female (6–10 cm long, 5–8 mm wide). The 5–8 mm wide flowers have four small sepals and petals (2 and 3 mm long, respectively) and eight to nine stamens or a pistil with a bifid, white stigma. The flowers appear at the beginning of the rainy season (Coppens d'Eeckenbrugge and Paull, 2008).

The polygamous tree is generally dioecious with male and female plants but some plants can have hermaphrodite flowers (Martin *et al.*, 1987). Flowering takes place once a year, from April to June in Puerto Rico, in April in Florida and June to August in Trinidad. In central Honduras, it flowers around May. Only a small proportion of the flowers develop into fruit that takes about 100 days, sometimes up to 150 days, to mature (Martin *et al.*, 1987).

Pollination and fruit set

The anthers of many bisexual flowers are non-functional or pollen-sterile so that fruit do not develop unless cross-pollination occurs. It is therefore necessary to

Fig. 6.1. Spanish lime (*Melicoccus bijugatus*) showing leaves and fruit (used with permission from León, J. (2000) *Botánica de los Cultivos Tropicales*. Editorial Agroamérica, Instituto Interamericano de Cooperación para la Agricultura (IICA), San José, Costa Rica).

interplant male plants with female or hermaphroditic plants to ensure adequate pollination and fruit set. Some trees have sufficient flowers of both sexes to yield fruit without interplanting. No information has been found about pollination agents but the flowers are rich in nectar and it is considered a plant suitable for the production of honey (Vargas *et al.*, 1999). This could be an indication that honey bees and other flying insects are involved in pollination.

Fruit

Fruit are borne in clusters and are globose drupes, 2–3.5 cm in diameter and 10–25 g in weight (Fig. 6.1), and green to yellowish when mature with a thin, leathery, brittle skin. The fruit have a short, sharp terminal protrusion. The pulp is a gelatinous aril that is juicy and tart or acid-sweet, translucent, somewhat astringent, orange to salmon pink, and contains one relatively large seed to which the aril is firmly attached. The single seed is large, yellowish-white or white, hard-shelled, starchy and astringent. Some trees produce a proportion of fruit with two hemispherical seeds.

Cultivar development

Little attention has been given to this fruit. Sporadically horticulturists have collected seeds and this has led to better selections. Cultivars have been developed by selection followed by vegetative propagation. Four selections, numbered 'Puerto Rico #1' to 'Puerto Rico #4', were obtained from several seedlings grown by George Jackson in the 1960s from 54 trees in Puerto Rico. In the 1970s to 1980s selections made in Puerto Rico included 'Mayagüez', 'Añasco', 'Sasa', 'Fana', 'Isabel', 'Gloria', 'Yinto' 'Santiago'. Four cultivars were selected in South Florida including 'Queen', 'Montgomery', 'Number 1' and 'Number 3', with 'Montgomery' and 'Number 3' being recommended (Vargas *et al.*, 1999). Cabrera and Brunner (2007) describe in detail 30 varieties in their germplasm collection in Puerto Rico. Eleven germplasm collections of Spanish lime are listed by Knudsen (2000). In many countries people sow seeds from the best fruit but their variability due to outcrossing often negates these efforts.

Cultural practices

Propagation

SEXUAL The species is usually propagated from seeds that take 15–30 days to germinate and start bearing after 5–6 years (Vargas *et al.*, 1999). Germination is not difficult but seeds should be sown as soon as possible after they are extracted from the fruit. The progeny from seed propagation are highly variable due to cross-pollination and the sex of the seedling is unknown until flowering occurs (Alix and Duarte, 1999).

ASEXUAL Only a small percentage of herbaceous and hardwood cuttings develop roots, irrespective of whether girdling or synthetic auxin treatment are used.

Grafting, as with litchi, is difficult and generally has been unsuccessful though approach grafting is possible. Air layering on vigorously growing, upright shoots is successful, with adequate root development in 5–6 weeks (Larson *et al.*, 1991). Asexually propagated material bears fruit in about 3–4 years.

Orchard establishment

In the absence of commercial orchards, regular orchard land preparation should be followed. Tree spacing should be at least 8–10 m. Normally, 1-year-old plants are used that are 40–50 cm tall. Isolated trees frequently do not produce fruit due to self-incompatibility, so that some male plants should be interplanted for optimal yields (Vargas *et al.*, 1999).

Irrigation

Young trees receive watering and fertilizer when planted out. Occasional deep watering in summer for established plants is recommended.

Pruning

Pruning and training is not normally practiced, but height needs to be controlled for ease of harvesting. Topping young plants at about 90–100 cm and selecting three to four well-distributed branches will ensure a low well-formed canopy that is manageable. Periodic pruning to remove dead or diseased branches, or those growing in the wrong direction, is helpful.

Fertilization

Recommendations as to fertilizer rates are unavailable. It is suggested to follow fertilization recommendations for similar fruit trees in the area.

Pest management

DISEASES The tree does not have many serious disease problems. In Florida, *Armillariella (Clitocybe) tabescens* causes mushroom root rot. Fungal leaf spot diseases are caused by *Fusarium* sp. and *Phyllosticta* sp.; *Cephaleuros virescens* causes algal green spot and scurf.

INSECTS The tree is a host for the citrus black fly, *Aleurocanthus woglumi*, which can be controlled with parasites. Mealy bugs can also cause some damage (Morton, 1987).

Weed control

Weed control in a commercial orchard should be the same as with other fruit trees.

Orchard protection

No information is available on the need for windbreaks, though the tree is fairly sturdy.

Harvesting and postharvest handling

An adult tree can produce up to 200 kg year^{-1} (Vargas *et al.*, 1999).

Harvesting

The entire fruit cluster is clipped from the trees when mature. A change in fruit peel is the best indicator of harvest maturity. The fruit rind becomes brittle but does not change color. Fruit mature in Puerto Rico between June and September, and between July and December in the Dominican Republic.

Postharvest treatment

Normally the fruit is packed in 6–10 kg boxes. The skin protects the fruit. Storage at 10–12°C is recommended for up to 14 days (Pérez *et al.*, 2008).

Utilization

The fruit are cut and sold mostly along roadsides and streets to be eaten out of hand as a snack. The aril of Spanish limes has a sweet taste when fully ripe that is a pleasant mixture of sweetness with acidic overtones. The tight, thin skin is easily cracked by the teeth or thumbnails. It is made into jam, marmalade, jellies or beverages and the peeled fruit can be boiled to make a cold drink. The fruit is particularly rich in iron and phosphorus (Table 6.1) and the percentage of edible aril is up to 56% in the best clones. The seeds are said to be edible after roasting. Indians of the Orinoco River basin use the cooked seeds as a substitute for cassava.

The abundant flower nectar allows the production of an appreciated dark, flavored honey. The wood is valued for rafters, indoor framing and cabinetwork. Pulverized, roasted seed kernels and a decoction of bark or leaves are used for dysentery and intestinal disorders. In Panama, the leaves are used to repel flies in houses (Coppens d'Eeckenbrugge and Paull, 2008).

Table 6.1. Composition of Spanish lime fruit per 100 g edible portion (Morton, 1987).

Proximate	Spanish lime
Energy (kcal)	58.1–73
Moisture (g)	68.8–82.5
Protein (g)	0.5–1.0
Lipid (fat) (g)	0.08–0.2
Fiber (g)	0.07–2.60
Ash (g)	0.34–0.74
Minerals	
Calcium (mg)	3.4–15.0
Iron (mg)	0.47–1.19
Phosphorus (mg)	9.8–23.9
Vitamins	
Thiamine (mg)	0.03–0.21
Riboflavin (mg)	0.01–0.20
Niacin (mg)	0.15–0.90
Carotene (mg)	0.02–0.44
Ascorbic acid (mg)	0.80–10.00

Guarana

Guarana, *Paullinia cupana* (Mart.) Ducke., is a climbing plant of the mid- to upper Amazon basin that is grown not for its fruit but rather for its seeds that are used for preparing beverages that are very popular in Brazil. The seed extract is used to isolate a number of products, especially caffeine. *Paullinia cupana* H.B.K. was its first name and is a synonym of *Paullinia sorbilis* Mart. The Amerindian word *uaraná* means "eye-like" and hence the common Portuguese name is *guarana*, though it is also referred to as *guarana-sipo*, *guarana-uva* and *urana*. In English and French it is *guarana* while in Spanish it is called *guaraná* or *cupana*, this last referring more to var. *cupana* known with that name in Venezuela and Colombia (Cavalcante, 1979).

Important genera and species

The genus *Paullinia* comprises about 180 species of which about 40 are utilized by several Amerindian groups (Benlekehal *et al.*, 2001). Guarana belongs to the most important commercial species, of which two types are reported: *Paullinia cupana* var. *sorbilis* (Mart) Ducke, which is the type commercially grown and has a high degree of variability; and *Paullinia cupana* var. *cupana* with no tendrils and larger flowers and fruit, and whose leaflets are more strongly lobed. *Paullinia yoco* Schultes & Killip is another species found along the Putumayo River in the vicinity of the border between Peru and Colombia is also used as a stimulant but it is not cultivated (Erickson, 2008).

Origin and distribution

Guarana was first described by Humboldt, Bonpland and Kunth in 1824 when found near the Orinoco River in Venezuela. It seems to have originated in the mid- to upper Amazon basin. Since no wild plants are found, it is difficult to establish the exact place of origin. The region of Manaus in the state of Amazonas in Brazil has been considered as its center of origin, since that is where the var. *sorbilis* was domesticated, while *P. cupana* f. *typica* could have its center of variability in the region of the upper Orinoco and upper Negro rivers in the confluence of Colombia, Ecuador and Brazil (Cavalcante, 1979; Villachica *et al.*, 1996).

Guarana was used in pre-Columbian times and the modern plant seems to have been derived by selection from tall lianas for shortness by the people growing it in Amazonia. The cultivated area of guarana was not very large until fairly recently. Closely related *Paullinia* species and biotypes, all native to the Brazilian Amazon, are well described but it is unlikely that a wild counterpart to the commercial guarana will be found (Erickson, 2008).

Guarana is grown mainly for its caffeine content and its stimulatory effects but the plant and its products were little known outside Brazil until fairly recently. According to Erickson (2008), commercial orchards in Brazil were established at the beginning of the 20th century and Empresa Brasileira de Pesquisa

Agropecuária (EMBRAPA) in Brazil started active research about 30 years ago. Most commercial plantations were along the Amazon River, but since the 1990s production has expanded to other areas such as the cacao-producing area in the southern state of Bahia that now has more area planted than Amazonia. This increase was a response to a growing demand in the USA and several other countries. Production has extended little to other tropical regions with similar climatic conditions to those of Brazil. In addition, Brazil has imposed a control on exports of planting material to avoid its spread outside of Latin America and avoid potential competition.

Ecology

Soil

The soils where it is grown in Brazil are red-yellow acidic oxysols with a pH of about 4.0. Traditional cultivation of guarana is carried out on fairly acid soils (pH 3.5–5.0) with a low fertility and with a high concentration of aluminum (Erickson, 2008). It is also cultivated in eutrophic soils where yields are much better. The best soils must be deep, medium to heavy in texture, well drained, porous and with a high content of organic matter (Villachica *et al.*, 1996).

Rainfall

The mean annual rainfall of the areas where it is grown is 2,200–2,500 mm. Annual precipitation must exceed 1,400 mm, with rain well distributed during the year. A maximum water deficiency of 250 mm is tolerated (Villachica *et al.*, 1996).

Temperature

Recommendations from EMBRAPA include growing guarana in areas with similar climates to its region of origin, with a mean annual temperature between 28 and 29°C. The minimum temperature tolerated is 12°C (Erickson, 2008). It will grow well in areas with a mean annual temperature of 23–26.5°C, an average annual minimum of 20.0–23.0°C and an average minimum temperature of the coldest month of 16.5–23.0°C (Villachica *et al.*, 1996).

Light and photoperiod

In the nursery and during the initial 6 months in the field, it is recommended to grow the plants under partial shade; afterwards they are grown in full sun. It flowers with no problems in the areas where it is grown so no photoperiod influence seems to be involved with the flowering process under those conditions.

Wind

Being a climbing plant attached to supports such as a tree or trellis, tolerance to wind will depend on the resistance of the support to wind (Erickson, 2008). In its native area, there are normally no strong winds so that this is not a concern. It is recommended that the trellis runs in the same direction as the main winds to reduce the impact.

General characteristics

Plant

Guarana is a climbing perennial evergreen shrub or woody liana 2–3 m tall under full sun and prostrate when grown under semi-shade. The brown-yellowish main stem is smooth, while the long new shoots have four to five longitudinal furrows. The wide alternate leaves are 27–33 cm long and 10–15 cm wide and are pinnately compound having five leaflets of which four are in pairs and the fifth is at the distal end (Fig. 6.2). The leaves have a 7–15 cm petiole that has a channel in its upper side as well as a developed sheath and a deciduous bract at each side (Fig. 6.2). In young seedlings, growth is very slow with about one new leaf appearing every month. These seedlings normally have entire leaves that become compound after the seventh to tenth leaf. Tendrils arise from the leaf axils in var. *sorbilis*, while var. *cupana* has no tendrils. These tendrils can bear inflorescences that normally arise at the nodes.

Flowers

This monoecious vine has both male and female flowers in the same racemes. Each raceme is 10–30 cm long with several hundred flower buds. Pistillate flowers are about one fifth or one sixth of the staminate flowers. Female flowers have apparently normal stamens but anthers are indehiscent; the ovary has three carpels and three locules with one ovule each that can be fertilized. The male flowers have eight stamens of three different sizes, a rudimentary ovary and poorly developed stylus and stigma. The calyx has two outer sepals surrounding three inner sepals

Fig. 6.2. Guarana (*Paullinia cupana*) showing (A) leaves on stem with fruit, (B) fruit and (C) seed (used with permission from León, J. (2000) *Botánica de los Cultivos Tropicales*. Editorial Agroamérica, Instituto Interamericano de Cooperación para la Agricultura (IICA), San José, Costa Rica).

that look like the four or five white, 5 cm long petals forming a cap-like structure. The internal part of the petals has coriaceous scales.

The juvenile stage lasts about a year with inflorescences appearing on new stems initiated early in the rainy season and that have become semi-hardened. More recent growth can also bear flowers, on canes that are still soft and succulent.

Rainfall patterns appear to dictate the onset of flowering in the Amazonian region where the driest months are June–September. Flowering is correlated with this dry period and normally peaks between July and early September and fruit matures in January–February.

Pollination and fruit set

Guarana is an allogamous (cross-fertilized) species. In a raceme, flowers of the same sex open at a certain day in unison with the other racemes on that branch. On other branches of the same plant, flowers of the opposite sex can be open. These flower openings come in waves that can last 1 month or more of open male or female flowers followed by a wave of opening of the opposite sex flowers in a given raceme (Erickson, 2008).

Bees of the genera *Melipona* and *Apis* are common pollinators. Wasps also seem to be involved, though ants are especially abundant on guarana plants, continually moving from branch to branch, and may also assist in pollination. Flowers open in the morning and most pollen has been released by midday (Erickson, 2008).

Fruit

The 2–3 cm long and 1–1.5 cm wide fruit is a dehiscent capsule with a pointed apical end that dehisces along its three septae (Fig. 6.2). The epicarp is a deep yellow to reddish-orange or reddish-yellow when ripe and can contain from one to three seeds depending on the number of fertilized ovules. The 1.2–2.0 cm seeds weigh about 1 g and comprise around 84% of the fruit mass. The seeds are ovoid and cream colored at their base, with the rest being dark brown, black or greenish (Benlekehal *et al.*, 2001). The seeds are partially surrounded by a waxy white ariloid that is thicker at the seed base and does not cover its distal portion (Donadio *et al.*, 2002). The fruit develop and mature with the onset of the new rainy season. A fruit will have an average of 1.3 seeds (Cavalcante, 1979). The seeds have two fleshy cotyledons surrounded by a thin, woody coat. The embryo is immature at the time of fruit ripening and has no differentiated radicle. Birds are the main dispersers of the seeds (Erickson, 2008).

Cultivar development

Cytogenetics, genetics and breeding

Guarana is very diverse in plant characteristics and growth, amount and time of production, caffeine content, ability to root by cuttings, germination speed and disease resistance (Villachica *et al.*, 1996). The major limitation is low productivity, since average dry kernel yield per plant ranges between 250 g with traditional cultivation and 520 g under improved management. This variation gives considerable scope for improvement particularly in better cultivation practices and plant care (Erickson, 2008).

Selection and evaluation

The selection of more productive, early maturing lines with resistance to disease and stress began in Manaus as early as 1980. Plants have been selected in commercial and experimental plantings with yields of between 4 and 6 kg of dry seed per plant. The production of hybrids, either through traditional methods or genetic engineering, especially in conjunction with the production of clonal material, will allow more uniform production and management to be achieved. Leadership was provided by the Brazilian research service (EMBRAPA) at its station near Manaus that studied cultural methods, genetics, pathology and other aspects of guarana growing (Erickson, 2008). CPATU/EMBRAPA at Belém, Pará has a germplasm bank with around 200 accessions and cloned selections. The CPAA/EMBRAPA in Manaus has more than 150 clones. The Comissão Executiva de Planejamento da Lavoura Cacaueira (CEPLAC) in Bahia selected 22 progenies of half-brother plants. The Instituto Agronômico de Campinas (IAC) in the state of São Paulo has some accessions. The Peruvian Instituto Nacional de Innovación Agraria (INIA) has a collection of the best ecotypes cultivated at Pucallpa, Peru (Villachica *et al.*, 1996).

Cultivars

The EMBRAPA breeding work in Manaus focused on higher yields and resistance to diseases such as anthracnose and stress. Selections that produce 4–6 kg per plant per year versus 0.8–4.4 kg by traditional material, and that are normally propagated by seed, have been developed and released (Erickson, 2008). The selections do show significant genotype by environment interaction when grown in different parts of the Amazon (Nascimento Filho *et al.*, 2011). EMBRAPA-released clones in 2005 include 'BRS-Amazonas', 'BRS-CG648', 'BRS-CG612', 'BRS-CG882', 'BRS-CG611' and 'BRS-Maués' (Tavares *et al.*, 2005).

Cultural practices

Propagation

SEXUAL Seed propagation due to the outcrossing nature of the crop gives very heterogeneous plants. The seeds are normally taken from completely mature fruit and washed to eliminate the ariliod. Immediate sowing is necessary as the seeds are very recalcitrant, losing their germination capacity in 3 days (Villachica *et al.*, 1996). Seeds can be mixed with a moist substrate like sawdust, peat moss or clean sand, and held in a plastic bag to prevent germination loss for 30–45 days. Radicle emergence starts after 60–70 days and germination slows after 180 days when about 85% of the seeds have germinated. In the nursery, the seedlings are kept under semi-shade (60%) until they are taken to the field. For more efficient germination, seeds can be put in bags with moist substrate and kept there until radicles start to emerge; at this time they are sown in nursery bags with the radicle tip facing down.

ASEXUAL Clonal propagation is recommended to propagate superior plants. This propagation is achieved by cuttings or tissue culture and subsequently planted as for seedlings. Superior plants are selected as parents in traditional breeding programs and from these, cuttings can be obtained. In a trial done in Brazil herbaceous and hardwood cuttings with their bases etiolated with a 3 cm wide black plastic tape, 40 days before being detached from the mother plant, were put to root in sand under mist and 80% shade after being treated with 50 ppm indole butyric aid (IBA). The herbaceous cuttings rooted 91.7% and 29.2% without and with IBA, respectively, against 70.8% and 33.3% for non-etiolated cuttings. The best rooting of hardwood cuttings was 5%. The number of roots per cutting was higher with the IBA treatment (Rodrigues and Lucchesi, 1987). At present, it is recommended that leafy cuttings treated with 5,000–6,000 ppm IBA be used and put to root under an intermittent mist system. Depending on the genotype rooting will start after 45–100 days. Air layers can also be used but they are more cumbersome and do not lend themselves for mass production since there is limited material available on a mother plant. Grafting or budding is possible but the presence of the longitudinal furrows in the branches makes it difficult to obtain the buds and to match the tissues.

Orchard establishment

Typically, guarana is planted on newly cleared virgin forest. A common observation is that the plants are intolerant of soil compaction. Plantations prepared by bulldozing trees and brush fail to grow as well as those cleared by felling and burning the trees, avoiding the use of heavy soil compacting machinery (Erickson, 2008).

Transplanting is usually done in January and February at the height of the rainy season. Plants are then established before the dry season. Planting distances usually vary from 3 × 3 m, to 5 × 5 m and 6 × 3 m. Commercial production usually begins when the plants are 3–4 years old. Traditional cultivation of guarana was carried out with full sun exposure from transplanting; now shading is provided during at least the first 6 months then slowly removed. Shading can be provided by palm leaves, wood pieces, or shade plants such as castor bean, pigeon peas, and cassava.

Irrigation

Normally this crop is not irrigated, as cultivation takes place under high rainfall conditions. It could probably be cultivated with irrigation in places with less rainfall but few studies are available.

Training

Some attention has been paid to training plants on trellises, much as with grapes. Plant populations can be substantially increased and pruning done to maximize the amount of new, productive growth. Where this has been done on an experimental basis they have been interplanted with passionfruit (*Passiflora* sp.) to temporarily furnish revenue until guarana comes into heavy production. Both crops appear to thrive under similar conditions (Erickson, 2008).

Pruning

In the 1980s, crop management changed with the adoption of fertilization and pruning to direct the branches along supports. Since flowering occurs in the new shoot, production pruning is done after harvest to eliminate the branches that carried fruit, as well as to control plant size by shortening or eliminating branches that are too long or lying on the ground. At the same time, a sanitary pruning is carried out to remove old, broken and diseased branches.

Fertilization

It is recommended that at planting time 3 kg aged cow manure and 120 g triple superphosphate be applied to the planting holes. Three months later 20 g urea and 90 g potassium and magnesium sulfate should be applied in a circle 20 cm from the stem, after 6 months 40 g urea, 10 g borax and 10 g zinc sulfate should be applied and at 9 months after transplanting another 40 g urea. Normally fertilization is made during the rainy season between January and March in Brazil.

Research has led to the recommendation that each vine from the second to the fifth year after transplanting should receive 110, 150, 180 and 210 g urea; 120, 150, 180 and 200 g triple superphosphate; 200, 250, 300 and 350 g potassium chloride plus 30 g magnesium sulfate, 10 g zinc sulfate and 1 g boron. The fertilizer is applied within 1 m of the stem (EMBRAPA, 1998). Nitrogen should be split into three applications in January, February and March, phosphorus in one application in January and potassium in two applications in January and March. The micronutrients magnesium, zinc and boron are applied once in March, which coincides with the peak of the rainy season. When vines are 5 years old, yearly fertilizer applications are the same.

Pest management

DISEASES Several fungi and bacteria can cause severe damage. No virus has been reported. The most severe disease attacking flowers and foliage is anthracnose caused by *Colletotrichum graminicola*. This disease is present in all orchards with different degrees of severity according to genotype and year. *Septoria paullinae* causes a black crust. A disease that results in a proliferation of new shoots or in masses of tumor-like growths is caused by *Fusarium decemcellulare* and can be very severe in nurseries, resulting in seedling mortality; heavily infected mature plants need to be removed. *Ganoderma philippii* causes the so-called red root-rot resulting in yellowing of the foliage, followed by gradual decline and death; infected plants should be eliminated. Other root rots include those caused by *Phytophthora cactorum* and *Cylindrocladium clavatum*. Leaf spot, a bacterial disease caused by *Xanthomonas campestris*, can be controlled with copper fungicides.

INSECTS The thrips *Lotaris adisi* attacks young growth and leaves as well as floral parts. It is also believed that it transmits the shoot proliferation disease (Villachica *et al.*, 1996).

Weed control

The natural growth is not that of a typical liana, as the shoots are only a few meters in length. The result is a dense sprawling shrub. Ground covers are sometimes planted. These may include broad beans (*Vicia faba*) and kudzu (*Pueraria lobata*). The latter is so aggressive that great care must be taken to prevent it from overgrowing the orchard (Erickson, 2008). Otherwise weeding will be necessary, especially during the first years after planting in order to avoid competition. Mulching around the plant can help to reduce weed pressure.

Orchard protection

Normally the plant is cultivated in areas that are not prone to strong winds. If this is not the case the trellises have to be solid and probably oriented in the same direction as the main winds in order to present the least resistance.

Harvesting and postharvest handling

Harvesting

A raceme can have 50–115 fruit at different maturity stages due to its sequential flowering. Production from individual plants varies widely. Average production for non-improved types ranges from 0.1 to 4.0 kg per plant per year with new selections producing 4–6 kg per plant per year. Year to year yield variation is equally great with changes in rainfall pattern being mainly responsible for these yield fluctuations.

The fruit must be harvested before complete maturity. Harvesting is done by hand over a period of several weeks. Normally the whole raceme is cut using shears when about half of the fruit have started dehiscence.

Postharvest treatment

After the harvest, the fruit are put in heaps in the shade for 2 or 3 days to allow for ventilation and apparently some fermentation. Seeds are then extracted by removing the husk of the fruit capsule either by machine or manually. After that, the seeds are washed to remove the aril or ariloid tissue, and are left to dry in the open or oven dried. This drying is followed by roasting in a clay or metal oven in 10 cm deep layers for 4–5 h with periodic turning of the layer until they achieve 8–10% moisture content. Solar driers can be used with the seeds covered by transparent plastic film; drying times of 4–5 days are normal in full sun.

Utilization

Commercial guarana is produced only from the seeds; all other parts of the fruit are discarded. A common method of commercialization is to take the fermented, roasted and cleaned kernels and sell them (*em rama* in Brazil). Alternatively, the dried seeds without the integument are ground to a powder to which some water is added to form a dough or paste that is molded into a stick-like or loaf-like form (called *bastao* in Brazil). These 3–4 by 18–20 cm sticks or loaves are then dried and smoked for about 30–60 days in an oven on moderate fire until they become stone-hard and change color (Cavalcante, 1979). These sticks or loaves, weighing 120–150 g, are sent to local markets and consumers will shave off part of them to mix with water to prepare their drinks.

Another preparation is to use the pure powder for preparing guarana-flavored drinks from either previously fermented (10 days) (gauarana Luzela) or non-fermented seeds (guarana Marau). In both cases, sticks are made initially from the seeds but later ground to produce a powder from fermented seeds that is bitter and has a neutral flavor if non-fermented (Benlekehal *et al.*, 2001). These two forms are usually sold in stores in the cities; each is about 200 g and can be used to make 100 l of beverage.

Guarana is also consumed very much as a hot drink, like an instant coffee, prepared at home especially at breakfast where it is said to serve as a stimulant to start the day in proper shape and mood. Syrups are prepared for making beverages, mainly for industrial use to make carbonated drinks. For this product, the fruit is fermented for about 10 days, dried at 40°C to bring the moisture content down from 83% to 2% and then the fruit parts are removed to leave the clean kernels. The kernels are roasted at 200°C for 12 min. All of these preparation steps make it easy for the guarana powder to dissolve in water. An alcoholic extract is obtained by macerating the clean dry kernels for 24 h in a 70% alcohol solution and can be used directly to prepare beverages.

Guarana has become a very popular soft drink, especially in Brazil, where it is sold as a carbonated drink and in a number of natural health products. The dried product has a caffeine content of 4.3–4.7%, sometimes even up to 10.0%, which is much higher than coffee (0.8–1.3%), cocoa (0.4%) and kola nut (2.8%). The seeds, in addition to their high caffeine content, contain guaranin, tannin, theobromine (1.2%) and other components in small amounts. They also have a high protein and starch content (Table 6.2). The appearance of the fruit, which resembles the human eye, and its stimulatory properties have given it a special place among plants of the region.

In popular medicine, it is used as an elixir. It is also used for treating headaches, menstrual problems, fever, heat stress and intestinal disorders. Some people claim it has aphrodisiac properties. It is said it helps in weight loss and fatigue reduction. Nowadays guarana is readily found in health food stores as

Table 6.2. Composition of guarana seeds per 100g edible portion (from Villachica *et al.*, 1996, after Vasconcelos *et al.*, 1976).

Proximate	Guarana (g)
Moisture	7.65
Fiber	49.12
Red resin	8.8
Starch	8.35
Guarana tannic-acid	5.90
Caffeine	5.39
Yellow oil	2.95
Piro-guarana acid	2.75
Red dye	1.52
Saponin	0.06
Yellow dye	0.60
Pectin, malic acid, mucilage, dextrose	7.47

a component of dietary medications, "rejuvenation" agents and other products. Recent studies suggest potential for the treatment of oily skin and cellulitis. Guarana products are distributed internationally, most commonly as a carbonated soft drink in the USA and other industrialized countries although the restrictions of the US government to products with high caffeine content have limited somewhat the consumption of these beverages.

References and Further Reading

Acevedo-Rodríguez, P., Welzen, P.C., Adema, F. and Ham, R.W.J.M. (2011) Sapindaceae. In: Kubitzki, K. (ed.) *Flowering Plants. Eudicots, vol 10. The Families and Genera of Vascular Plants.* Springer Berlin, Heidelberg, Germany, pp 357–407.

Spanish lime

Alix, C. and Duarte, O. (1999) *Propagación de Especies Frutales Tropicales*. CURLA; PDBL; AFE/COHDEFOR; DICTA; SETCO/PROFORFITH. La Ceiba, Honduras, p. 80.

Cabrera, I. and Brunner, B.R. (2007) *Variedades de Quenepa: Evaluación del germoplasma de Melicoccus bijugatus* Jacq. *en la Estación Experimental Agrícola de Juana Díaz*. Universidad de Puerto Rico, Recinto Universitario de Mayagüez, Colegio de Ciencias Agrícolas, Estación Experimental Agrícola, San Juan, Puerto Rico, Boletín No. 30.

Campbell, C.W. (1976) Selection and propagation of the Spanish lime in Florida. *Proceedings of the Florida State Horticultural Society* 89, 227–229.

Coppens d'Eeckenbrugge, G. and Paull, R.E. (2008) *Melicoccus bijugatus*, Mamoncillo. In: Janick, J. and Paull, R.E. (eds) *Encyclopedia of Fruit and Nuts*. CAB International, Wallingford, UK, pp. 808–809.

Crane, J., Balerdi, C., Campbell, R., Campbell, C. and Goldweber, S. (1994) Managing fruit orchards to minimize hurricane damage. *HortTechnology* 4, 21–27.

Fouque, A. (1973) Espèces fruitières d'Amerique tropicale. *Fruits* 28, 290–299.

Francis, J.K. (1992) *Melicoccus bijugatus* Jacq. *Quenepa*. US Department of Agriculture (USDA) Forest Service, Southern Forest Experiment Station SO-ITF-SM-48, New Orleans, Louisiana, 4 pp.

Knudsen, H. (2000) *Directorio de Colecciones de Germoplasma en América Latina y el Caribe*. Primera edición. International Plant Genetic Resources Institute (IPGRI), Rome.

Larson, K.D., Schaffer, B. and Lera, S.R. (1991) Vegetative propagation of Spanish lime and jaboticaba. *Proceedings of the Florida State Horticultural Society* 104, 54–56.

León, J. (1987) *Botánica de los Cultivos Tropicales*. Instituto Interamericano de Cooperación para la Agricultura, Servicio Editorial, San José, Costa Rica, 445 pp.

Little, E.L. and Wadsworth, F.H. (1964) *Common Trees of Puerto Rico and the Virgin Islands*. Agriculture Handbook No. 249. US Department of Agriculture (USDA), Agriculture Marketing Service (AMS), Washington, DC, 548 pp.

Martin, F.W., Campbell, C.W. and Ruberté, R.M. (1987) *Perennial Edible Fruits of the Tropics: An Inventory.* Agriculture Handbook No. 642, US Department of Agriculture, p. 57.

Morton, J.F. (1987) *Fruits of Warm Climates*. Creative Resource Systems Inc., Winterville, North Carolina, pp. 267–269.

Pérez, C.H.C., Gómez, P.M. and Vila, J. (2008) Evaluación de los parámetros de calidad en frutos de mamoncillo (*Melicoccus bijuga* L.). Características químicas. *Revista Iberoamericana de Tecnología Postcosecha* 9, 7–15.

Pérez, C.H.C., Gómez, P.M. and Vila, J. (2009) Physical characteristics of mammon fruit (*Melicoccus bijugatus* Jacq.) according to its location in the tree and the storage temperature. *Bioagro* 21, 189–194.
Popenoe, W. (1974) *Manual of Tropical and Subtropical Fruits*. Hafner Press, New York, 474 pp.
Vargas, O., Alix, C., Lobo, A.D. (authors), Duarte, O. and Sanchez, J. (technical reviewers) (1999) *Frutales y Condimentarias del Trópico Húmedo*. CURLA; PDBL; AFE/COHDEFOR; DICTA; SETCO; PROFORFITH, La Ceiba, Honduras, pp. 290–291.

Guarana

Bauman, T.W., Schulthess, B.H. and Hanni, K. (1995) Guarana (*Paullinia cupana*) rewards seed dispersers without intoxicating them by caffeine. *Phytochemistry* 39, 1063–1070.
Benlekehal, H., Clotteau, M., Dornier, M. and Reynes, M. (2001) Un produit amazonien particulierement riche en caféine: la graine de guaraná (*Paullinia cupana* H.B.K. var. *sorbilis* (Mart.) Ducke). *Fruits* 56(6), 423–435.
Cavalcante, P.B. (1979) *Frutas Comestiveis da Amazonia III.* Publicaciones Avulsas No. 33. Instituto Nacional de Pesquisas da Amazonia, Museo Paraense Emilio Goeldi, Belém, Pará, Brazil, pp. 43–54.
Donadio, L.C., Moro, F.V. and Servidone, A.A. (2002) *Frutas Brasileiras*. Editora Novos Talentos, Jaboticabal, São Paulo, Brazil, pp. 174–177.
EMBRAPA (1998) *Sistema de Produçao para Guarana*. EMBRAPA, Documentos #13. Centro de Pesquisa Agroforestal da Amazonia, Manaus, Brazil.
Erickson, H.T. (2008) *Paullinia cupana* Guarana. In: Janick, J. and Paull, R.E. (eds) *Encyclopedia of Fruit and Nuts*. CAB International, Wallingford, UK, pp. 817–820.
Erickson, H.T., Correa, M. and Escobar, J. (1984) Guarana (*Paullinia cupana*) as a commercial crop in Brazilian Amazonia. *Economic Botany* 38, 273–286.
Escobar, J.R., Corrêa, M.P.F. and Motta, A.S. (1984) *Seleção de clones de guaraná (*Paullinia cupana *var.* sorbilis *(Mart.) Ducke) baseada em vigor e adaptação ao campo*. In: Anales I Simpósio del Trópico Húmido (Cultivos perennes), Belém, Brazil, EMBRAPA/CPATU 4, 295–304.
Lleras, E. (1994) Species of *Paullinia* with economic potential. In: Hernando Bermejo, J.E. and León, J. (eds) *Neglected Crops: 1492 from a Different Perspective*. Plant Production and Protection Series No. 26. Food and Agriculture Organization (FAO), Rome, pp. 223–228.
Nascimento Filho, F.J., Atroch, A.L., Cruz, C.D. and Carneiro, P.C.S. (2011) Comportamento de clones de guaraná em três localidades. *Revista de Ciências Agrárias (Journal of Agricultural Sciences)* 53, 38–45.
Poltronieri, M.C., Durante, M. de L.R., Rodrigues, J.E.L., Nazare, R. de F.R., Kato, A.K. and de Oliveira, A.F.F. (1995) *A Cultura do Guarana*. EMBRAPA - SPI - Colecao Plantar #29. Centro de Pesquisa Agroforestal da Amazonia, Manaus, Brazil.
Rodrigues, J.E.L.F. and Lucchesi, A.A. (1987) Propagaçao vegetativa do guaranazoiro (*Paullinia cupana* (Mart.) Ducke) a través de estacas induzidas (capeadas) e com ácido indolbutírico. *Anais da Escuela Superior de Agronomía Luiz de Queiroz* XLIV, 1–17.
Tavares, A.M., Atroch, A.L., do Nascimento Filho, F.J., Pereira, J.C.R., Abreu de Araújo, J.C., Cardoso Moraes, L.A., Pereira Santos, L., Rodrigues de Arruda, M., Sousa, N.R. and Angelo, P.C. da S. (2005) *Cultura do Guaranazeiro no Amazonas* (4th Edition). Pereira, J.C.R. (Technical ed.), Manaus, Embrapa Amazônia Ocidental, 40 pp. Available at: http://www.infoteca.cnptia.embrapa.br/bitstream/doc/678408/1/SistemaProdGuarana.pdf (accessed 16 July 2013).
Villachica, H., Carvalho, J.E.U., Muller, C.H., Diaz, S.C. and Almanza, M. (1996) Guaraná, In: *Frutales y Hortalizas Promisorios de la Amazonía*. Tratado de Cooperación Amazónica, Lima, Perú, pp. 147–150.

7 Passifloraceae and Caricaceae

PASSIFLORACEAE

The Passifloraceae include about 530 species divided into 27 genera. The family Passifloraceae has now been absorbed into the large order Malpighiales, which is very diverse except at the molecular level. The family includes mostly tropical trees, shrubs, lianas and climbing vines. The genus of most interest in the Passifloraceae is *Passiflora*, which includes the passionfruit amongst its 500 or so species that are mostly vines with a few shrubs and herbs. The genus is absent from Africa, though found in the Americas, Asia and Australasia. Most species have edible fruit with the most common commercial species being the purple passionfruit or maracujá (*Passiflora edulis*) for the fresh market and the more acid yellow passionfruit (*P. edulis* f. *flavicarpa*) for the juice industry.

The sweet granadilla (*Passiflora ligularis* A. Juss) is a lesser-known member of this genus also grown for its fruit. Other members of this genus grown for their edible fruit include the giant passionfruit (*Passiflora quadrangularis*), the banana passionfruit (*Passiflora mollisima*) and several Andean lesser-known species. The banana passionfruit (*P. mollisima*) belongs to a different subgenus called *Tacsonia* that includes the so-called *curuba india* from Colombia (*Passiflora mixta*), sometimes classified as a different form of the same species. *Passiflora pinnatistipula* from the same area of the South American Andes has an almost spherical fruit. In total, there are more than 60 *Passiflora* species with edible fruit worldwide with the ones from the Andes including *P. antioquiensis*, curubejo *P. popenovii*, galupa *P. pinnatistipula*, chulupa *P. maliformis*, rosy passionfruit *P. cumbalensis*, *P. schlimiana*, *P. ampullacea*, *P. tripartite*, *P. ambigua* and *P. mandonii*. A number of species are grown for their ornamental value, such as *P. coccinea* and *P. caerulea*, and others are grown for their fruit and as ornamentals.

Sweet Granadilla

This species is native to middle to higher elevations from Mexico to Central America down to the Andes of Peru and western Bolivia (Romero Castañeda, 1961). It is known as *granadilla* in Spanish probably because of its inside resemblance to the pomegranate (*granada* in Spanish). It is also known as *fruta moco* (mucus fruit) for its pulp, and in English it is the sweet granadilla. Colombia is the main producer with more than 3,000 ha, both for local demand and a good portion that is exported to the EU and the USA. Peru has about 800 ha that satisfies a large local market and limited exports. These estimates do not include backyard plants that are often seen growing climbing on trees and fences (Mamani-Quispe, 2000). Ecuador also exports part of its production while Bolivia and Venezuela produce mainly for the local market. Other Latin American countries also produce this fruit for the local market where growing conditions exist. Outside the Americas, producing countries include South Africa, Kenya, Rwanda and Australia.

Ecology

Soil

Sweet granadilla prefers deep light soils from sandy loams to loamy clays with a medium to high content of organic matter, good drainage, a depth of 30–40 cm, and a pH around 6–6.5 although it grows well with pH above 7 in the Peruvian coast. Locations with risk of flooding are avoided. In many places, they are grown in hilly or mountain areas with 25% up to 75% slope.

Rainfall

The ideal range is between 1,500 and 2,500 mm uniformly distributed throughout the year, otherwise irrigation is required during the dry months to avoid reduction in yields. Irrigation is a must in the dry Peruvian coast and in the valleys running down from the western Andes slopes where a good part of the Peruvian production occurs and rains are absent. In the Cusco area, with rainfall less than 500 mm year^{-1} that falls between October and April, irrigation is essential.

Temperature

The ideal range is between 16 and 24°C, with an optimum of 16–18°C or even 12–15°C in the Cusco area of Peru (Mamani-Quispe, 2000). It produces at between 1,500 and 2,500 m in the Andes; above this elevation it flowers but sets fruit poorly, in part because of the lower insect populations. It can also produce at sea level in the subtropical Peruvian central and southern coast where winters never reach freezing temperatures and summers rarely exceed 30°C. The vine does poorly in the subtropics where the summers are too hot or winters too cold. It will not withstand frost. In Central America, it produces above 1,000 m (Barbeau, 1990). In southern Peru when planted above 2,200 m, the rainfall is less than further to the north and temperatures cooler, growth is slow and production low. At higher temperatures, the plant requires more water and fertilizer; yields improve but the chances of *Nectria* infection increases, especially if above 20°C, while below 10°C lower flowering and higher fruit abscision rates occur. Excessive daily temperature fluctuations result in peel cracking (Fischer *et al.*, 2009).

Light and photoperiod

The plant does not seem to be photoperiodic, since it flowers almost year round at different latitudes. The longer the day the more the plants will grow and photosynthesize. Ideally, the plant needs a daily average of 8 h of sunshine or the fruit will become brownish in color.

Wind

Its large soft leaves and delicate flowers make it prone to damage by strong winds, especially when planted vertically in the fence type of trellis. Strong winds can interfere with insect pollination, branch breakage occurs, stigma and pollen can dehydrate and trellises, especially those of the fence type, can be damaged. Scarring of fruit occurs due to rubbing against other fruit and stems.

General characteristics

Plant

It is a vine with simple axillary tendrils and a cylindrical softwood glabrous stem. The 8–17 by 6–14 cm entire simple alternate leaves (Fig. 7.1) are ovate with

Fig. 7.1. Sweet grandilla (*Passiflora ligularis*) showing leaves, tendrils and flowers (used with kind permission from Isa Degener, appeared in Otto Degener's book *Flora Hawaiiensis* in 1938). Fruit by Odilo Duarte.

a cordiform base, and glabrous with a dark green almost blue upper side and a green-grayish underside that can have a slight or strong violet tone and shows the prominent leaf veins. The petiole has three to five pairs of green-colored filamentous glands (León, 1964). The roots are fasciculate and shallow.

Flowers, pollination and fruit set

The flowers have a peduncle that can be up to 5 cm long. The involucre has three green ovate bracts with an acute apex (Fig. 7.1). There are five triangular sepals and petals measuring 3–5 cm with a central protruding line that ends in the acumen; they are pale green on the outer side and white in the inner side. The crown is formed by filaments with transversal alternating bands colored purple, white and deep purple, the external filaments being longer than the petals. Anthers show the yellow-colored pollen and the stigma is trifid. The plant seems to be allogamous and will start flowering about 9–10 months after planting and 75–85 days later fruit are ready for harvest (Bernal, 1988). Pollination is performed by bumble bees (*Epicharis*), honey bees (*Apis mellifera*) and a large wasp. *Trigona* bees are sporadically found. The flower opens for only 1 day and the pollen is not viable early in the morning or late in the afternoon. Manual pollination may be required when there is poor insect activity.

Fruit

The indehiscent capsular fruit is 6.5–8 cm long and 5–7 cm wide with a 6–12 cm peduncle and can be ovoid to spherical to slightly flat at the poles with the tip pointing towards the stem (Fig. 7.1). It has a 1 mm epicarp, which is yellow to orange, sometimes purple, green or a combination, with white freckles is most varieties. This smooth epicarp shows six faint longitudinal lines. The epicarp is a hard sclerophyllous tissue that gives the fruit a firm texture and brittle nature and will crack rather than wrinkle as yellow passionfruit does when it ripens or under pressure. The 4–5 mm thick underlying white mesocarp is soft and spongy and the endocarp is a white film that separates from the mesocarp at maturity. The black elliptic and flat seeds are arranged on three longitudinal placentae and each seed is surrounded by a transparent jelly-like pulp that is very sweet and aromatic. There can be 250–350 seeds per fruit.

Cultivar development

Cytogenetics, genetics and breeding

The chromosome number is $2n = 18$, as for most horticulturally important Passiflora. Characterization using RAPD markers indicates that there is a high degree of genetic diversity within *P. ligularis* with high intra-specific polymorphism (Sánchez *et al.*, 1998).

Selection and evaluation

Selections have been made mainly in Colombia and to a lesser extent in Peru and Ecuador. One of the main problems is that cross-pollination results in fairly large plant to plant variability so that stable varieties are difficult to maintain.

Cultivars

Selections are named by the location where they are cultivated or by fruit characteristics. Fruit are most often characterized by: (i) fruit size: large (> 100 g), medium (70–100 g) and small (< 100 g); (ii) fruit shape: round, round-flattened at poles, ovate and pear-shaped; and (iii) peel thickness: thick, medium and thin. In Colombia, some selections exist with superior characteristics like: 'Criolla', a large round 124 g fruit, with thick exo- and mesocarp, but a low pulp content; 'Pecosa', a medium-sized round to flattened fruit, with large white freckles and good pulp content; 'Valluna', a medium oval to elongated fruit, weighing 120 g with thin exo- and mesocarp and a high pulp content; 'Urrao', a large round to flattened fruit, thick peel and less pulp than 'Valluna'. Two recent selections are: 'Cundinamarca' with a large round-elongated fruit, with reddish thin peel with very pronounced white freckles but low pulp content; the other is 'Huila' with a large oval fruit, intense yellow peel, 130 g and high pulp content (Zuleta-Ospina *et al.*, 2011).

Peru has made some selections and finally introduced some of the better Colombian materials that are more attractive, productive and have a higher fruit weight that the traditional local types. Ecuador has also introduced some of these selections.

Cultural practices

Propagation

SEXUAL Seed propagation is the most used method for this species. Seeds should be taken from good quality completely ripe fruit of superior plants. The pulp should be fermented for 2–3 days and then rubbed to remove the arils and then put in a container that is filled with water so the seeds sink and the arils float and are discarded by a repetitive decanting. All floating seeds should be discarded. Clean seeds are put on paper towels or newspaper and dried in the shade for 2–3 days. After that, they are ready to be sown either in seedbeds, germination boxes, nursery bags or better in cell trays. The advantage of cell trays is that seedlings can be extracted with a root ball and transplanted directly, bypassing the nursery bag stage. Germination should start within 15–20 days; if sown in seedbeds transplanting to bags should be done when seedlings are 5–8 cm tall and after a further 30–45 days plants will be ready to be taken to the field. The total time from seed sowing to planting in the field is 65–80 days (Bernal, 1988).

ASEXUAL The vine can be propagated by cuttings using mature semi-lignified branches cut into 30–40 cm pieces containing three to four buds; the upper cut should be slanting and made 3–4 cm above a bud and the lower transversal cut 3–4 cm below a bud. After this the cuttings should be treated with 1,000–2,000 ppm indole butyric acid and their 15–20 cm basal part buried in the substrate, normally in a nursery bag. This propagation is not a very popular method and the plants have a shorter life.

Grafting can also be used. In Peru, sweet granadillas are grafted onto yellow passionfruit for better adaptation to slightly alkaline or saline soils of the coast. In Colombia, they tried grafting on *P. maliformis* var. *pubescens*, which is very resistant to *Nectria* but very susceptible to *Meloidogyne* nematode. *P. ambigua* has also been tried. This will become more important as soil-borne diseases increase in importance and several wild *Passifloras* will have to be tested for resistance. Cleft graft is the most used type. Tissue culture is also successful (Miranda, 2009).

Field preparation, transplanting and plant spacing

Orchards can be installed on flat lands that are more prone to frost; however, the majority of orchards occur on slopes. If flat land is used the usual preparation includes plowing and harrowing. In other cases, the land in cleared of bushes and tall weeds and only the zone where the plants are to be transplanted will be prepared; the same occurs on hilly land. Normally 40 × 40 × 40 cm holes are dug and amendments added according to soil analysis. Recommendations include the addition of 4–5 kg decomposed manure, 100–150 g complete formula fertilizer, 60 g nematicide and 0.5 g lime if pH is low in the planting hole (Bernal, 1998). This addition can include fungal inoculum, such as *Metarhizium* and *Paecilomyces* sp., to prevent possible attacks by soil-borne pathogens. Planting distances vary from 3 × 4 m, 3 × 6 m, 4 × 4 m, 6 × 6 m to 8 × 8 m with plants in the middle of the squares or rectangles formed by the trellis posts or along the post row in the case of fence-type trellis. In some instances, a mound 20–30 cm high and 80–100 cm in diameter is prepared so the plants are in the middle of it to insure water will not stand at their base. After planting, the plants have to be directed towards the trellis using tutors or strings.

Trellises

Since the plant is a vine it is grown on trellises to obtain good yields and quality. Several types of trellises can be used. One type is the "pergola" called *emparrado* where the canopy grows horizontally on top of an intercrossed wire system forming a roof. Posts with a diameter of 12–18 cm and a length of 2.60 m are used. They are buried 50–60 cm every 4 m forming squares. To economize sometimes a sturdy post is installed every 8 or 12 m and thinner wood or bamboo posts at 4 and 8 m form the trellis. The peripheral posts have to be very sturdy, and to prevent them from inclining towards the center of the field a cable or a No. 10 guy wire is attached to the post top that goes to an anchor about 2.0 m from the post base. The anchor can be a concrete plate buried 80–100 cm or a well–buried short post. In other cases, a second post is planted in a slanting position in the inside of the post with its top making close contact at about 1.8 m on the standing post so that it prevents it from inclining. A third method is to install the peripheral posts in an inclined position with a 60–65° angle to the external side of the plot. On top of the posts following their lines either barbed or No. 10 or 12 wires are fixed to make large squares. On these thick wires, thinner No. 16 or 18 wires are installed every 50–100 cm in both directions forming a mesh.

The other trellis type is the "fence" trellis where 2.6–3.0 m by 15–20 cm posts are buried 50–60 cm along the established line that can follow the contour, the main wind direction or run down slopes with up to 70% steepness. Posts are placed every 6–8 m with thinner wood or bamboo posts between (3–4 m from the main posts); after this four horizontal rows of wire are installed: the first at 0.80 m from the ground, the others every 40 cm. The rows are ~3.0 m apart.

This "fence" system is cheaper to install and allows for more uniform distribution of the plants along the lines, but implies adequate orientation to obtain the best radiation and requires more work in arranging new growth along the fence. However, less productive branch development occurs, it costs more to hang the branches and it is more difficult to prune and overall is less productive. It can also result in higher incidence of burned fruit because of higher exposure to the sun (Miranda, 2009).

Irrigation

Irrigation will be necessary if rainfall is too little or the dry periods are too long. The ideal system is drip irrigation, since it saves water, does not wet the foliage, does not create high humidity and reduces weed proliferation because it does not wet the whole field.

Pruning

FORMATION PRUNING The purpose of formation pruning is to induce the plant to produce a functional canopy. In the pergola trellis, this consists of leaving a single main stem and once it has reached the wires remove the top by cutting at 1.0–1.5 m to induce laterals and if necessary pinch the new shoots to get four to eight branches; the number will depend on plant density and relative humidity, and if these are high fewer branches will be left. These branches will have to be directed in the proper direction. After the plant reaches the top wires and is attached to them, the guiding strings have to be removed to avoid strangling of the branches. The older leaves will also have to be removed from the base when the canopy reaches the pergola top. In the fence-type trellis, the stem is tipped to make sure two branches go along each of the four lines of wire, one in each direction.

MAINTENANCE PRUNING After the first harvest all unproductive, thin, sick or damaged branches are removed redirecting and/or topping some of the tertiary and quaternary remaining branches to stimulate reproductive structure. Branchlets crossing with those of neighbor plants should be cut before they get entangled.

RENOVATION PRUNING Renovation pruning consists in cutting off all main branches at 1.0 m from the stem leaving two to three per plant to start a new cycle. This can be done after two or three seasons or when the canopy is overcrowded assuming the plants are still healthy and vigorous.

Fertilization

Fertilization is based upon soil analysis. This species has low requirements of P and medium to high of N and K. Some recommendations are to apply four times per year 300 g of 17-6-18-2 the first year and 450 g the second year, plus 1 kg dolomite, 5 kg decomposed chicken manure and 50 g minor elements per plant twice a year (Bernal, 1998). Other authors recommend applying a 10-30-10 or 13-26-6 monthly according to soil analysis results, adding microelements and manure.

Pest management

DISEASES Probably the most important disease in young and adult plants is called *secadera* derived from the Spanish word "seco" for dry, because there is a dry rotting of the plant base caused by the fungus *Nectria haematococca* and its anamorphic state *Fusarium*, which penetrates through the roots and clogs the vascular bundles. It is favored by excess moisture and wounds and it can be transmitted by infected plant material and soil, even shoes (Zuleta-Ospina *et al.*, 2011). It starts with yellowing and dropping of leaves followed by necrosis that ascends the stem and the whole plant wilts, the fruits shrivel, and finally the plant dies. The use of clean seed and clean or disinfected seedbed is important and incorporating *Trichoderma harzianum* into the planting hole is also useful to protect the plant.

Other diseases of the seedbeds include "damping off" caused by *Pythium*, *Rhizoctonia solani* and *Phytophthora*. Scab caused by *Cladosporium herbarum* appears as necrotic circular spots and fruit deformation. Anthracnose caused by *Colletotrichum gloeosporioides* affects leaves, buds and small fruit producing dry, sunken, brown-colored lesions. The *Phomopsis* fungus produces "chicken eye" spots. Powdery mildew can be a problem under high humidity and *Botrytis* can also attack the flower buttons, young peduncles and fruit.

INSECTS Seedlings can be attacked by the larvae of the moth *Agrotis ipsilon* that feed on the roots and later cut the hypocotyl. Insects attacking the foliage include: aphids (*Aphis gossypii* and *Myzus persicae*), *Empoasca*, *Agrotis*, *Thrips*, *Agraulis juno* and *A. vanillae*, and leaf cutting ants (*Atta cephalotes*). Stem attacking insects include the scales *Ceroplastes* and *Chrysomphalus*. Other insects like the stingless bee *Trigona testacea* and the moth *Scybalista bifascialis* attack flower buttons and branch terminals. The larvae of the flower button flies *Dasiops inedulis* and *Lonchaea* sp. feed on the anthers and the ovary. The fruit fly (*Anastrepha* spp.) lays its eggs in immature fruit causing its shriveling and drop. Other problems include mites (*Tetranychus mexicanus* and *Brevipalpus* sp.), snails and slugs. Root knot nematodes (*Meloidogyne* spp.) can also be serious pests.

Weed control

Herbicides are not recommended or should be used very carefully because the plant is very susceptible to these chemicals. Normally four weedings are done using machete or hoe very carefully to avoid wounding the roots or stem. In many cases, hoes are used to mound some additional soil around the stem to improve its anchorage. After the canopy develops at the top of the pergola, weeds tend to diminish significantly due to the shading. Associated crops like beans or vegetable crops are also used successfully and help keep the soil weed free.

Orchard protection

Wind damage can be serious so that fence-type trellis lines should be installed in the direction of main winds or wind barriers have to be installed, ideally before planting the crop.

Harvesting and postharvest handling

Yields

An adult plant can produce 500–600 fruit year^{-1}, equivalent to 50–60 kg year^{-1} (Bernal, 1998). Other Colombian authors indicate that yields can go from 2 t ha^{-1} the first year, to 16 and 15 t the second and third years, 12 and 10 t the fourth and fifth years, and 6 t the sixth year (Miranda, 2009). Estimated yields in Peru are 7.7 t ha^{-1} in the second year and 14 t ha^{-1} in the third year (Mamani-Quispe, 2000).

Harvesting

Flowering starts about 9–10 months after planting and harvesting 75–85 days later. Fruit should be harvested when 75% of the skin is yellow and dry. Pruning shears should be used to pick the fruit or pressure is applied at the third node above the fruit. Fruit should not be touched with the hands to avoid wiping the natural waxy cover. Harvesters should have their nails cut and are sometimes required to wear wool gloves (Miranda, 2009). Fruit is put in cardboard boxes lined with paper and all peduncles are directed in one direction; no more than three layers of fruit separated by paper should be packed. Some authors recommend spraying with fruit protecting fungicides and sodium hypochlorite the day before harvest to reduce fungal attacks. Harvest is almost continuous with peaks during the year following temperature and rainfall patterns; normally harvest will last 8–9 months. The orchard can produce acceptable yields for 6–8 years.

Postharvest treatment

The fruit can be kept on the plant up to 45 days. It generates ethylene and can be kept for 3–4 days in a closed space to allow complete skin yellowing. The fruit can be stored for up to 30 days at 6–7°C and 85–90% relative humidity.

Utilization

The fruit is normally eaten out of hand by taking away a piece of its brittle peel or splitting it in two and scooping the sweet and aromatic pulp with a spoon; the seeds are swallowed. It can also be mixed with water or milk in a blender to make a very good tasting juice after separating the broken seeds. Marmalade and jelly can also be made. It contains vitamins A, C, and K, phosphorus, iron, and calcium (Table 7.1) and it is one of the first juices pediatricians recommend to feed infants, separating it from the seeds with a colander. There is little industrial use of this very good tasting fruit. The juice has digestive and diuretic properties and is recommended for patients with ulcers or hiatal hernia since it contains wound-healing compounds and helps counter reflux in adults and babies. It has an antispasmodic effect and induces sleepiness. The beautiful flowers are used in the perfume industry and the peel can be used as cattle feed.

Table 7.1. Proximate fruit composition for 100 g of flesh of sweet granadilla, banana passionfruit, mountain papaya and babaco.

Proximate	Sweet granadilla[a]	Banana passionfruit[a]	Mountain papaya[b]	Babaco[c]
Edible pulp (%)	73	50		81–87
Water (g)	76.3	92.6	93.2	93–94
Energy (kcal)	94	25		20–22
Energy (kJ)				82–93
Protein (g)	2.4	0.5	1.0	0.7–1.3
Lipid (g)	2.8	0.1	0.3	0.02–0.3
Carbohydrate (g)	17.3	6.2	3.3	4.6–6.0
Fiber (g)	4.2	0.6	1.4	0.5–1.0
Ash (g)	1.2	0.6	0.8	0.2–0.3
Minerals				
Calcium (mg)	10.0	8	36.0	8.0–12.0
Iron (mg)	0.9	0.4	1.3	0.3–4.0
Magnesium (mg)				6.0–15.3
Phosphorus (mg)	64.0	18	28.0	7.7–17.0
Potassium (mg)				132–220
Sodium (mg)				1.3–3.0
Vitamins				
Ascorbic acid (mg)	20	52	26.0	23–31
Thiamin (mg)	0.0	0.0	0.06	0.02–0.03
Riboflavin (mg)	0.11	0.04	0.05	0.02–0.06
Niacin (mg)	1.6	1.5	0.6	0.5–1.0
Vitamin A	5 mcg Activity	20 mcg Activity		0.16–2.0 (mg)
Juice				
pH				4.0–4.3
Titratable acidity				7.8–10.4

[a]Leung and Flores, 1966.
[b]Undurraga, 1989.
[c]Romero-Rodríguez *et al.*, 1994.

Banana Passionfruit

Some authors like Quintero (2009) use the name *Passiflora tripartita* var. *mollisima* as a synonym for *Passiflora mollisima* L.H. Bailey. Other names given to this species are: *P. tomentosa* Lam., *Tacsonia mollisima* HBK, *Murucuja mollisima* Spreng., *Tacsonia mixta* subsp. *tomentosa* Mast., *Tacsonia mollisima* var. *glabrescens* Mast, *P. tomentosa* var. *mollisima*. There is some confusion in the nomenclature of the species because Coppens d'Eckenbrugge *et al.* (2001) consider *P. tarminiana*, the so-called *curuba india*, as a different species, though Lobo and Medina (2009) and Fischer *et al.* (2009) indicate that *P. mollisima* and *P tarminiana* produce similar fruit; given that the two can be crossed without difficulty it suggests that they might rather be botanical varieties and not different species. The name in English is banana passionfruit because of the fruit shape. In Spanish, it is called *tacso* or *tacso de Castilla* in Ecuador, *curuba* or *curuba de Castilla* in Colombia, *tumbo* or *tumbo serrano* in Peru and Bolivia, and *parcha* in Venezuela. There is also a species called *curuba roja* (red banana passionfruit) that is *P. cumbalensis.*

Origin and distribution

The subgenus *Tacsonia* is restricted to the Andes of South America with about 50 species that grow above 2,000 m altitude (Escobar, 1992). The banana passionfruit comes from higher elevations (2,000–3,000 m) mainly in the Andes of Colombia but can also be found in Ecuador, Peru, Venezuela and Bolivia. It is little known outside these countries. It is cultivated commercially mainly in Colombia with Boyaca being the principal producing area with a harvested area of 663 ha and a production of 6,868 t, contributing 48% of the national production of the banana passionfruit (Lizarazo *et al.*, 2013). In Ecuador and the other Andean countries, no large orchards exist and fruit sold in the markets comes from small areas or backyard plants. Occasionally, it can be found in the markets of San Salvador or Mexico City. Limited volumes are exported to the USA or the EU mainly from Colombia and Ecuador. There are some plantings in southern Australia, Madras in India and New Guinea. New Zealand seems to have a climate that is suitable for it and it is grown in small amounts. In Hawaii, it was introduced as an ornamental and for the fruit, but it has become a very aggressive forest-destroying weed on the Big Island and Kaua'i.

Ecology

Soil

The plant needs a soil profile that is at least 50–60 cm deep, medium textured (loam to sandy loam) and rich in organic matter (MAG-INCCA, 1991). The soil should have a good water retention capacity since the vine does not stand long dry periods. The area should not be subjected to flooding and requires good drainage. The ideal pH should be 5.5–6.5 (Campos, 1992).

Rainfall

The optimal rainfall is between 1,500 and 2,000 mm (Quintero, 2009) uniformly distributed, otherwise irrigation is necessary for continuous fruit production. Dry periods occur in most of the Peruvian and Bolivian Andes where rain seldom reaches these amounts and normally occurs during only 6 months of the year.

Temperature

This is a cool climate crop with ideal average temperatures of 14–16°C, conditions found between 1,800 and 3,200 m in the Andes (Fischer *et al.*, 2009), with an optimum of 2,200–2,400 m altitude (Bonnet, 1988; Schoeniger, n.d.). It grows at 3,400 m in Cusco, Peru and has adapted well to altitudes of 1,200–1,800 m in Hawaii and New Zealand. It can stand light frosts (Munier, 1961) and temperatures of −5°C for a short time. At higher altitudes, fewer anthracnose problems are experienced (Campos, 1992).

Light and photoperiod

Cloudy areas should be avoided; the ideal is to have 1,200–1,500 h of sunshine per year (Campos, 1992) Photoperiod does not seem to have an effect on flowering since it does flower for long periods of the year at different latitudes.

Wind

Wind is very damaging to this crop, breaking young shoots and causing flower drop (MAG-INCCA, 1991). Fruit can also be damaged by rubbing with other plant parts. Strong winds can also destroy a "fence type" trellis if weakly built and not properly anchored. Pollinating insects are also affected by wind. Areas with strong winds should be avoided.

General characteristics

Plant

This is a vigorous vine with cylindrical stems coated with yellow hairs that has tendrils like most of its relatives and can grow up to 8–10 m. It has three-lobed shiny green leaves 7–10 cm long that are pubescent, especially grayish or yellowish-velvety on the underside and downy above, with clearly defined veins and finely toothed margins (Fig. 7.2). The root system is shallow with 70% of the roots in the first 30 cm of soil (Bonnet, 1988). The stipules are short, slender and curved.

Flowers, pollination and fruit set

The length of the floral cup is the most conspicuous morphological difference of the *Tacsonia* group where the cylindrical 7–10 cm hypanthium is longer than the sepals (Fig. 7.2), gray-green, frequently blushed with red, and rarely downy. The flowers are very showy and the corolla opens with 2–3 cm pink to reddish oblong petals. The calyx or corona has very short and reduced sepals (León, 1964) described by some authors as a single row of short tubercules (Segura *et al.*, 1998). Self-incompatibility has not been observed in the *Tacsonia* group (Coppens d'Eeckenbrugge *et al.*, 1997). There is no stigmatic movement during anthesis as in the *Passiflora* group. The flower morphology is adapted to cross-pollination by hummingbirds. According to Quintero (2009) honey bees are important pollinizers helping to improve fruit size and quality so that beehives should be present in the production areas. Flowers can be found at all months of the year but in more abundance in the dry season and fruit is copiously produced. The abundant fruit set observed in Hawaii seems to be due to a mixture of spontaneous self-pollination and pollination by insects. The newly opened flowers have exposed stamens, favorable to cross-pollination by insects; if cross-pollination does not occur, each flower later pollinates itself through movement of the stigmas to touch the stamens.

Fruit

The fruit is an oblong or oblong-ovoid berry 5–12 cm long and 3–4 cm wide (Fig. 7.2) that has a soft but thick and protecting leathery peel with a pale yellow or yellow-orange color, sometimes pale green at maturity, covered by a fine pubescence. The epicarp is hard but flexible and the mesocarp very thin. The orange shiny pulp, making up 60% of the fruit weight, contains small, black, flat,

Fig. 7.2. Banana passionfruit (*Passiflora mollisima*) showing leaves, flower and fruit (used with permission of Wagner *et al.* (1990) *Manual of Flowering Plants of Hawai'i*, Vol. II. Bishop Museum Press/University of Hawai'i Press, Honolulu). Fruit drawing by Odilo Duarte.

elliptic seeds that are surrounded by a sweet-sour, sometimes almost sweet, other times fairly sour aril, with a typical fine and special aroma. The fruit from Colombia and Ecuador are larger than those from Peru (León, 1964) and they have more than 100 seeds.

Cultivar development

Cytogenetics, genetics and breeding

The chromosome number has been reported as $2n = 18$. Molecular markers used to study the *Tacsonia* subgenus found that similarity among different accessions was high indicating inter-species gene flow, which is important for traditional and molecular breeding studies for these fruit (Sánchez *et al.*, 1998). It has also been established that there is a closer relationship between *P. mollisima* and *P. mixta* than suggested by present taxonomical classification (Segura *et al.*, 1998). In the same region, there are more than 37 species of the *Tacsonia* group so there is a tremendous potential to improve banana passionfruit types, especially in Colombia where 15 of these exist including red types that can be as sweet as a sweet granadilla (Quintero, 2009). Several crosses have been tried in Colombia between species like *P. mollisima*, *P. mixta*, *P. cumbalensis*, *P. antioquiensis*, *P. tarminiana* and *P. manicata*.

Selection and evaluation

Some selections have been made, especially in Colombia and Ecuador, that are being vegetatively propagated to maintain their phenotype. The main aim in these programs is good fruit size, high yield, tolerance to anthracnose, long postharvest life and that the fruit can be eaten out of hand.

Cultivars

Due to cross-pollination, selections are propagated by grafting or tissue culture resulting in stable cultivars. In Colombia, they have the most promising cultivar 'Momix', apparently a cross of *P. mollisima* by *P. tripartita*, which is highly productive, fairly tolerant to anthracnose and with a good flavor; it is propagated in some cases by grafting on regular banana passionfruit. Other selections propagated by tissue culture are 'Ruizquin 1' and 'Ruizquin 2', both very productive but susceptible to anthracnose and not popular for export. A third interspecific cross is 'Tintin', which is tolerant to anthracnose, has good sized fruit but low yields and has been used only for preparing juices, not for export to be eaten out of hand, so it is not recommended to be extensively planted (Quintero, 2009).

Cultural practices

Propagation

SEXUAL Seed propagation is the most widely used system where local types are grown (Munier, 1961). For this the pulp of healthy large and completely ripe or overripe fruit are left to ferment for 1–3 days then rubbed against a wire mesh or the bottom of a colander. Rubbing the pulp containing the seeds with ash can also be done. The rubbed seeds and aril are repeatedly washed thoroughly with water in a container allowing the good seeds to sink until the arils are discarded and the seeds are cleaned. The cleaned seeds are allowed to dry on paper for 3–5 days under shade. Normally seeds are sown in seedbeds or germination boxes and later transplanted to nursery bags (Delgado, 1982). Recently cell trays have begun to be used and transplanting done directly to the field bypassing the bag phase. Seeds should be buried about 1 cm and the substrate can be covered with plastic to speed up the germination that can last about 6 weeks.

ASEXUAL Grafting can be done especially when there are soil problems, but it is seldom used. The cultivar 'Momix' is propagated by grafting. According to Quintero (2009) in very poor soils *P. manicata* can be used as a rootstock. Cuttings can be used but it is not common.

Field preparation, transplanting and plant spacing

Field preparation can include plowing and harrowing the whole field in the case of a new orchard on flat land. Only the planting holes are prepared where the plants will be grown with the trellises already in place or in steep areas where mechanization is not possible. Transplanting is done when plants are 30–45 cm tall (Schoeniger, n.d.) or when the seedlings have three to four true leaves (Delgado, 1982). In cell trays, it takes about 2–3 months for a seedling to be ready for transplanting. Planting holes should be cubes with 30 cm sides (MAG-INCCA, 1991) or up to 60 cm sides (Schoeniger, n.d.) depending on soil characteristics. In poor soils, the upper portion of the soil from the hole should be separated from the lower portion and at planting the upper portion is placed at the bottom of the hole. Decomposed manure or compost should be added to the planting hole at a rate of 5–10 kg per hole or mixed with the soil.

Plants should be placed 4–5 m apart and about 1,000 plants per ha are recommended (Quintero, 2009). In heavy soils, a mound 30 cm high and 80–100 cm diameter should be made for each plant. If no irrigation is available transplanting should coincide with the start of the rainy season (Bonnet, 1988).

Trellis

Since this is a vine it has to be grown on trellises of several types but one of the two most widely used is the "fence" trellis, where 2.50 m posts are buried 50 cm every 3–5 m or more, depending on their sturdiness, and plants every 4–6 m. The distance between trellises should be 2–3 m but if associated with feijoa 400 plants per ha will be used and plants and trellises will be separated 5 m with feijoa lines running in the middle of two trellis lines. The orientation should be the same as the prevailing winds (Schoeniger, n.d.) while other authors indicate to use contour curves or N–S orientation (Quintero, 2009). Each trellis will have lines of No. 14 wire at 0.5, 1.0, 1.5 and 2.0 m from the ground (Delgado, 1982) or at 1.0 and 2.0 m only (Schoeniger, n.d.). This trellis has the disadvantage that the fruit skin will be damaged by the contact with the adjacent leaves that have serrated margins. The other trellis is the "pergola" described for sweet granadilla, the main differences being that the posts are installed every 5–6 m, the plant rows every 4 m and the cross wires are 1.0 m apart instead of 50 cm. This trellis, although the fruit hangs free without damage risk, should not be used when moisture is too high since diseases can become a problem due to reduced aeration and high relative humidity under the trellis.

On steep land, a third type of trellis called the "half-roof" trellis is recommended where one line of posts 2.1 m high runs parallel 1.2 m from another line that is 1.20 m high. Each high post is connected to its adjacent low post by a No. 12 wire that will support two lines of No. 16 wire that run parallel to the post lines and support the canopies. Plants are guided to reach the top of the 2.1 m line and then pinched to induce two branches to form that will grow along the No. 12 wire; when they meet the branch from the neighboring plant their tips are cut. This will induce lateral growth and these branches will grow towards the lower line so the fruit will hang freely and will not be damaged by rubbing.

Irrigation

If too little rain falls or rains are scarce or absent for a reasonable time, irrigation is necessary for continued production. In hilly conditions, drip irrigation is a good choice. On flat land drip or furrow irrigation can be used.

Pruning

FORMATION PRUNING Formation pruning consists of guiding the plant with strings to the trellis and cutting off the stem to induce branching. In the "fence" trellis two branches should be running in opposite directions along each wire line until they meet that of a neighboring plant where they are topped; this means four or eight branches will be selected according to the number of wires used. In the "pergola" trellis four to eight branches should be induced to form to fill the upper part.

In the "half-roof" trellis, once the stem reaches the top of the 2.10 m line it is pinched to get two branches that will run in opposite directions on the wire until they meet the branch of the neighbor and here they are tipped which will cause them to produce laterals that will fill the half roof. All undesired shoots on the main stem and basal shoots should be eliminated.

PRODUCTION PRUNING Normally sweet granadilla and banana passionfruit branches form flowers in the axils of the leaves but the first nodes do not flower so that once a branch has produced it will not flower again and should be discarded by pruning. In the "half-roof" trellis type once the branches reach the lower part of the roof they should be eliminated as well as all tertiary branches. This removal is required to avoid them from becoming entangled and facilitate removal of pruned branches in the future. This pruning will induce the formation of new shoots from the primary branches and it should be made about 4 months before the desired harvest time. This means 2.5–3 pruning cycles per year. It is important to prune the branchlets where the canopies of neighbor plants meet so that they do not get entangled.

During the last harvest the lower wires' primary branches on the fence trellis can be partially cut in a slanting form leaving three to four buds and then the third and fourth wire branches will be cut. Sometimes the branches in the upper wire are left in place until the new growth on the lower wires is attached in order to keep the vine in position.

RENOVATION PRUNING This is done after 7 or 8 years to renovate the plants by cutting all primary branches at 20 cm from the main stem and the regrowth is given the formation pruning.

ELIMINATION OF THE SHRIVELED PETALS This should be done as soon as they start to senesce to avoid the attack of the peel by spider mites and insects that will cause small wounds that enlarge with fruit growth and result in defects and a low fruit quality.

Fertilization

Some recommendations include four applications per year of 2 kg poultry manure, humus or compost, plus 200 g dolomite, 100 g DAP or 10-30-10 and twice a year, 50 g Agrimins (minor elements). Schoeniger (n.d.) recommends a 10-30-10 formula to be applied twice a year, once at the beginning of flowering and the second at the end of fruiting. Delgado (1982) recommends 100 g per plant of 10-30-10 every 3 months during the first 18 months and 250 g after that, with 15 g Agrimins added in all applications. MAG-INCCA (1991) recommend applying 2 kg of decomposed manure and 60 g of 10-30-12 twice a year. Fertilizer should be applied in a circle 30–60 cm away from the stem and better if a 5–10 cm deep furrow is made to bury the product. In steep land, apply in a half circle in the uphill part of the plant. Boron deficiency causes peel cracking. The deficiency of N, K and Mg in banana passionfruit reduces the accumulation of dry matter, demonstrating their importance for the growth process (Lizarazo *et al.*, 2013).

Pest management

DISEASES The main disease is anthracnose (Munier, 1961; MAG-INCCA, 1991) that causes round black spots on the fruit peel; in early attacks, fruit remain small and deformed, and the peel can crack open. Under some climatic conditions, powdery mildew can be a problem (Schoeniger, n.d.). *Botrytis* can also be a problem with flowers and young fruit.

INSECTS The main problems are leaf eaters (*Dione* or *Agraulis juno*, Cram) that are found in groups. Leaves and shoots can be attacked by leafhoppers (*Empoasca* sp.). Fruit flies (*Anastrepha* sp.) can be sporadically found. Foliage can be infested by mites (*Tetranychus* sp.). In humid and poorly drained situations, nematodes (*Meloidogyne* sp.) can be a problem.

Weed control

The soil has to be kept weed free. This can be achieved with machete or hoe, making sure that the stem base is not wounded. Herbicides should be used very carefully after consulting with experienced people and if a new product is to be used a test in a small area should be made before applying to the whole orchard.

Orchard protection

In windy situations if the fence type of trellis is used, the lines should go in the same direction as the wind. With other trellis types, their anchorage will have to be very solid. The ideal is to install wind barriers before the plantation so they will protect it from the beginning.

Harvesting and postharvest handling

Yields

The individual fruit weigh from 50 to 150 g. Yields are around 10 t ha^{-1} $year^{-1}$ (Lizarazo *et al.*, 2013). In dense planting with good weed control, adequate fertilization and sanitary control the annual harvest in Colombia can be 150–250 fruit per vine, equal to 150,000–250,000 fruit per ha or around 20–25 t ha^{-1} (Campos, 1992).

Harvesting

Harvest starts 10–12 months after transplanting (MAG-INCCA, 1991). For immediate use, the fruit is harvested as it starts to turn yellow. For fruit to be shipped to distant markets, fruit should be harvested while still green outside but orange inside or when the yellow just starts to appear on the peel (Delgado, 1982) or as the peel develops a lighter green. The fruit should be cut with a piece of peduncle attached. Care has to be taken not to bruise the fruit or subject it to rough handling. There is more or less continuous fruiting year round in Colombia, Ecuador, Peru and Bolivia while in New Zealand, the crop ripens from late March or early April to September or October (Morton, 1987).

Postharvest treatment

The fruit should be packed in wood or cardboard boxes lined with paper with a maximum of three to four layers of fruit separated by paper with the last layer being covered with paper. The fruit withstands shipment well and will keep for about 10 days to 2 weeks when stored under dry and not too cold conditions. During shipping, the fruit will turn completely yellow and gives off a delicious aroma. Storage at 4–7°C and 90% relative humidity allows fruit to be held for a month (Botia-Nino *et al.*, 2008).

Utilization

The fruit of the sweeter types can be eaten out of hand but normally it is used to make juice. The juice can be made in a blender mixing it with water or milk and sugar, and straining out the seeds. It is also used in gelatin desserts. In Ecuador, ice cream is made with the pulp. It is also used to prepare drinks as a replacement of the lemon or lime flavors. In Bolivia, the juice is combined with *aguardiente* and sugar, and served as a pre-dinner cocktail. In New Zealand, the Department of Agriculture has developed recipes to encourage the growing and utilization of the pulp for pie fillings, and for making meringue pie, sauce, spiced relish, jelly, jam and other preserves. It is also advocated as an ingredient in fruit salad, especially with pineapple, and for blending with whipped cream as a pudding, and for cooking and preserving as an ice-cream topping (Morton, 1987).

CARICACEAE

The family Caricaceae is a small basal family in the order Brassicales, the same order that contains *Arabidopsis* and the Cruciferae (cabbages). The most widely recognized and grown member worldwide is papaya (*Carica papaya*). Caricaceae species are normally evergreen shrubs or small trees divided into six genera. Four genera are of tropical American origin (*Carica, Jarilla, Jacaratia, Vasconcellea*) and one, *Cylicomorpha*, from equatorial Africa. Approximately 71 species have been described, though Badillo (2000, 2001) reduced the number of species with the following distribution: *Carica*, one species; *Cylicomorpha*, two species; *Jacaratia*, eight species; *Jarilla*, three species; *Vasconcellea*, 20 species; and *Horovitzia*, one species.

Many of the species that were formerly in the genus *Carica* have been reassigned based on molecular evidence to the genus *Vasconcellea*. *Vasconcellea* species are dioecious, except for *V. monoica* Desf. and some *V. pubescens*. Most species are herbaceous, single-stemmed and erect. The edible species are *V. pubescens*, *V. monoica* Desf. (Col de Monte), *V. goudotiana* Solms-Laubach (papayuelo), and *V. quercifolia* Benth. and Hook. These are mostly eaten cooked as the fruit are low in sugars and lack palatability being dry without the juicy flesh of *C. papaya*.

The edible species *V. pentagona* (syn. *V. heilbornii*-Badillo var. *pentagona*), "babaco", is of subtropical origin found at 2,000–3,000 m in Ecuador (Oosten, 1986). It is probably a natural hybrid between *V. stipulata* Badillo and *V. pubescens*

A. DC. Kuntze; they are known in Ecuador by their respective names of *toronche* and *chamburo*. It has also been indicated that *V. chrysopetala* is the result of hybridization between *V. pubescens* and *V. stipulata* (Badillo, 1967).

Mountain Papaya

The name *Vasconcellea pubescens* A. DC. was given after all the cold climate *Carica* species were reclassified as *Vasconcellea*. Synonyms for this species include *Carica candamarcensis* Hooker, f., *C. cestriflora* (A. DC.) Solms, *C. cundinamarcensis* Linden, *C. pubescens* (A. DC.) Solms, *C. pubescens* Lenné & Koch, *V. cestriflora* A. DC., and *V. cundinamarcensis* V. M. Badillo. The species is consumed for its fruit and seen as a source for disease and cold tolerance in *C. papaya* breeding though crossing is difficult and the progeny are frequently sterile or with poor fruit quality and yield (O'Brien and Drew, 2009). It is one of the parents of the hybrid babaco that resulted from a natural cross with *V. stipulata*.

In English, the tree and fruit are known as mountain papaw or mountain papaya while in Spanish a number of regional names exist: *chilhuacán*, *chiglacón*, *chamburo* (Ecuador), *chamburo*, *huanarpu hembra*, *papaya de monte*, *papaya arequipeña*, *papaya de altura* (Peru, Bolivia) and *papayuela* (Colombia, Chile). In French, it is *papayer de montagne* and in German *Berg papaya*.

Origin and distribution

The species is native to Panama, Venezuela, Bolivia, Colombia, Ecuador and Peru at elevations of 1,500–3,000 m in areas ranging from dry, windy, open plateaus to humid, shaded forests. It was removed from the wild and cultivated in gardens for the fruit and as an ornamental plant. It was utilized there before the tropical papaya was introduced. The species has been distributed worldwide but is not grown as a significant commercial crop. It is widely seen growing in family gardens from Colombia to Bolivia, and some commercial orchards exist in Chile totaling about 220 ha.

Ecology

Soil

The species demands good soils, preferring loams to silty loams that are rich in organic matter and 80–100 cm deep. In sandy soils, nematodes can become a problem although it has some resistance to them. Flooding can easily cause plant death due to root rotting.

Rainfall

Mountain papaya grows well with rainfall above 500 mm and better with 1,000 mm evenly distributed. The plant does not tolerate drought, which results in profuse leaf fall. Under seasonal rainfall, it has to be irrigated to obtain good yields.

Temperature

This tree is found in low dry mountain forest areas between 1,000 and 3,000 m, depending on the latitude. Mean temperatures range between 12 and 18°C (22°C in winter at midday). It is sensitive to lower temperatures that occur early in the morning to noon, especially with intense sunlight levels, and is also sensitive to intense sun during the colder months. Severe frost will kill the plant, though it does survive −3°C without serious injury except that the fruit will be damaged at 0°C. It does not need cold temperatures in winter like deciduous trees do, preferring moderate temperatures with high humidity, even in summer, with little fluctuation between day and night or between seasons. Temperature extremes affect the foliage and normal ripening of the fruit.

Light and photoperiod

There is no information on photoperiodic response, but it does not seem to be affected as it flowers at the equator and at very high latitudes in Chile.

Wind

High winds are a problem with this herbaceous stemmed plant with large leaves and a shallow root system.

General characteristics

Plant

The tree, actually a herbaceous and succulent plant, is 3–6 m tall, sometimes up to 10 m (León, 1987), and has succulent, medullose and branched stems with a very broad basal part (pachycaul). The stem is marked by conspicuous leaf scars. The broad leaves are clustered at the terminal parts of the branches, forming a dense crown; they have 17–34 cm long petioles with a leaf blade that is 20–26 cm long and 34–40 cm wide (Fig. 7.3). The leaves are normally five-lobed with two lateral lobes at each side; in some cases there are three or seven lobes. Growth is slow and leaves continuously appear at the apex while the basal leaves now self-shaded abscise. The species is distinguished by the coating of hairs on the underside of the leaves. The root system is fairly shallow. The juvenile period is 10–12 months.

Flowers, pollination and fruit set

Most plants are dioecious and, as in *C. papaya*, there are three sexual forms: pistillate, staminate and hermaphrodite plants. Pistillate and staminate specimens do not respond to seasonal climatic changes while the hermaphrodite plants form female, male and perfect (hermaphrodite) flowers in different proportions, depending on the season; normally hermaphrodite plants form inflorescences where terminal flowers are pistillate and the basal staminate (Sánchez-Vega, 1994). The male plants are very leafy and have an abundance of inflorescences or panicles that are 1–15 cm long, and flowers are slender with a long pubescent peduncle and flower tube, having five sepals and five yellowish-white or greenish petals, five stamens and five sessile staminodes inserted in the corolla throat; they can produce some fruit that are larger than the female's but irregular in shape. Female plants are less

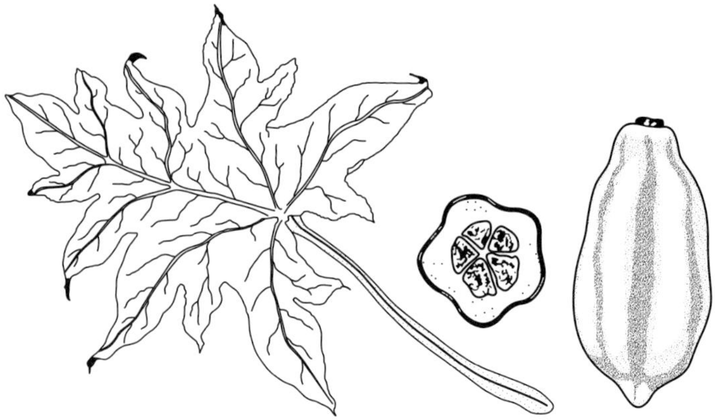

Fig. 7.3. Mountain papaya (*Vasconcellea pubescens*) showing leaves and fruit with a cross-section of a fruit (used with permission from FAO Plant Production and Protection Series No. 26, Neglected Crops: 1492 from a different perspective, 1994 (A001/2013)).

leafy, and have larger flowers with yellow to white petals that are solitary or rarely in short cymes, arising in the leaf axils with a very short peduncle. They are pentamerous, have a five-lobed stigma and produce the typical regular shaped fruit; these are the ones the industry uses. Hermaphrodite plants produce fruit similar to the female but generally deformed and so with less usable flesh (Undurraga, 1989).

Fruit

The obovoid or ellipsoidal five-sided fruit is 6–15 cm long and 3–8 cm wide (Fig. 7.3), weighing about 130 g, and changes from green to yellow or orange-yellow when ripe; the firm pulp is about 1 cm thick, tart, fragrant and yellow when ripe. The numerous seeds (100–200) in the interior cavity are spiky, brown-reddish and covered by a juicy sarcotesta. The fruit of hermaphrodite flowers has more pronounced lobules and the pulp is thinner while that from female flowers is less lobulated and smoother. The pericarp makes up to 70–75% of total weight, placenta tissue is 20–25% and seeds around 5% (Luza *et al.*, 1990).

Cultivar development

Cytogenetics, genetics and breeding

The diverse sexual forms of *Vasconcellea* species have created some confusion in their classification. This sexual variability added to the fact of possible interspecific crosses opens the possibility of creating hybrids of different combinations and adding further variability.

Selection and evaluation

Some selection has been made by using asexual propagation to obtain a majority of female plants in the orchard and a few hermaphrodites as pollen sources. No systematic work has been done to get improved material for commercial uses.

Cultivars

No cultivars are recognized in its native range with the greatest diversity in Ecuador and northern Peru.

Cultural practices

Propagation

SEXUAL Seeds are the normal method of propagation with female fruit being the source. After extraction from the ripe fruit, they should be washed to eliminate the mucilage and normally superficially dried under shade. Seeds begin to germinate at 30 days and a 60% germination rate has been noted. They can be germinated in beds or trays and transplanted to bags to attain a larger size; this is used normally when no irrigation is available. Seeds can also be sown in trays with cells and plants taken directly to the field when about 25–35 cm tall if irrigation is available. The seeds apparently have no dormancy although a short postharvest storage period seems to improve germination (Alarcón *et al.*, 1997). Seeds stored for 4 months at room temperature (18–22°C) had ~80% germination compared to those stored for one year and freshly harvested seeds that had 2.23% and 36% germination, respectively. Germination was best at 30°C and started after 7 days at 30–35°C and after 14 days at 15–20°C. Sometimes seeds are pre-germinated by putting them in sand at 20–28°C that is kept under high moisture. A problem with sexual propagation is the presence of male or monoecious plants. Normally 60% female, 12–20% male and 20% hermaphrodite plants are obtained (Undurraga, 1989).

ASEXUAL Occasionally cuttings are used but the percentage of takes is low unless treated with 4,000–5,000 ppm indole butyric acid, 20–24°C bottom temperature and mist whereby 90–100% rooting can be obtained (Undurraga, 1989). Cleft graft has been used successfully and could be a solution to root diseases using resistant species as in the case of babaco.

Field preparation, transplanting and plant spacing

In the Andes, there is very little agricultural knowledge concerning *V. pubescens*. Its cultivation is traditional and it is grown in rural home gardens as a decorative plant and for fruit for household consumption. One to three plants are grown in each garden and these receive the same management as other species in the plot, so there are no specific cultivation techniques for this species.

Chile is one of the few places where technical cultivation is performed. The seedlings are planted out when they are 10–15 cm high (two to four leaves). It is recommend to plant at 2 × 1 m or 2 × 3 m with the shorter distances used with formation of two axes after topping while the larger distances are for a four axis formation that involves two toppings (Undurraga, 1989). Plantings can also be done using 1.5 m between plants and 2 m between rows, resulting in 3,000–3,500 plants per ha (CAF, 1992), although distances of 3 × 3 m can be used, reducing plant density to 1,111 plants per ha. It is recommended to eliminate male plants, leaving about one male for ten female plants to ensure pollination (CTIFL, 1992).

Irrigation

In the absence or deficit of rainfall, irrigation is needed. In Chile, the estimated need is for 10,000–12,000 m^3 ha^{-1} $year^{-1}$ in areas where it does not rain.

Pruning

Plants of *V. pubescens* tend to branch naturally from the base; under intense cultivation they can be formed with one, two or four branches. To form two branches, formation pruning is required by topping the stem to force branching and then selecting the best two branches, while for four branches the two initial branches are topped to get them to branch again and the best two of each branch are retained. This species showing higher and multi-branched trees needs more severe pruning in order to ease harvest activities and plant management, and reduce phytosanitary problems, while heavy fruiting sometimes requires fruit to be thinned, leaving two fruit per peduncle (CAF, 1992).

Fertilization

In Chile where commercial plantations exist, they recommend heavy fertilization with 400–600 kg N, 100–200 kg P_2O_5 and 70–80 kg K_2O ha^{-1} (Undurraga, 1989).

Pest management

DISEASES The main diseases are root rots caused by *Fusarium*, *Pythium* and *Phytophthora* that are all difficult to control. Field rotation and careful water management avoiding an excess are the best preventive measures. This species is tolerant to ring spot virus and this could be used to transfer genes to papaya (Bernal and Correa, 1990), though so far with limited success (O'Brien and Drew, 2009).

INSECTS The main problems are aphids that transmit virus diseases that cause leaf mottling or netting.

Root nematodes such as *Meloidogyne arenaria* or *M. incognita* can do severe damage, especially in sandy soils low in organic matter. Other nematodes that sometimes attack are *Pratylenchus thornei*, *Helicotylenchus* spp. and *Aphelenchoides* sp., although the species has shown some tolerance to some of them (Bernal and Correa, 1990).

Weed control

Normally contact herbicides are used as well as mechanical or weeding with a hoe especially around the stem. Chemicals are preferred since any wounding done by a weeding tool can be the entrance point for fungal diseases.

Orchard protection

In areas with strong winds, windbreaks have to be planted and the rows should follow the main wind direction.

Harvesting and postharvest handling

Yields

Yields per unit of area are not known, but counting fruit on single trees, you can expect between 15–20 fruit per month. Yields of 18–200 fruit per plant per year can be expected (Luza *et al.*, 1990). Information from Chile indicates around 15–30 t ha^{-1} $year^{-1}$. The useful life of a plant is between 5 and 8 years.

Harvest

Fruit need to be harvested at least on a weekly basis. The harvester detaches fruit that has started to change color.

Postharvest treatment

This is a climacteric fruit where the maximum respiration rise coincides not with the ripe stage but with the overripe stage, so that the fruit has to be used quickly. Fruit resistance to pressure goes from 25.5 N for green-skinned fruit to 10.9 N for fruit with yellow skin. Ideal storage temperature seems to be around 7°C that will prolong life of green-skinned fruit for 30 days and only 10 days for yellow-skinned fruit (Undurraga, 1989).

Utilization

The fruit pulp varies greatly in sweetness and some can be eaten fresh or prepared as a juice with the blender, but most must be cooked and sweetened to be eaten (Luza *et al.*, 1990). In Chile, they estimate that 80–85% is used by industry and the rest goes to the fresh product market although some sort of processing will be needed at home since its low sugar content and high papain content do not allow for fresh fruit consumption (Undurraga, 1989). Some people do like to eat it without processing. The fruit yield a clear juice and are excellent in pies, ice cream, marmalades, or sweets. The boiled or baked green fruit can be eaten as a vegetable. Canned preserves are marketed in Chile. This species produces latex with a high level of papain that is fivefold stronger than that of papaya (Scheldeman, 2002). The latex can be used as a meat tenderizer or in the pharmacological industry; it can also be used against skin mycosis, verruca plana and as an antihelminthic. The blended pulp helps alleviate dispepsy. Other parts of *V. pubescens* have medicinal uses to treat arterial sclerosis.

Babaco

Babaco (*Vasconcellea* × *heilbornii* V.M. Badillo) is a natural hybrid that occurred in Ecuador between *V. pubescens* and *V. stipulata*, both formerly in the genus *Carica*. This is the cool climate papaya-type fruit grown commercially in the inter-Andean valleys of Ecuador and Colombia. Commercial interest has increased since the 1970s. Babaco has been cultivated in New Zealand since 1973 and was introduced to Europe in 1985, notably in Italy and Spain. More recently, its cultivation has spread to California, Australia, Israel and Greece. Babaco production occurs mostly in Ecuador and New Zealand. In Ecuador, between 1996 and 1999 annual production was about 600 t on about 100 ha, of which most was domestically consumed. The high yield and the unusual flavor have led to this expansion in production.

Ecology

Soils

It prefers light and well-drained deep loamy soils or sandy loams. Sandy soils that are rich in organic matter (> 3%) are also recommended (Dornier *et al.*, 2008). Too much clay in the soil should be avoided. Soils should be slightly acidic or neutral with a pH of 6.5–7.0, but it can stand pH from 5.8 to 8.2 (Viteri, 1992). It is very susceptible to standing water.

Temperature

A growing temperature of 13–22°C is ideal. This can be found between 800 and 3,000 m altitude, depending on the particular spot in the Andes. It can also produce near sea level in subtropical climates such as New Zealand, the Peruvian central and southern coast and other places. It tolerates 2°C but it does not stand frosts or prolonged periods above 30°C (Merino, 1989). In places with very cold winter temperatures, it is grown in greenhouses.

Rainfall

For optimal growth 600–1,300 mm of rainfall are needed. If not uniformly distributed irrigation will have to be used during the dry season. A high relative humidity favors fungal diseases.

Light and photoperiod

The babaco needs plenty of sunshine with at least 4.5 h of sun exposure daily, otherwise the plant will get slender and flowers will drop in larger numbers; if sunlight is too intense leaf sunburn can occur.

Wind

The plant is sensitive to wind due to its shallow root system, and the potential for leaf damage and defoliation due to their large size.

General characteristics

Plant

This semi-ligneous single and rather thin-trunked shrub grows rapidly and can attain 3 m in one year and 25 cm diameter at its base. The stem is spongy-fibrous, gray or brown and shows prominently the scars of the fallen leaves. It has a tuberous main taproot that can attain a diameter of up to 40 cm and bears many lateral fleshy roots that are fairly long with many thin absorbent rootlets.

The alternate glabrous leaves have between five and seven lobes that can be entire or slightly divided with prominent veins in the underside and a darker green on the upper side and a long petiole (Fig. 7.4).

Flowers, pollination and fruit set

It produces only bell-shaped pistillate flowers (Fig. 7.4) that measure between 3.5 and 4 cm long and grow out of the leaf axil 2–3 months after planting either singly or in groups of two or three. They have five white-yellowish or greenish petals joined at their bases, and thin and reflexed at their tips (León, 1964). The sepals are dark green and rudimentary. The yellow stigma is star-shaped. Since there are no staminate flowers, the plant can only be propagated asexually. No pollinators are needed to produce fruit (Dornier *et al.*, 2008).

Fruit

The parthenocarpic fruit is a berry with no apparent ovarian cavity. The pentagonal fruit has five longitudinal well-defined sides, can measure 15 and up to 30 cm in length and 6–15 cm in diameter (Fig. 7.4), and weighs between 0.5 and 2.0 kg. It is rounded at its base with a slight concavity at the union with the peduncle and is pointed at its apex. The skin is green and turns

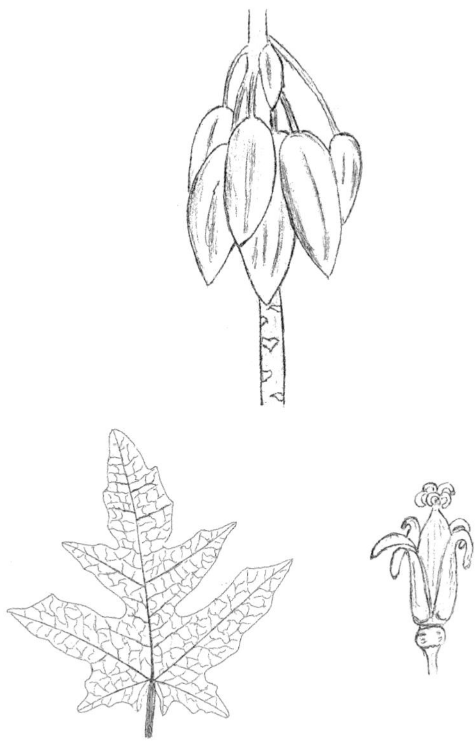

Fig. 7.4. Babaco (*Vasconcellea × heilbornii*) showing fruit and stem scars, leaf and flower. Drawings by Odilo Duarte.

yellow upon ripening. The edible ripe ivory or almost white and watery flesh is localized 1–2 mm under the epicarp and is rather acid and aromatic. The central cavity of the fruit is filled with a spongy structure. Eventually some aborted yellow-colored seed is present. It takes about 7 months from anthesis to harvest.

Cultivar development

Cytogenetics, genetics and breeding

Babaco has a chromosome number of $2n = 18$ (Delgado-Ortiz, 1988). One problem with babaco in the developed countries markets is that the first fruit the plant produces are too large and are not very well accepted. With the idea of obtaining a plant with smaller and more uniform fruit, Soria and Soria (1992) produced some hybrids between 1963 and 1965 using babaco as the female plant and *V. pubescens* (locally called chamburo) as the male parent. The first two hybrids were produced using male chamburo pollen; one of them was called "Chambaco". "Chambaco" is a very precocious plant that grows rapidly, has only female flowers, has fruit 14–16 × 6–8 cm (a little larger than half a babaco), has a thicker peel, a pinkish-cream pulp and an aroma that is a mixture of babaco and chamburo, and propagates very easily from cuttings. This hybrid has greater commercial potential. The other hybrid called "Modesbaco", with female flowers only and a smaller fruit (7–9 × 5 cm), has fruit and pulp similar to babaco. The third hybrid between babaco and a hermaphrodite chamburo is called "babaquillo", which is a hermaphrodite plant with leaves and flowers similar to babaco, and a fruit 11–12 × 7 cm, with 20 cm peduncles and 10–50 seeds. These authors indicate that the fact that babaco can be crossed with *V. pubescens* and viable seeds are produced is evidence that this species was one of its parents. Badillo in 1967 proposed that *Vasconcellea* × *heilbornii*, *V. chrysopetala* and *V. fructifragrans* (all formerly *Carica*) are all hybrids originating from the cross of *V. pubescens* and *V. stipulata*.

Selection and evaluation

The above mentioned "Chambaco" is vegetatively propagated and is being planted in some areas of Ecuador under different names.

Cultivars

There are no defined cultivars but it is possible to see in the field ecotypes with some differences in internode length, fruit size, form and soluble solids content and leaf form. Some nurseries promote what they call superior clones with higher sugar content. In Spain, they have promoted 'Dulce Babaco Hortex' No. 2 that has higher sugar content than No. 1.

In its native range, people cultivate what they call the "baby babaco" as a backyard plant. This is actually the jigacho or toronche (*V. stipulata* or *V. chrysopetala*) with a fruit similar to babaco but smaller (14 cm long and 450 g). It is more compact, has a sweet aroma and an intense yellow flesh. It could be an alternative for export.

Cultural practices

Propagation

Babaco is propagated vegetatively by cuttings, typically from 2-year-old plants. The trunk is divided into sections of 25–30 cm in length and 4–6 cm in diameter, which root in 10 weeks. For uniformity, the middle portion of the stem should be used. Rooting occurs sooner, in 6–7 weeks, when 10 × 2 cm cuttings are used. Propagation using cuttings from young plants is very frequent in greenhouse production. Another system is to cut a mother plant to a 50 cm stump and when the new shoots reach 40 cm they are pulled not cut so that the cutting comes with a "heel" at the base. Cuttings are cured by leaving them in a dry and shaded place so the wound starts healing and after 4–6 days they are planted in the nursery bags. Sometimes cuttings are cured and after 4–5 days planted directly in the field.

Grafting can done with chamburo (*V. pubescens*) or toronche (*V. stipulata*) as rootstocks. Plants at the end of their fruiting cycle or young nursery plants are used as rootstock. The graft is made at a height of 10–15 cm. This type of propagation generates more vigorous plants, with better cold and pest resistance (Dornier *et al.*, 2008). Cleft is the most used type of grafting. Micro-propagation from shoots and axillary buds is relatively easy and enables the rapid production of large numbers of healthy plants.

Field preparation, transplanting and plant spacing

Cultivation occurs in both the field and the greenhouse. The land must be deeply plowed (40 cm) to facilitate root development. Organic matter in the form of decomposed cattle or chicken manure should be added to the planting hole at a rate of 5–7 kg. The use of mounds for planting is convenient. Triangles or squares with 1.5 m sides result in 4,444–5,128 plants ha^{-1}; sometimes 1.5 × 1.2 m is used, resulting in 5,555 plants ha^{-1}.

Irrigation

Frequent regular water supply is required for plant growth and yield. Irrigation should be adapted to soil type and season; for clay soils a 12-day irrigation cycle is sufficient. Irrigation frequency should be increased for sandy soils (every 8 days during the wet season and every 4 days during the dry season). Irrigation after planting, at the start of flowering and during fruit development and maturation is crucial (Viteri, 1992). Drip irrigation is well suited with one or two lines of drippers per row. Soil should be kept near field capacity.

Pruning

Side shoots must be removed to obtain plants that are sufficiently tall and to produce high quality fruit. The side shoots can be used for propagation by preparing cuttings. Elimination of old leaves using pruning shears is also done periodically. After the plant finishes its productive cycle of 18–24 months, the trunk is cut back to about 30 cm above the soil level (coppiced). Following 8 months of regrowth, the plant is ready for another cycle of fruit production. Yield decreases with plant age and the plant is generally only exploited for four production cycles (Villareal *et al.*, 2003).

Babaco plants can be pruned leaving two or three branches depending on nutrition, plant distances, greenhouse or field conditions and desired fruit sizes. To obtain smaller fruit sometimes two branched plants are left. Single-stemmed plants yield larger, heavier fruit, especially the fruit at their base that are often not suited for marketing (Soria and Viteri, 1999). Cossio (1988) confirms the preference of V-shaped plants to obtain suitable fruit sizes for European markets.

Fertilization

During plant growth, nitrogen, potassium and magnesium supplements every 3 months are recommended. A biannual phosphorus supplement can be added. Fertilization depends on the nutritional status of the soil. Recommended average applications for babaco are in kg ha^{-1}: N, 100–250; P_2O_5, 100–200; K_2O, 200–400; Ca, 150–200; Mg, 100–150; and S, 30–50. Cacioppo (1987) recommends the use of: 150 g N; 90 g P_2O_5; 150 g K_2O; and 20 g MgO per plant per year during the 2 years the plant will be productive. Boron can be important in certain soils.

Additionally 60 t ha^{-1} of organic matter should be incorporated at soil preparation time. This must be higher in greenhouse conditions due to a faster development and more intense production (Soria and Viteri, 1999). Organic fertilization is recommended every 6 months (CTIFL, 1992).

Pest management

DISEASES Disease and pests limit production of babaco (Villareal *et al.*, 2003). Fusarium (*Fusarium* sp.) causes root rot, wilting and defoliation. *Phytophthora* sp., *Pythium* sp. and *Rhizoctonia* sp. also attack the roots and result in plant death. Small yellow then brown spots that appear on the leaves, are often due to alternaria leaf spot (*Alternaria* sp.), particularly on younger leaves, and can lead to leaf abscission. Powdery mildew (*Oidium* sp.) causes a white down to develop on the underside of the leaves and on the flowers sometimes preventing fruit set. Anthracnose (*Colletotrichum* sp., *Mycosphaerella* sp.) results in large brown spots on the leaves (up to 3 cm in diameter) and can cause defoliation. *Agrobacterium* can produce the so-called "Elephant foot".

Some viruses such as the "Babaco Mosaic" and "Babaco Roughness" can attack the plants and the viruses are propagated by insects, tools, nematodes and infected material.

INSECTS Aphids (*Myzus persicae*) deform young shoots. Mites (*Tetranychus bimaculatus* and *Hemitarsonemus latus*) form colonies on the underside of the leaves, which turn yellow. They may also attack the fruit (Villareal *et al.*, 2003). *Meloidogyne incognita* and *Meloidogyne javanica* nematodes cause root nodules to form and slow growth and development. The trunk becomes weak and the leaves turn yellow and abscise; they are one of the most important production problems.

Weed control

Can be done using plastic mulch or with tools used so they do not damage the shallow roots. A hoe is a popular tool for this and in most cases, weeding is combined with mounding the soil around the stem to get better root development and plant anchorage.

Orchard protection

In windy conditions, a windbreak will have to be established and rows should run along the main wind direction.

Harvesting and postharvest handling

Yields

Annual yield is between 40 and 60 t ha^{-1} and can range from 10 to 100 t with each plant producing between 25 and 80 fruit. Yields of 200 t ha^{-1} have been reported for a 2-year productive life period (Viteri, 1992). The highest yields are obtained under greenhouse conditions in Ecuador.

Harvest

Harvest starts about 10–12 months after planting. The fruit is harvested as soon as a yellow appears at one side of the fruit (Mencarelli *et al.*, 1990). At this time soluble solids are about 4.1%. Fruit at this yellow stage have around 20.4 N resistance. It is important to cut the fruit with the peduncle to avoid fungal disease infection through the wound and to slow ripening. Fruit should be put into wood or plastic boxes and protected from the sun.

Postharvest handling

After harvesting, the climacteric fruit reaches commercial maturity (75% yellow) in 2–4 weeks at 20–25°C and in 4–6 weeks at 15°C. The fruit suffers chilling injury below 5°C. The respiratory rate and ethylene production increase from 4–7 to 12–14 ml CO_2 kg^{-1} h^{-1} and from 0.5 to 7 l C_2H_4 kg^{-1} h^{-1} (at 18°C) and peak when the skin is 50–70% yellow (Harman, 1983). Ethylene or holding the fruit at 25°C can be used to hasten ripening.

Utilization

The fruit is mainly consumed fresh when completely ripe. The flesh is juicy and somewhat acid in taste. It can be processed to make pasteurized and concentrated fruit juice, dehydrated fruit powder, dried fruit, marmalade and pieces in syrups. The fruit has more water, protein and organic acids than papaya (Table 7.1) and its sugar and vitamin content is lower. The two fruit have similar mineral composition. Up to 119 fruit aroma compounds have been identified (Barbeni *et al.*, 1990); the most important are 1-butanol, 1-hexanol, ethyl hexanoate and butanoate. Green babaco fruit latex contains proteolytic and lipolytic enzymes; the activities are similar to those of papaya latex (Dhuique-Mayer *et al.*, 2001).

References and Further Reading

Sweet granadilla

Barbeau, G. (1990) *Frutas Tropicales en Nicaragua.* Dirección General de Técnicas Agropecuarias, Ministerio de Desarrollo Agropecuario y Reforma Agraria, Editorial Ciencias Sociales, Managua, Nicaragua, pp. 285.

Bernal, J.A. (1988) *El Cultivo de la Granadilla.* Curso de Frutales. Gerencia Regional UNO. Instituto Colombiano Agropecuario, Colombia, pp. 104–116.

Castaño-Zapata, J. (2009) Enfermedades importantes de las pasifloráceas en Colombia. In: Miranda, D., Fischer, G., Carranza, C., Magnitskiy, S., Cassierra, F., Piedrahita, W. and Flores, L.E. (eds) *Cultivo, Poscosecha y Comercialización de las Pasifloráceas en Colombia: Maracuyá, Granadilla, Gulupa y Curuba.* Sociedad Colombiana de Ciencias Hortícolas, Bogotá, Colombia, pp. 223–243.

Fischer, G., Casierra-Posada, F. and Piedrahita, W. (2009) Ecofisiología de las especies pasifloráceas cultivadas en Colombia. In: Miranda, D., Fischer, G., Carranza, C., Magnitskiy, S., Cassierra, F., Piedrahita, W. and Flores, L.E. (eds) *Cultivo, Poscosecha y Comercialización de las Pasifloráceas en Colombia: Maracuyá, Granadilla, Gulupa y Curuba.* Sociedad Colombiana de Ciencias Hortícolas, Bogotá, Colombia, pp. 45–68.

León, J. (1964) *Plantas Alimenticias Andinas.* Boletín No. 6. Instituto Interamericano de Ciencias Agrícolas, Zona Andina, Lima, Perú, 112 pp.

Leung Woot-Tsuen, W. and Flores, M. (1966) *Tabla de Composición de Alimentos para Uso en América Latina.* Instituto de Nutrición de Centro América y Panamá (INCAP), Editorial Interamericana S.A., 58 pp.

Lobo, M. and Medina, C.I. (2009) Recursos genéticos de pasifloráceas en Colombia. In: Miranda, D., Fischer, G., Carranza, C., Magnitskiy, S., Cassierra, F., Piedrahita, W. and Flores, L.E. (eds) *Cultivo, Poscosecha y Comercialización de las Pasifloráceas en Colombia: Maracuyá, Granadilla, Gulupa y Curuba.* Sociedad Colombiana de Ciencias Hortícolas, Bogotá, Colombia, pp. 7–17.

Mamani-Quispe, P. (2000) *Frutales Nativos de Exportación. Cultivo de Granadilla en Sistema Tipo Parrilla.* Informe Técnico No. 02-00 Instituto Nacional de Investigación Agraria, Lima, Perú.

Miranda, D. (2009) Manejo integral del cultivo de la granadilla (*Passiflora ligularis* Juss.) (Integrated crop management of sweet granadilla). In: Miranda, D., Fischer, G., Carranza, C., Magnitskiy, S., Cassierra, F., Piedrahita, W. and Flores, L.E. (eds) *Cultivo, Poscosecha y Comercialización de las Pasifloráceas en Colombia: Maracuyá, Granadilla, Gulupa y Curuba.* Sociedad Colombiana de Ciencias Hortícolas, Bogotá, Colombia, pp. 121–157.

Mora, H. and Benavides, M. (2009) Plagas de importancia económica asociadas a las pasifloráceas y su manejo en Colombia. In: Miranda, D., Fischer, G., Carranza, C., Magnitskiy, S., Cassierra, F., Piedrahita, W. and Flores, L.E. (eds) *Cultivo, Poscosecha y Comercialización de las Pasifloráceas en Colombia: Maracuyá, Granadilla, Gulupa y Curuba.* Sociedad Colombiana de Ciencias Hortícolas, Bogotá, Colombia, pp. 245–265.

Romero Castañeda, R. (1961) *Frutas Silvestres de Colombia, Volumen 1.* Editorial San Juan Eudes, Bogotá, Colombia, pp. 169–172.

Sánchez, I., Angel, F., Fajardo, D., Castillo, M.F., Lobo, M., Tohme, J. and Roca, W. (1998) Uso de marcadores moleculares como base en el mejoramiento genético del género *Passiflora* L. *Proceedings of the Interamerican Society for Tropical Horticulture* 42, 267–271.

Zuleta-Ospina, J., Rodríguez-Aguilar, M.H. and Miranda-Lasprilla, D. (2011) *Plan de transferencia en el manejo de la secadera (Fusarium sp.) en los cultivos de granadilla y maracuyá, y perla tierra (Eurhizococcus sp.) en el cultivo de mora de Castilla, en cuatro localidades del Departamento del Huila*. Consejo Departamental de Ciencia, Tecnología e Innovación del Huila, Codecti, Huila, Neiva (Huila), Colombia.

Banana passionfruit

Bonnet, J.G. (1988) *El Cultivo de la Curuba*. Curso de Frutales. Gerencia Regional UNO. Instituto Colombiano Agropecuario, Colombia, pp. 34–46.

Botia-Nino, Y.C., Almanza-Merchan, P. and Balaguera-López, H.E. (2008) Efecto de la temperatura sobre la maduración complementaria en curuba (*Passiflora mollissima* Bailey). *Revista U.D.C.A. Actualidad y Divulgación Científica* [online] 11(2), 187–196.

Campos, T. (1992) El cultivo de la curuba (*Passiflora mollisima* (H.B.K. Bailey)) en Colombia. *Acta Horticulturae* 310, 215–232.

Castaño-Zapata, J. (2009) Enfermedades importantes de las pasifloráceas en Colombia. In: Miranda, D., Fischer, G., Carranza, C., Magnitskiy, S., Cassierra, F., Piedrahita, W. and Flores, L.E. (eds) *Cultivo, Poscosecha y Comercialización de las Pasifloráceas en Colombia: Maracuyá, Granadilla, Gulupa y Curuba*. Sociedad Colombiana de Ciencias Hortícolas, Bogotá, Colombia, pp. 223–243.

Coppens d'Eeckenbrugge, G., Segura, D.S., Hodson de Jaramillo, E. and Góngora, G.A. (1997) Les fruit de la passion. In: Charrier, A., Jaquot, M., Hamon, S. and Nicolas, D. (eds) *L´amelioration des Plantes Tropicales*. CIRAD, Montpellier, France, 623 pp.

Coppens d'Eeckenbrugge, G., Barney, V.E., Jorgensen, P.M. and MacDougal, J.M. (2001) *Passiflora tarminiana*, a new cultivated species of *Passiflora* subgenus *Tacsonia*. *Novon* 11, 8–15.

Delgado, C.H. (1982) *El Cultivo de la Curuba*. Cartilla de Divulgación CCH No. 12. Instituto Colombiano Agropecuario, Bogotá, Colombia.

Escobar, L.A. (1992) *La Sistemática y Evolución de las Passifloras*. Primer Simposio Internacional de Passifloras, Palmira, Colombia.

Fischer, G., Casierra-Posada, F. and Piedrahita, W. (2009) Ecofisiología de las especies pasifloráceas cultivadas en Colombia. In: Miranda, D., Fischer, G., Carranza, C., Magnitskiy, S., Cassierra, F., Piedrahita, W. and Flores, L.E. (eds) *Cultivo, Poscosecha y Comercialización de las Pasifloráceas en Colombia: Maracuyá, Granadilla, Gulupa y Curuba*. Sociedad Colombiana de Ciencias Hortícolas, Bogotá, Colombia, pp. 45–68.

León, J. (1964) *Plantas Alimenticias Andinas*. Boletín No. 6. Instituto Interamericano de Ciencias Agrícolas, Zona Andina, Lima, Perú, 112 pp.

Leung Woot-Tsuen, W. and Flores, M. (1966) *Tabla de Composición de Alimentos para Uso en América Latina*. Instituto de Nutrición de Centro América y Panamá (INCAP), Editorial Interamericana S.A., 58 pp.

Lizarazo, M.A., Hernández, C.A., Fischer, G. and Gómez, M.I. (2013) Response of the banana passion fruit (*Passiflora tripartita* var. *mollissima*) to different levels of nitrogen, potassium and magnesium. *Agronomía Colombiana* 31(2), 184–194.

Lobo, M. and Medina, C.I. (2009) Recursos genéticos de pasifloráceas en Colombia. In: Miranda, D., Fischer, G., Carranza, C., Magnitskiy, S., Cassierra, F., Piedrahita, W. and Flores, L.E. (eds) *Cultivo, Poscosecha y Comercialización de las Pasifloráceas en Colombia: Maracuyá, Granadilla, Gulupa y Curuba*. Sociedad Colombiana de Ciencias Hortícolas, Bogotá, Colombia, pp. 7–17.

MAG-INCCA (1991) *El Cultivo del Taxo*. Serie Agrícola, Documento No. 2. Ministerio de Agricultura y Ganadería (MAG), Instituto Nacional de Capacitación Campesina (INCCA), Quito, Ecuador.

Mora, H. and Benavides, M. (2009) Plagas de importancia económica asociadas a las pasifloráceas y su manejo en Colombia. In: Miranda, D., Fischer, G., Carranza, C., Magnitskiy, S.,

Cassierra, F., Piedrahita, W. and Flores, L.E. (eds) *Cultivo, Poscosecha y Comercialización de las Pasifloráceas en Colombia: Maracuyá, Granadilla, Gulupa y Curuba*. Sociedad Colombiana de Ciencias Hortícolas, Bogotá, Colombia, pp. 245–265.

Morton, J. (1987) *Fruits of Warm Climates*. Creative Resource Systems Inc., Winterville, North Carolina, pp. 332–333.

Munier, P. (1961) La curuba, passifloracée frutière cultivée en Colombie. *Fruits* 16(8), 403–404.

Quintero, O.C. (2009) Manejo integrado del cultivo de la curuba (*Passiflora tripartita* var. *mollisima*) (Integrated crop management of banana passionfruit). In: Miranda, D., Fischer, G., Carranza, C., Magnitskiy, S., Cassierra, F., Piedrahita, W. and Flores, L.E. (eds) *Cultivo, Poscosecha y Comercialización de las Pasifloráceas en Colombia: Maracuyá, Granadilla, Gulupa y Curuba*. Sociedad Colombiana de Ciencias Hortícolas, Bogotá, Colombia, pp. 191–209.

Romero Castañeda, R. (1961) *Frutas Silvestres de Colombia, Volumen 1*. Editorial San Juan Eudes, Bogotá, Colombia, pp. 176–179.

Sánchez, I., Angel, F., Fajardo, D., Castilo, M.F., Lobo, M., Tohme, J. and Roca, W. (1998) Uso de marcadores moleculares como base en el mejoramiento genético del género *Passiflora* L. *Proceedings of the Interamerican Society for Tropical Horticulture* 42, 267–271.

Segura, S.D., Coppens d'Eeckenbrugge, G. and Ollitrault, P. (1998) Isozyme variation in five species of *Passiflora* subgenus *Tacsonia* and *P. manicata*. *Proceedings of the Interamerican Society for Tropical Horticulture* 42, 260–266.

Schoeniger, G. (nd.) *Cultivo de la Curuba*. Mimeograph. Universidad de Nariño, Colombia.

Caricaceae

Badillo, V.M. (1967) Acerca de la naturaleza híbrida de *Carica pentagona*, *C. chrysopetala* y *C. fructifragrans*, frutales de Ecuador y Colombia. *Revista de la Facultad de Agronomía, Maracay* 4, 92–103.

Badillo, V.M. (2000) *Carica* L. vs. *Vasconcella* St. Hil. (Caricaceae) con la rehabilitación de este último. *Ernstia* 10(2), 74–79.

Badillo, V.M. (2001) Nota correctiva *Vasconcellea* St. Hill. y no *Vasconcella* (Caricaceae). *Ernstia* 11(1), 75–76.

Oosten, H.J. van (1986) A meadow orchard with *Carica pentagona* (Babaco). *Acta Horticulturae* 160, 341.

Mountain papaya

Alarcón, L., Benavides, L. and Parodi, G. (1997) Efecto de la temperatura y el almacenamiento sobre la germinación de papayuela (*Carica pubescens* Lenne et Koch). *Proceedings of the Interamerican Society for Tropical Horticulture* 41, 242–245.

Bernal, H. and Correa, J. (1990) *Especies Promisorias de los Países de Convenio Andrés Bello*. Tomo IV. Editorial Guadalupe, Bogota, Colombia, pp. 268–300.

CAF (1992) *Manual técnico del cultivo de chamburo*. Centro Agrícola de Quito, Corporación Andina de Fomento, Quito, Ecuador, 36 pp.

Calzada Benza, J. (1993) *143 Frutales Nativos*. Ediciones Universidad Nacional Agraria – La Molina, Lima, Perú, 366 pp.

CTIFL (1992) *Nuevas Especies Frutales*. Ediciones Mundi-Prensa, Madrid, Spain, 164 pp.

León, J. (1964) *Plantas Alimenticias Andinas*. Boletín No. 6. Instituto Interamericano de Ciencias Agrícolas, Zona Andina, Lima, Perú, 112 pp.

León, J. (1987) *Botánica de los Cultivos Tropicales*. Instituto Interamericano de Cooperación para la Agricultura, San José, Costa Rica, p. 380.

Luza, F., Lizana, A.L. and Fichet T. (1990) Comparison of fruit and flower of female and hermaphroditic papaya plants (*Carica pubescens* Lenne & Koch). *Proceedings of the Interamerican Society for Tropical Horticulture* 34, 131–138.

National Research Council (1989) *Lost Crops of the Incas. Little known plants of the Andes with promise for worldwide cultivation*. National Academy Press, Washington, DC.

O'Brien, C.M. and Drew, R.A. (2009) Potential for using *Vasconcellea parviflora* as a bridging species in intergeneric hybridisation between *V. pubescens* and *Carica papaya*. *Australian Journal of Botany* 57, 592–601.

Sánchez-Vega, I. (1994) Andean Fruits. In: Hernando Bermejo, J.E. and León, J. (eds) *Neglected Crops: 1492 from a Different Perspective*. Plant Production and Protection Series No. 26. FAO, Rome, pp. 181–191.

Scheer, C. and Lüdders, P. (1997) Cultivation of *Carica pubescens* in Chile. *Erwerbsobstbau* 39, 54–57.

Scheldeman, X. (2002) Distribution and Potential of Cherimoya (*Annona cherimola* Mill.) and Highland papayas (*Vasconcellea* spp.) in Ecuador. PhD Thesis, Faculty of Agricultural and Applied Biological Sciences, Department Plant Production, Laboratory of Tropical and Subtropical Agronomy and Ethnobotany, Coupure links 653 B-9000 Ghent, Belgium.

Undurraga, P.L. (1989) El cultivo de la papayuela (*Carica candamarcensis* Hook) in Chile. In: Cuellar de García, S. (comp.) *Producción, Transformación y Comercialización de Frutas Tropicales*. Reunión Técnica de la Red Latinoamericana de Agroindustria de Frutas Tropicales, 3, Manizales (Colombia), 19–23 Junio 1989. Federación Nacional de Cafeteros de Colombia, Bogotá (Colombia); FAO, Oficina Regional para América Latina y el Caribe, Santiago (Chile), pp. 143–149.

Babaco

Badillo, V.M. (1967) Acerca de la naturaleza híbrida de *Carica pentagona*, *C. chrysopetala* y *C. fructifragrans*, frutales de Ecuador y Colombia. *Revista de la Facultad de Agronomía, Maracay* 4, 92–103.

Badillo, V.M. (1983) *La Flore d'Equateur*. Vol. 20. Swedish Research Council Pub., Stockholm.

Badillo, V.M. (2000) *Carica* L. vs. *Vasconcellea* St. Hil. (Caricaceae) con la rehabilitación de este último. *Ernstia* 10, 74–79.

Barbeni, M., Guarda, P.A., Villa, M., Cabella, P., Pivetti, F. and Ciaccio, F. (1990) Identification and sensory analysis of volatile constituents of babaco fruit. *Flavour and Fragrance Journal* 5, 27–32.

Bartley, J.P. (1988) Volatile flavor components in the headspace of the babaco fruit. *Journal of Food Science* 53, 138–140.

Bois, D. (1928) *Les Plantes Utiles*. Tome II. Paul Lechevalier, Paris.

Cacioppo, O. (1987) *Il Babaco Manuale Pratico*. REDA Edizioni per l'Agricoltura, Italia.

Castaldo, D., Lo Voi, A., Palmeri, L., Dall'Aglio, G., Carpi, G. and Rillo, L. (1989)Transformatione del babaco in prodotti semi finiti. *Industria Conserve* 64, 117–120.

Cossio, F. (1985) Babaco, frutto prossimo futuro. *Informatore Agrario* XLI (40), 63–67.

Cossio, F. (1988) *Il Babaco*. Edizioni Calderini Edagricole, Bologna, Italy, 60 pp.

Cossio, F. and Bassi, G. (1987) Babaco, dopo il boom tiriamo le somme. *Terra e Vita* (Cagliari), 7, 88–93.

CTIFL (1992) *Nuevas Especies Frutales*. Ediciones Mundi-Prensa, Madrid, Spain, 164 pp.

Delgado-Ortiz, C.H. (1988) *El Cultivo del Babaco* (*Carica pentagona*). Curso de Frutales. Gerencia Regional UNO. Instituto Colombiano Agropecuario, Colombia, pp. 133–139.

Dhuique-Mayer, C., Caro, Y., Pina, M., Ruales, J., Dornier, M. and Graille, J. (2001) Biocatalytic properties of lipase in crude latex from babaco fruit. *Biotechnology Letters* 23, 1021–1024.

Dornier, M., Ruales, J. and Paull, R.E. (2008) Babaco *Vasconcellea pentagona* Heilb. Caricaceae. In: Janick, J. and Paull, R.E. (eds) *Encyclopedia of Fruit and Nuts*. CAB International, Wallingford, UK, pp. 247–249.

Harman, J.E. (1983) Preliminary studies on the postharvest physiology and storage of babaco fruit. *New Zealand Journal of Agricultural Research* 26, 237–243.

Jordan, M. and Piwanski, D. (1997) Regeneration of babaco using leaf explants and shoot-tip culture. *Revista Internacional de Botanica Experimental* 61, 109–115.

Kempler, C., Kabaluk, T. and Nelson, M. (1993) Greenhouse cultivation of babaco: effect of media, container size, stem number, and plant density. *New Zealand J. Crop Hort. Sci.* 21, 273–277.

Little, C. (1982) Babacos – New goldmine or Ecuadorean curiosity? *N.S. Farmer*, February 11, 1982, 16–17.

León, J. (1964) *Plantas Alimenticias Andinas*. Boletín No. 6. Instituto Interamericano de Ciencias Agrícolas, Zona Andina, Lima, Perú, 112 pp.

León, J. (1987) *Botánica de los Cultivos Tropicales*. Instituto Interamericano de Cooperación para la Agricultura, San José, Costa Rica, p. 38.

Mencarelli, F., Fontana, F. and Massantini, R. (1990) Postharvest behaviour and chilling injury in babaco fruit. *Acta Horticulturae* 269, 223–231.

Merino, D. (1989) *El Cultivo del Babaco*. Agroguías Mundi-Prensa, Mundi-Prensa, Madrid, Spain, 86 pp.

Oosten, H.J. van (1986) A meadow orchard with *Carica pentagona* (Babaco). *Acta Horticulturae* 160, 341.

Romero-Rodríguez, M.A., Vázquez-Oderis, M.L., López-Hernández, J. and Simal-Lozano, J. (1994) Composition of babaco, feijoa, passionfruit and tamarillo produced in Galicia (North-West Spain). *Food Chemistry* 49, 23–27.

Scheldeman, X. (2002) Distribution and Potential of Cherimoya (*Annona cherimola* Mill.) and Highland papayas (*Vasconcellea* spp.) in Ecuador. PhD Thesis, Faculty of Agricultural and Applied Biological Sciences, Department Plant Production, Laboratory of Tropical and Subtropical Agronomy and Ethnobotany, Coupure links 653 B-9000 Ghent, Belgium.

Shaw, G.J., Allen, J.M. and Visser, F.R. (1985) Volatile flavor components of babaco fruit (*Carica pentagona* Heilborn). *J. Agric. Food Chem.* 33, 795–797.

Soria, J.V. and Soria, M.V. (1992) Nuevos cultivares híbridos originados de babaco (*Carica pentagona* Heilb.). *Rumipamba* VI(2), 111–118.

Soria, N.I. (1994) Frutales andinos, perspectivas de desarrollo y exportación en el Ecuador. *Proceedings of the Interamerican Society for Tropical Horticulture* 38, 18–22.

Soria, N.I. and Viteri, P.D. (1999) *Guía para el Cultivo de Babaco en el Ecuador*. Proyecto de Fruticultura, INIAP-COSUDE, Quito, Ecuador, 48 pp.

Vega de Rojas, R. and Kitto, S.L. (1991) Regeneration of babaco (*Carica pentagona*) from ovular callus. *J. Amer. Soc. Hort. Sci.* 116, 747–752.

Villareal, L., Dhuique-Mayer, C., Dornier, M., Ruales, J. and Reynes, M. (2003) Evaluation de l'intérêt du babaco (*Carica pentagona* Heilb.). *Fruits* 58, 39–52.

Viteri, P. (1992) *El Cultivo del Babaco en el Ecuador*. Manual N19, Programa de frutales, Instituto Nacional de Investigaciones Agropecuarias INIAP, Quito, Ecuador.

8 Arecaceae

The Arecaceae (Palmae or Palmaceae) are monocots that include perennial lianas, shrubs, and trees, commonly called palms. The family has 202 genera and about 2,600 species divided into subfamilies and tribes. The palms separated early from the other monocot families and developed unique specializations and diversity. Palms are found in warm temperate and subtropical regions though more commonly in tropical areas from deserts to rainforests. About two-thirds of the species are found in tropical forests. The species are readily identified by their compound evergreen leaves at the top of normally unbranched solitary or clustered stems. Palms are cultivated as food and for food products derived from their fruit and hearts of palm. Heart-of-palm or palm heart is the growing apical bud including unexpanded leaves of certain palms, used as a vegetable.

Palms have been important to man since antiquity being mentioned numerous times in the Bible, Quran and other ancient texts. The most frequently mentioned is the date (*Phoenix dactylifera* L.), others include coconut (*Cocos nucifera* L.), rattans (*Calamus* spp.), oil palm (*Elaeis guineensis* Jacq.), betel nut (*Areca catechu* L.) and numerous species used to make toddy (ex. *Borassus flabellifer* L.). Palms are widely used as ornamentals in tropical landscapes given their wide appeal.

Acai or Assai Palm

Important genera and species

Acai, *Euterpe oleracea* Mart. is a slow-growing, clustering palm of the Amazonian rainforest. The genus contains eight species of either solitary or clustering palms found from Central America to northern South America and Trinidad in the West Indies. A single-stemmed (solitary) species *E. precatoria* Mart. has similar floral biology, reproductive system and uses for palm hearts to that of acai. *E. precatoria* occurs almost exclusively in the non-inundated forests of Peru and Brazil and grows to a maximum of 20–22 m in height. Another single-stemmed species *E. edulis* is found in the so-called "Mata Atlantica" of Brazil; it is also used for palm hearts. Because of exploitation for palm hearts its use has been prohibited following the rapid decline in population numbers (Donadio *et al.*, 2002). This same species is also being exploited for the palm hearts in Bolivia with noticeable reductions in wild population numbers although some replanting is being done by the canning industries.

Acai or assai is also known by the common names Pará palm (English), *açaí, açaí do Pará, açaí de touceira, açaí do baixo Amazonas, juçara* (Portuguese), *murrap, naidi* (Colombia), *palmier pinot* (French Guiana), *manaka* (Surinam), *manac* (Trinidad), *manaca,* and *palmiche de Rio Negro* (Spanish). *E. precatoria* Mart. is known as *açai do terra firma* or *açaí do Amazonas* in Brazil and as *huasai* or *chonta* in Peru, and *E. edulis* is called *palmito, palmito dulce, palmitero* in Spanish and *sarova* or *jussara* in Portuguese.

Origin and distribution

Acai grows wild on inundated soils that are often close to the coast or along rivers. In Brazil, it is found in the states of Amazonia, Pará, Maranhão, Amapá, Tocantins and Mato Grosso. Only recently has there been interest in developing plantations of acai in several parts of Brazil because of the nutritive value of the pulp and for the palm hearts (Nogueira *et al.*, 1995). The pulpy fruit mesocarp of *E. oleracea* is an important food resource for native peoples in tropical America. In recent years, the fruit and its juice have taken on a mystique in developed countries that acai possess some "magical" health benefits. Little data is available on production of acai. It is estimated that fruit production in Brazil in 2006 was 500,000 t and juice consumption was 298,000 t with 12,000 t exported (Bichara and Rogez, 2011) and that 150,000 t year^{-1} of its palm hearts are processed, representing more than 90% of the Brazilian palm heart production. About half of the acai comes from Pará State, with much being processed in Belém, the capital about 100 km from the Atlantic on the Pará River, part of the Amazon river system.

Ecology

Soil

The plant prefers sandy soils in marshy areas and is most productive on deep flooded organic acidic soils (Martin *et al.*, 1987; Donadio *et al.*, 2002). It does well in areas subjected to periodic flooding but where water does not stand still (Vargas *et al.*, 1999). Hydromorphic soils with a high content of partially decomposed organic matter are ideal although latosols also allow for good growth. Acai can form extensive stands in swampy forest and along river courses in the rainforest. In the Amazon estuary, it occurs in extensive pure stands. The palm is intolerant of dry situations.

Rainfall

Acai adapts to areas with rainfall greater than 2,000 mm year^{-1}, ideally uninterrupted or with only short dry periods. It also needs a high relative humidity from 80 to 90% (Donadio *et al.*, 2002).

Temperature

The palm grows in hot wet tropical lowlands (Martin *et al.*, 1987; Donadio *et al.*, 2002). It is most adaptable to humid, tropical climates where the mean annual temperatures are around 26°C and rarely drop below 10°C. It does produce well in places with average temperatures of 21°C in Brazil. In Central America, it can be established from 0 to 500 m altitude (Vargas *et al.*, 1999).

Light and photoperiod

The plant needs plenty of sunshine for growth with about 1,500–2,500 h of sunshine a year. It is not sensitive to photoperiod and produces at several latitudes as long as temperature, sunshine and moisture are adequate.

Wind

Wind does not seem to be a problem because of the flexible stems and its well-ramified root system. In its native range, strong winds are generally not a problem.

General characteristics

Plant

This clustering palm produces several stems (5–25) that reach up to 18–20 m (León, 1987); sometimes even 30–35 stems occur under partial shade (Donadio *et al.*, 2002). The stems are slender and gray at maturity, and 7–20 cm in diameter. At the base of the stems, a skirt of visible roots occurs and in swampy conditions, upward growing pneumatophores may be formed. The crown consists of 8–14 pinnated, olive-green leaves, each 2–4 m long (Fig. 8.1). The leaf bases are tightly sheathed and form an attractive, smooth crown shaft that can be green or variously yellow, red or purple. The 50–62 widely spaced leaflets are pendulous in orientation on the rachis, and can be up to 1 m in length with parallel veins (Meerow, 2008).

Flowers, pollination and fruit set

The many-branched axillary flower stems are about 1 m in length, protected by a large bract (prophyll) and emerge from below the crown shaft (Fig. 8.1). The species is monoecious with each inflorescence producing numerous, small, sessile staminate and pistillate flowers. The inflorescence is composed of a central stiff rachis with about 54 lateral branches (rachillae), each of which bears clusters of two lateral staminate flowers and one central pistillate flower. The exception is at the terminus where only staminate flowers occur. The flowers are directly attached to the rachillae. The purplish male flowers are 4.5 × 2.7 mm with six free equally long stamens while the female flowers are purple to light brown and 3.2 × 2.6 mm

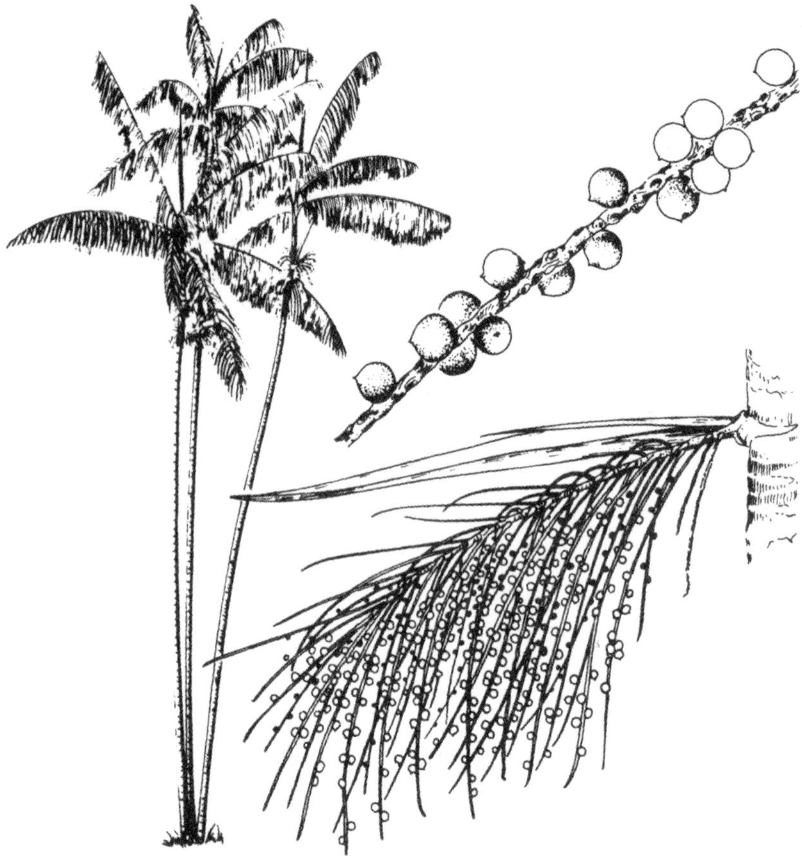

Fig. 8.1. Acai (*Euterpe oleracea*) showing plant, fronds, inflorescence and fruit on rachis (used with permission from León, J. (2000) *Botánica de los Cultivos Tropicales*. Editorial Agroamérica, Instituto Interamericano de Cooperación para la Agricultura (IICA), San José, Costa Rica).

with a trilocular ovary and three stigmas. The sepals are pink. There are from three to eight inflorescences in different stages per adult plant.

The juvenile period is 3–4 years though plants growing inside a tree canopy take longer to flower. This palm can flower and produce continuously under certain climatic conditions such as the coastal area near São Paulo; drought can potentially cause inflorescence abortion.

Pollination and fruit set

Male flowers mature and shed pollen before the female flowers become receptive. A variable amount of self-pollination occurs depending on the synchronization between different inflorescences on the same or different stems. The primary pollinators are small bees and flies, as well as beetles. Seed dispersal over short distances is by rodents while birds and water are involved in long distance dispersal.

Fruit

The fruit is a drupe that comes in clusters of about eight with a diameter of 1.5–2.0 cm, and weighs 0.8–2.3 g. At maturity, the fruit can range in color from green to deep purple. The populations with dark green fruit at maturity are known locally as green or white and there is another type that also matures green called *tinga* in Brazil. Fruiting occurs year round with the peak between July and December. The edible mesocarp is sweet and pulpy comprising about 5–15% of fruit volume. There are 900–1,000 seeds per kg, which are very rich in oils.

Cultivar development

Cytogenetics, genetics and breeding

No formal breeding effort has been carried out with *E. oleracea*, though interspecific hybrids between *E. oleracea* and *E. edulis* have been produced in Brazil (Bovi *et al.*, 1987). The Centro de Pesquisa Agropecuária do Trópico Úmido (CPATU) in Brazil has a collection of around 150 accessions with the main aim being to obtain thicker stems and larger fruit bunches. The Instituto Nacional de Pesquisas da Amazônia (INPA) at Manaus has also some accessions of native types.

Selection and evaluation

There is great variability among this palm with some specimens having green fruit at maturity ("white" type) while others are purple and the latter are preferred for fruit. Both types are used for palm hearts with the purple-fruited types showing great variability in fruit number, stems precocity, stem diameter and height of the first bunch. The yield of pulp is more dependent on the environment than on the genetic make-up. One of the main goals of the selection program is to harvest between January and June when fruit production is at its lowest (Villachica *et al.*, 1996).

Cultivars

No defined cultivars have been developed although there is great variability among the population.

Cultural practices

Propagation

SEXUAL Slow initial growth and considerable mortality of seedlings have been the main problem in successful field establishment (Bovi *et al.*, 1987). Acai seeds are recalcitrant and sensitive to both dehydration and low temperature during storage (Araujo and Silva, 1994). The seeds lose their viability in a month if stored below 15°C or more than 20% seed dehydration occurs (Villachica *et al.*, 1996). If seeds are air-dried for several days after removal from the fruit, then held in plastic bags at room temperature, they can be stored for a few months. Seed stored for 15 days had 33.3% moisture content and 79% germination percentage (Araujo and Silva, 1994) while after 2 months germination is reduced to 28%, and 8% after 7 months.

The ideal is to sow the cleaned seeds as soon as extracted from the fruit into a seedbed or bags at 25–30°C. Seeds set horizontally and buried 3 cm usually germinate in 4–8 weeks and germination may continue for a year (Bovi and de Castro, 1993). To reduce germination variability, seeds can be pre-germinated by holding them in tightly sealed plastic bags with moist washed sand at about 25°C; as soon as germination is observed, the seeds are transplanted to the nursery bags. The nursery bags (20 cm wide × 30 cm height) contain a substrate of organic soil, manure and decomposed sawdust. In about 6 months, plants will be about 30 cm tall and ready to be taken to the field. Chup (1999) found beneficial effects on acai growth by inoculating seedlings with arbuscular mycorrhizal fungi.

ASEXUAL Asexual propagation using the small sprouts arising from the clump can be used but it is labor demanding and not very practical, especially for large-scale propagation. Successful somatic embryogenesis is also possible and has potential utility in the rapid propagation of selected clones.

Orchard establishment

The land is prepared as for any orchard crop. If the land was previously used for grazing or for growing crops, plowing and harrowing are necessary. Converting forest land requires the removal of large trees and shrubs leaving the lower vegetation to prevent erosion. Recommendations include planting this clumping palm on a 5 by 5 m square giving about 400 clumps per ha when fruit production is the objective. For palm heart production planting distances used are 2 × 2 m. Holes 40 × 40 × 40 cm are normally dug and 5 l of decomposed manure added to each hole at planting time.

Pruning

For fruit production, each clump should be left with three to four stems, while for palm hearts four to five stems are left. Palm heart stems are harvested when they reach an adequate size with more than one stem removed per plant (clump) during each year. Several stems will arise at different times and palm hearts can be harvested by selecting some for harvest and removing older stems to manage the clump size.

Irrigation and fertilization

Little data is available on mineral nutrition and fertilization. Villachica *et al.* (1996) recommend 50 g ammonium sulphate, 50 g triple superphosphate and 50 g potassium chloride applied during the first 2 years. From the third year on these amounts are doubled and 5 l of decomposed manure is applied per plant clump every second year.

Pest management

No major pests or diseases of acai have been reported. Zorzenon and Bergmann (1995) described fruit and seed predation by *Xyleborus ferrugineus* (Coleoptera: Scolytidae) with seed germination reduced by 80%. An aphid *Cerataphis lataniae* damages leaves, young stems and inflorescences, and the moth *Brassolis astyra* damages the leaves, while the beetle *Cocotrypes* sp. attacks the seeds after the fruit fall. Pest control is very difficult due to the height of the palm (Villachica *et al.*, 1996).

Weed control

As most palms are susceptible to many systemic herbicides, contact herbicides can be used around the clump along with manual weeding. Between the rows, mechanical or manual weeding can be done. The use of cover crops that fix nitrogen can be useful.

Orchard protection

Wind protection does not seem to be needed in current planting. Windbreaks may be needed with strong prevailing winds.

Harvesting and postharvest handling

Yields

One stem on average produces four to eight inflorescences annually beginning in the third year, each yielding approximately 4 kg of fruit. One stem can produce 16–32 kg of fruit, with a mean of 24 kg $year^{-1}$ (Bovi and de Castro, 1993). Villachica *et al.* (1996) indicate that production starts 4 years after transplanting with each stem producing 15 kg of fruit. This production can result in 12–20 t ha^{-1} $year^{-1}$ in non-flooded areas and up to 25 t in alluvial flooded soils.

Harvesting

Inflorescences are at different stages of development on a single stem, from flower stems enclosed in their prophylls to clusters of ripe fruit. Regular harvesting is required to minimize fruit loss. Fruit is ready for harvest of the more common black variety when it turns an almost purple/violet and is covered by waxy-white coloration (Bichara and Rogez, 2011; Rogez *et al.*, 2011). This maturity stage is called *tuira* in Brazil. In its native area, harvesting is more intense from September to December with some fruit available for harvest year round. People harvest the fruit by climbing the palms, cutting the inflorescences and lowering them gently to the ground using a rope. The fruit are detached and put in baskets lined with leaves and taken to the point of sale or processing, often by small boats, where the pulp is extracted mechanically or by hand.

Postharvest treatment

The fruit is non-climacteric and the ideal temperature for transport and storage is 10–15°C (Bichara and Rogez, 2011). In the very warm region where acai is produced, no cooling facilities are readily available to remove field heat. The fruit has a high bacterial load of lactic and acetic acid bacteria (Aguiar *et al.*, 2013). The use of closed polystyrene boxes during transport at about 30°C leads to significant nutritional and functional changes compared to when transported in open baskets. In closed boxes, fermentation was detected in 3 h and it is recommended to transport in more aerobic open baskets even though weight loss is four times greater in 24 h at 1.6%. Ideally, the fruit should be processed within 12 h to avoid fermentation.

After being received at the processing site, acai is washed, preferably in chlorinated water, and the fruit pulp is separated either manually or mechanically from the seed. In both cases, before pulp separation from the seed, the fruit are held in water at about 25°C for about an hour to induce softening. Using warm water at 35–40°C soaking time can be reduced to 20 min. For manual processing, the fruit is squeezed by hand as water is added to make the separated pulp liquid; afterwards the seeds are washed to remove any residues. Machines with blades that promote rubbing of the fruit have been developed for mechanical pulp separation with water added during the process. Sometimes the pulp is pasteurized at 90°C for 10 min, but in spite of pasteurization's sanitary benefits, the process lowers the anthocyanin content. The juice can be treated at 80–85°C for 10–30 s and stored at −20°C for several months.

Utilization

Acai palm hearts that are collected from wild palms are by far the most important in Brazil in both the fresh market and in industrial processing in jars and cans. This importance is in spite of the expansion in recent years of peach palm plantings, which is also a multi-stemmed perennial. Peach palm can be harvested earlier and easier than acai since it has shorter stems but it has to be cultivated for that purpose.

Acai pulp is an extremely important indigenous resource for people in areas bordering the Amazon estuary (Strudwick and Sobel, 1988). It has been the focus of considerabe research oriented towards its commercial exploitation (Calzavara, 1972; Urdaneta, 1981; Anderson, 1988) of the highly nutritious pulp; locally the paste-like pulp is known as *acai*. A drink, locally called *vinho de açaí* though not alcoholic, is not as nutritious. The pulp has a high lipid content and caloric value and is rich in calcium, iron and Vitamin B1 (Table 8.1). It is also very rich in polyphenols, especially flavonoids, which explains its popularity in the developed world markets where it is sold in health food stores, and used as an ingredient in so-called "energy" beverages.

The extracted pulp is processed into beverages, ice cream and pastries. These products are sold in different concentrations at local and regional markets by the businesses that process the fruit. The businesses are known in Brazil as *açaílandias*. Mixed with cassava flour, rice, shrimp or meat, it is consumed in great quantities by the populations of the lower Amazon River. The flavor is somewhat nutty with a metallic aftertaste. The texture is creamy to slightly oily (Strudwick and Sobel, 1988).

Acai pulp is highly perishable and has to be consumed within 24 h unless refrigerated. This perishability has limited its exportation. The gathering, transportation and processing of the pulp frequently do not use a high level of sanitation or technology. Dehydrated pulp has a storage life of more than 100 days (Melo *et al.*, 1988). In the dehydrated form, as well as frozen, it has been exported to the USA and Europe in recent years.

Table 8.1. Mean chemical composition per 100 g of acai, peach palm mesocarp and heart, buriti mesocarp and ungurahui mesocarp.

Proximate	Acai[a]	Peach palm mesocarp[b]*	Peach palm heart[b]**	Buriti[c]	Ungurahui[d]
Calories (kcal)	62	273.5	47.6	283.0	317.2
Proteins (g)	0.8	3.3	1.5	3.0	3.3
Fats (g)	0.7	6	1.3	21.1	12.8
Saturated		2.2	0.73		
Monounsaturated		3.3	0.35		
Polyunsaturated		0.5	0.22		
Carbohydrates (g)	14.9	34.9	5.2	18.1	47.2
Fiber (g)	1.7	2.0	0.9	10.4	31.5
Ash (g)	0.4	0.8	NA	0.9	1.1
Minerals					
Calcium (mg)	6	18.9	42.4	74.0	
Iron (mg)	2.0–2.6	0.59	0.23	0.7	
Magnesium (mg)	9	17.1	3.4		
Phosphorus (mg)	20	NA	NA	27.0	
Potassium (mg)	128	240.5	193.6		
Sodium (mg)	Trace	4.3	0.1		
Vitamins					
Vitamin A, Carotene (mg)	1	1.1	NA	1062.0***	1.3
Vitamin C, A. Acid (mg)	10.3	18.7	3.2	26.0	
Thiamin B1 (mg)		0.045	NA	0.12	
Riboflavin B2 (mg)	0.04	0.135	NA	0.17	
Niacin (mg)	Trace	0.81	NA	0.30	

*Mesocarp of 3 fruits; **a 9 cm long by 2 cm diameter section; ***as mg retinol.
NA, not available.
[a]Anonymous, 2011.
[b]Yuyama *et al.*, 2003.
[c,d]Flores-Paytán, 1997.

Peach Palm

The peach palm, *Bactris gasipaes* Kunth, is the only domesticated palm in tropical America. Other names for the species were *Guilielma gasipaes* (HBK.) Bailey and *B. speciosa* (Mart.) Karst. It was probably first used for its wood and was fully domesticated for its starchy-oily fruit, and is now most important for its heart-of-palm (Clement, 2008). In English, it is known as peach palm, as *pejiballe* and *pejivalle* in Spanish, and *pupunha* in Portuguese (Brazil). It is also known as *tembe* (Bolivia), *chontaduro* (Columbia, Ecuador, Peru), *pejibaye* (Costa Rica, Nicaragua), *pibá* (Panama), *pijuayo*, *chonta* (Peru), *cachpay*, *chontadura*, *pirijao*, *chontaduro colorado*, *pichiguao* (Colombia) and *pijiguao* (Venezuela). The name in English and other European countries is a misnomer, as the fruit is more like cassava than a juicy peach (Clement *et al.*, 2004). The name peach palm was coined by Prussian naturalist and explorer Alexander van Humboldt (1769–1859) and is derived from the pleasant fruity aroma of the fermented pulp. The heart-of-palm is a vegetable composed of the tender unexpanded leaves, stem and apex of the palm's crown.

Origin and distribution

The origin of peach palm has been debated extensively and inconclusively for more than a century (Clement, 2008). Recent morpho-anatomical evidence suggested that peach palm's origin will probably be found in south-western Amazonia (Ferreira, 1999), in what is now northern Bolivia, south-eastern Peru and western Brazil. Recent allozyme and molecular evidence also pointed towards that same region (Rojas-Vargas *et al.*, 1999; Rodrigues *et al.*, 2004) with two dispersions from that region. One dispersion from this region was to the north-east and the other to the north-west, resulting in the complex of landraces in central and western Amazonia, and Central America up to Nicaragua. All cultivated populations and landraces are now *B. gasipaes* Kunth var. *gasipaes* and all wild populations are now *B. gasipaes* var. *chichagui* (H. Karsten) Henderson.

During the century immediately following European conquest, the fruit was reported to be used principally as a cooked starchy staple, fermented to make a drink, or ground and dried into flour. The wood was important for tools and weapons because of its straight grain and durability (the modern Columbian name, chontaduro, means the "tough palm"). Throughout western Amazonia and extending up to Costa Rica, the peach palm appears to have been a staple starch crop of Amerindians, perhaps as important as maize (*Zea mays* L.) and cassava in much of this region.

Peach palm is still almost exclusively a neotropical crop, with only experimental areas in Africa, Asia and Oceania. As a fruit crop, it is grown almost exclusively by smallholders, in home gardens and in slash-and-burn swiddens, with a few small orchards near major consumption areas (Clement, 2008). Hence, all production data are estimates. Brazil, Colombia, Bolivia, Ecuador, Peru and Venezuela produce moderate amounts of peach palm in their Amazonian lowlands, while Venezuela produces it in the Orinoco River basin lowlands, and Ecuador in its Pacific coastal lowlands. French Guiana, Guyana, Nicaragua, Panama and Surinam are minor producers. Extrapolation from this scant data set suggests that total production is probably about 120,000 t, of which only about 50% is commercialized as fresh fruit, while the other 50% is used for subsistence, either directly or as animal feed, or is wasted (Clement *et al.*, 2004). A typical fresh bunch weighs 2–5 kg, and is worth US\$0.50–1.00 at the farm gate and US\$1.00–3.00 in the market.

In contrast to the fruit, the heart-of-palm is grown in high-density (> 5,000 plants ha^{-1}), high-input commercial monocultures and some production statistics exist. Brazil had approximately 20,000 ha in production in 2002 and another 3,000–5,000 ha recently planted. Costa Rica had 8,000 ha in production in 2002, down from 12,500 in 1998 due to severe competition from Ecuador, which had 10,000 ha in production in 2002.

Ecology

Wild peach palm (var. *chichagui*) occurs in transitional natural ecosystems and where natural disturbances are frequent, principally along river beds and in primary forest gaps, while cultivated peach palm (var. *gasipaes*) only occurs in ecosystems created by humans. Extensive natural stands of wild peach palm have not been reported (Clement, 2008).

Soils

The peach palm is most productive on relatively deep, fertile, well-drained soils with a pH of 5.0–6.0. The palm is often reputed to be well adapted to tropical soils with few nutrients (NAS, 1975) based upon its use in traditional agroecosystems in the American tropics.

This palm produces small and overall low total yield on low-fertility soils, including highly eroded laterites and in slash-and-burn agricultural systems of primary or secondary forest. The burn releases potassium, calcium and magnesium that neutralize the acidity and aluminum toxicity. Fruit production in this system decreases in the long term without additional lime and nutrient inputs (Clement, 2008). The slash-and-burn system also involves the use of transplants with developed root systems and microbial associations (Fernandes and Sanford, 1995). Symbiotic associations with vesicular-arbuscular mycorrhizae improve growth and are often essential for normal development.

Rainfall

The palm prefers an abundant and well-distributed rainfall (2,000–5,000 mm year^{-1}). It can grow in places with up to 7,000 mm rain but needs good drainage. The palm tolerates relatively short dry seasons (3–4 months) though longer droughts reduce yields.

Temperature

Peach palm grows well at low to middle altitudes (< 1,000 m above sea level) and average temperatures between 24 and 28°C. It can be found at 1,300–1,800 m but growth is slow and only one harvest of fruit is obtained per year.

Light and photoperiod

It should be grown in full sun as it does not tolerate shade, except at the very young stages in the nursery.

Wind

Being a sturdy palm wind is not a problem, especially when producing hearts since the suckers are harvested when they are 1.5–2.0 m.

General characteristics

Plant

The peach palm is a multi-stemmed palm that may attain 20 m in height. Stem diameter varies from 15 to 30 cm and internode length varies from 2 to 30 cm and becomes shorter with age after 5 years. The internodes are armed with numerous black, brittle spines (Fig. 8.2), although spineless types do occur. The stem is topped by a crown of 15–25 pinnate fronds more than 2 m in length,

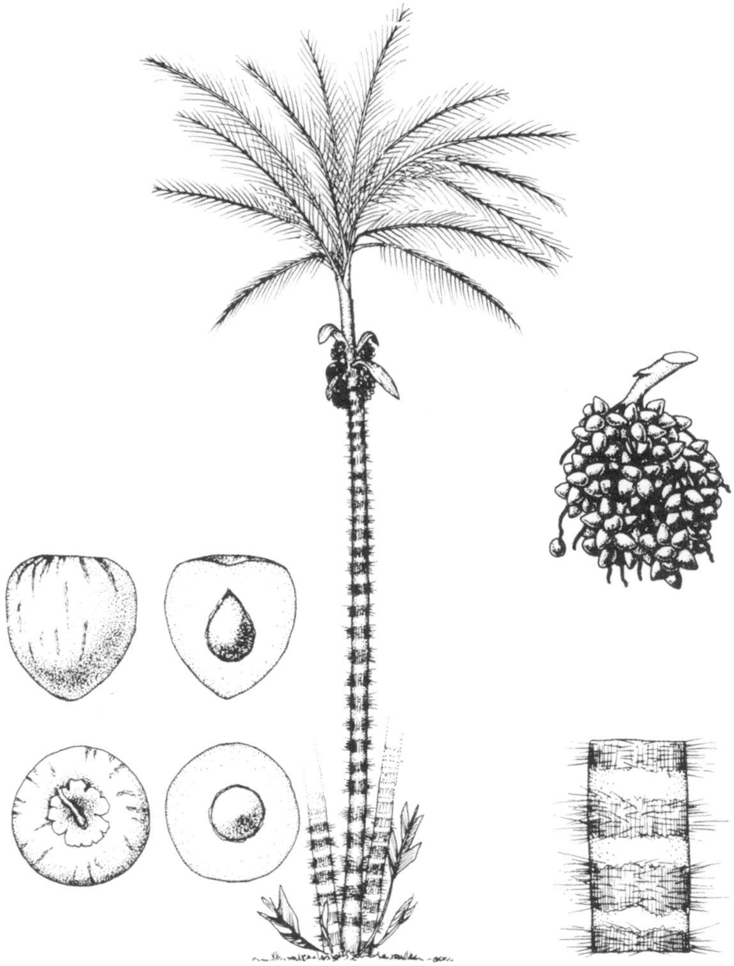

Fig. 8.2. The peach palm (*Bactris gasipaes*) showing clustered stem, fruit bunch and individual fruit and in cross-section, and spines in stem internode (used with permission from FAO Plant Production and Protection Series No. 26, Neglected Crops: 1492 from a different perspective, 1994 (A001/2013)).

with the leaflets inserted at different angles. Each frond has more than 200 leaflets and the petiole and rachis normally have spines (Romero Castañeda, 1961).

Flowers

The inflorescences appear among the axils of the senescent fronds protected by a spiny spathe; they are covered with two bracts, the outer short and thick, the inner surrounding the inflorescence until it matures. The inflorescence rachis is branched with thousands of male flowers intermixed with a few hundred slightly larger female flowers.

Phenology varies both within and among countries, and the environmental events that trigger flowering are not yet clear. In the central Brazilian Amazon, the main flowering season extends from the mid-dry season (August–September) to the beginning of the rainy season (November), and fruit ripen between late December and late March. In Costa Rica's Atlantic zone, the main flowering season extends from April to July and the fruiting season from September to December; this region has a much less pronounced dry season than central Amazonia. When the dry season is less pronounced, a second flowering period may occur in plants with good nutritional status. When this happens, the harvest seasons are separated by about 6 months. Costa Rican farmers have taken advantage of microclimatic, soil and altitudinal variability to supply their major San José market year round, effectively exploiting phenological variability (Clement *et al.*, 2004).

Pollination and fruit set

Both female and male flower opening occurs between 5 and 6 p.m. in a 3-day cycle for each inflorescence. The female flowers are fertile from spathe opening and are receptive for 24 h. Anthesis of the male flowers on the same inflorescence occurs on the second day of the 3-day cycle, 24 h after the females. Male flowers release their pollen in 15–30 minutes, showering the inflorescence, and then they abscise. Gravity also carries pollen to the inflorescences below and wind to adjacent plants on which female flowers may be receptive. The insects with their pollen load search for a recently opened inflorescence, attracted by a musky scent produced by glands on the male flowers. The reproductive biology of peach palm suggests a tight coevolutionary history with very small curculionid beetles, thousands of which are attracted by the musk to a single inflorescence. Where peach palm has been introduced recently, the lack of these curculionid beetles may severely limit fruit set, since the plant is basically allogamous.

Self-pollination is thought to be regulated by a genetic incompatibility mechanism, but there is considerable variation in self-fertility in peach palm. Self-pollination may occur: (i) within the same inflorescence; (ii) between inflorescences of the same stem; or (iii) between inflorescences on different stems of the same plant. The last event is probably much more common than the other two events, although the first event may be common at the beginning and end of the season when sufficient cross-pollen is not available (Clement, 2008).

Following pollination, the bunch contains between 50 and 1,000 fruit and weighs between 1 and 25 kg. Numerous factors cause premature fruit drop including poor pollination, poor plant nutrition, drought, crowding, insects and diseases.

Fruit

The fruit is a drupe with a fibrous red, orange or yellow exocarp, a starchy and oily mesocarp and a dark and hard endocarp with a fibrous/oily white kernel. Individual fruit of var. *gasipaes* weigh between 10 and 300 g, varying according to landrace. The seed weight is between 1 and 4 g, while the fruit of var. *chichagui* weigh between 0.5 and 5 g, and seeds between 0.3 and 1 g. Some varieties have very small or no seeds.

Fruit mature in 3–4 months. During the first 2 months of growth, the fruit expands in size and its composition is rich in starches and some protein. During the third month, starches are metabolized to produce oils and carotenes, and the mesocarp and exocarp gain color.

Cultivar development

Cytogenetics, genetics and breeding

The chromosome number is $2n = 28$. A complex hierarchy of landraces has been proposed on morphometric grounds (Rodrigues *et al.*, 2004). This analysis also suggests that the Andes were not a barrier to dispersion and the eastern/western distinction of landraces may not be important. This species is the result of the independent domestication of several wild populations leading to great local diversity and isolated populations. The recent exchange of germplasm has resulted in a considerable greater diversity, obscuring the characteristics of the local varieties.

Selection and evaluation

In situ collections can be found in Costa Rica, Brazil, Colombia, Peru and Ecuador with local and international funding. Spineless mutants have been selected in some areas.

Cultivars

There are spiny and spineless types but there is not much information about the superiority of either. No named cultivars have been released. The breeding effort aims at general population improvement to maintain the variability that helps control pests and diseases. This strategy is perfect for heart-of-palm production that is now based principally on the Pampa Hermosa landrace (Yurimaguas, Peru) because it is spineless, has rapid growth and has good quality (Mora Urpí *et al.*, 1999). This strategy is also desirable for the fruit market, since consumers desire uniformly high quality fruit that are not as common in open-pollinated landrace populations. The Benjamin Constant (BC) (Putumayo landrace) has a tendency for spineless plants to be superior in heart weight at one site and for spiny plants at another site. The BC spineless populations generally have more offshoots at both sites but superiority in terms of growth rates, heart weight and total edible weight depends upon location and seldom exceeds 10%. The use of spineless or spiny types depends more on the availability of germplasm and comparative costs, with spineless types having lower labor costs (Clement, 1996) and causing fewer labor accidents.

Cultural practices

Propagation

The peach palm is propagated by seed, as vegetative propagation of offshoots is difficult and a commercial tissue culture protocol has not yet been developed. The seed is considered to be recalcitrant. Seeds obtained from healthy productive trees at the beginning of the harvest season have greatest germination success

(> 80%). Seed preparation includes careful pulp removal and seed cleaning, with seeds often being left 24–48 h in water to be able to clean all pulp residues, sowing in appropriate irrigated substrates, moderate shade (75% of full sun) and frequent inspection for pests and diseases. Seeds should be sown 2–3 cm deep in an organic-matter rich, well-fertilized substrate. Fresh seeds take 30–90 days to germinate, depending upon temperature and humidity, and a transplant ready seedling takes another 4–6 months to produce.

Germination variability can be reduced by soaking the seeds for 24 h in water, then mixing them with moist moss or similar media in a plastic bag held in the shade until the first germination signs appear. The pre-germinated seeds can then be transplanted to nursery bags (20 × 30 cm) and kept until ready for transplanting. An alternative is to let the washed seeds dry out superficially and then put them in a double plastic bag with no substrate and leave them in the shade until germination starts. In both cases, there is no need to take care of seedbed preparation or maintenance (watering, weeding) in the bags. After transplanting the plants are left in the bags until they are 50–60 cm tall then they are taken to the field. Another approach to reduce costs is to transplant to seedbeds and then take bare root plants to the field.

Orchard establishment

Planting density, plant management and fertilization vary significantly between the fruit crop and the heart-of-palm crop. The fruit crop is generally grown in agroforestry systems of varying diversity and less frequently in monoculture. In monoculture, plantings should not exceed 333 plants per ha to avoid excessive shade (Villachica *et al.*, 1996). The heart-of-palm crop is grown at higher density (> 5,000 plants ha^{-1}) to as dense as 1 × 2 m. The plants should be put in holes that are larger than the nursery bags. The addition of organic matter and a chemical fertilizer to the planting hole at transplanting is recommended. If bare rooted plants are used they have to be transplanted into moist soil during an overcast day and after the start of the rainy season. The first heart-of-palm harvest is obtained within 18–24 months and the clumps are maintained by periodic harvesting of the larger offshoots if adequately fertilized and weeded.

Unlike for heart-of-palm production, for fruit the stem must fully develop for flowering to occur. The root system is therefore bigger and, because of this, greater drainage depth is required. The planting densities are 5 × 5 m and 4 × 4 × 8 m. The rectangular or triangular arrangements used allow sufficient light into the orchard and the wider spacings allow more intercropping during the early years. In most cases, multi-stemmed plants are left with one to three stems that are replaced after they become too old or tall in 10–15 years.

Harvesting leaves 90% of the biomass to mulch the field and many agro-industries compost the other 10% for use in the nursery or return it to the field, all of which makes the crop very sustainable. Both organic and conventional (mineral fertilizer, but almost no pesticides) heart-of-palm are available, although little research has been published on the organic alternative.

Irrigation

Commercial plantations are normally made in places with heavy rainfall, so that no irrigation is needed, although in some cases when the dry season is too long it may be desirable.

Pruning

In fruit production, offshoot management is critical to reduce within clump competition for light and nutrients, while maintaining enough offshoots to replace the fruiting stem when they grow too tall to be economically harvested (> 10 m, 10–15 years or less). To prepare for its renewal, a shoot is allowed to grow for 12–18 months before the removal of the taller stems after harvest. The new shoot will be produced in the following harvest period. In other cases, two to three stems are left in each clump and they are renewed in a similar way.

For palm heart production, the plant normally produces seven to ten shoots, of which only the largest are harvested about every 3–4 months. In some places the weakest shoots or those that are squeezed between other shoots are eliminated, but in other cases no pruning is done and only old leaves and weak shoots are removed.

Fertilization

Peach palm requires appropriate fertilization to remain productive. Dolomitic lime is recommended for managing soil acidity in tropical America's typically acid oxisols and ultisols, especially when aluminum concentration is high. Two tons per ha during field preparation and 1 t ha^{-1} at 3-year intervals for maintenance (Bovi, 1998) provide sufficient magnesium and calcium, while making phosphorous more available. Juvenile plants require abundant nitrogen and moderate phosphorous (90–120 kg N; 45 kg P; 80–90 kg K; all ha^{-1}), while fruiting plants require more potassium and nitrogen, with moderate phosphorous (140–190 kg N ha^{-1}; 90 kg P ha^{-1}; 150–180 kg K ha^{-1}).

For palm heart, the recommendation is to apply 35–40 g K_2O and 200 g N or 500 g of a 17-11-22-2.5 (Mg) formula per plant per year; these amounts can increase gradually six to ten times during the next 6–8 years. Nitrogen should be applied as ammonia since nitrates leach rapidly. Phosphorus is applied as triple superphosphate once a year, if possible in holes around the plant. Potassium should be applied as sulfate to help insure a constant sulfur presence in the soil. In most tropical soils, micronutrient deficiencies are common, but poorly studied with peach palm (Clement, 2008). Hence, animal manure is strongly recommended and mineral fertilization can be reduced proportionately. Leguminous groundcover crops are also highly recommended, as they suppress weeds, and provide organic matter and nitrogen to the system.

Pest management

DISEASES Anthracnose (*Colletotrichum*) often indicates inadequate phosphorous fertilization. Poor nursery management allows damping off (*Fusarium*) and other fungal diseases to become significant. The most serious fruit disease is *Monilia* sp. that is more prevalent during humid harvest seasons, especially if drainage is poor or there is very dense planting. Other fungal problems can include *Graphium* that causes fruit rotting and *Phytophthora* that attacks the growing point of older stems. The growing point can also be infected by the bacteria *Erwinia chrysanthemi*, which can be a problem where the drainage is poor or the plant is shaded in heart-of-palm stands. Most peach palm diseases can be managed with appropriate fertilization and field practices, and pesticides are not normally used.

INSECTS Beetles (Coleoptera) may be locally important as fruit or seed pests, but only in the Colombian Pacific have they seriously affected fruit yields. A seed-boring beetle has occasionally been reported in the south-west Amazon. The beetle *Rynchophorus palmarum* sometimes attacks older stems but more frequently stems that have been cut for heart extraction or the rhizomes under the soil surface of young plants while *Metamasius hemipterus* can be a problem in fruit production oriented plantings since it attacks the base of the inflorescence rachis and occasionally fresh cut stems in heart-producing plantations. Foliage mites (*Retracus johnstoni*) are indicative of poor plant nutrition and occasionally require chemical control on fruiting plants, but not in heart-of-palm orchards, where rapid growth and harvest keep their populations under control.

PESTS Parrots are the most destructive fruits pests. Often flocks of birds damage green fruit. Gophers, pigs, rabbits and rats are pests in some areas, and all can be controlled with fencing or poison baits.

Weed control

The peach palm does not like competition especially from grasses, therefore weeding is necessary. The use of herbicides tends to delay the development and reduce yields. In Hawaii, the use of woven black polypropylene mats resulted in effective weed control without negatively affecting yields (DeFrank and Clement, 1995).

Orchard protection

Nothing special since the plants are very sturdy and winds are normally not much of a problem in the places where it is grown, otherwise barriers have to be installed.

Harvesting and postharvest handling

Yields

Fruit production can be 3–5 t ha^{-1} year^{-1} in the fourth or fifth year and stabilize at 10–30 t ha^{-1} year^{-1} after 6–7 years, depending on density and management. For heart-of-palm the estimate is to harvest 10,000–14,000 hearts ha^{-1} year^{-1} with each heart weighing 100–200 g, resulting in 1.3–1.8 t ha^{-1} year^{-1}. About 46 hearts are needed to fill a carton of cans or jars so that approximately 50–60 ha of plantation are needed to fill a monthly container load of 1,500 cartons.

Harvesting

When the exocarp has 50% of its final color, the seed is mature and the fruit can be harvested. Full flavor and color come a week or so later, though the full flavor may be too strong for some consumers. Hence, fruit tend to be harvested and commercialized before full ripeness. Fruit from spineless peach palm is typically collected by climbing the stem and lowering the fruit bunches to the ground with a rope or dropping them into a net. Most peach palms, however, have very spiny stems so they are difficult to climb. Farmers use poles with a hook or curved knife at the end to detach the bunch, which is caught in a net or on a foam cushion. Harvesting from the ground is faster and safer than climbing the stem, but it causes more damage to the fruit.

The fruiting season typically extends over a 2–4-month period (Clement, 2008). The first harvest of the season normally yields the largest and best quality fruit, after which fruit quality gradually deteriorates due to increasing insect damage and fungal infection, favored by both presence of fruit and by gradual depletion of physiological reserves.

Heart-of-palm offshoots are harvested when they reach a commercial dimension that depends on market demands. Normally offshoots are harvested when they attain a diameter of 10 cm or more, 20–30 cm from the ground, and the hearts are about 2–3 cm in diameter. The first harvest is normally at 18 months after planting, although some precocious selections can start as soon as a year after planting. Harvesting is then continuous with three to four harvests of each individual. Normally the upper 80 cm is cut off and the leaves removed with a machete including the most external leaf sheaths leaving only two or three to protect the heart. In Brazil, there is a demand for three heart-of-palm dimensions: thin (1.5–2.5 cm) hearts to be canned for the export markets; medium (2–4 cm) hearts for both the bottled and fresh markets in Brazil; and thick (3–6 cm) for the Brazilian barbecued meat restaurants (*Churrascarias*) where they are served as garnish (Clement, 2008).

Postharvest treatment

After harvest, the fruit are very perishable due to their high fat content and must be handled carefully and marketed quickly (Clement, 2008). A ripe fresh bunch can be maintained in good condition without refrigeration for only 4–6 days if kept in the shade and well ventilated otherwise it will start spoiling after 2 days. With refrigeration (20°C, 70% relative humidity) and waxing, storage can be extended to 8 days, but few merchants have the necessary capital and facilities. The shelf life of fresh fruit can be extended by collecting well-developed fruit that are just starting to change from green. Frozen, dried or canned fruit can be conserved for months though with loss of final flavor and color.

Fresh fruit are commonly sold by the bunch, or they are minimally processed and packaged, especially in Costa Rica. The processing involves only removal of fruit from the bunch, washing, waxing, sorting and grading, and packaging in net bags of specified weight. Cooked fruit is handled like fresh fruit. In many countries, fruit are sold at market still attached to the bunch. Canned fruit is also produced in Costa Rica.

If fruit are destined for flour, they should be processed on the day of harvest or on the following day. The entire bunch is cooked to facilitate later removal of the fruit, to denature potential toxins, and improve starch quality. The whole fruit are cut into small pieces, the seed removed, and the pulp and peel dried before grinding and packaging.

The hearts should be processed rapidly in less than 48 h, ideally as soon as they reach the factory. After 2 days they start shriveling due to water loss.

Utilization

In pre-Columbian times the fruit was the most important product and used either cooked (boiled in water) or as a slightly fermented drink. In both forms, it constituted the basic food during the harvesting period for the indigenous communities (Clement, 2008). It was preserved in trenches between seasons and was dried, exposed it to heat and smoked. To be eaten, it just had to be boiled in water. It was also eaten in the form of tortillas made from its dough. The oil, which separates out when the fruit is boiled, was occasionally used for cooking other foods. Prolonged fermentation – lasting 1 week – enabled the alcoholic drink *chicha* to be made. The fruit was the basic source of energy, replacing the functions and uses of grain in other cultures.

The fruit is energy rich, due both to starches and to oils, while the heart-of-palm is essentially a dietary product (Table 8.1). The relative proportions of starch to oil vary inversely with the extent of domestication; the wild-type fruit being rich in oils while the domesticated types are rich in starches. Fruit protein quality is not exceptionally high, but fruit mesocarp oil is rich in mono-unsaturated oleic acid (Yuyama *et al.*, 2003). The fruit contain two anti-nutritional factors, a trypsin inhibitor and calcium oxalate crystals. The fruit chemical composition is similar to that of maize and better than that of potato or cassava on a fresh weight basis.

The heart-of-palm is normally used for canning and in this process the upper portion harvested contains the apex and the very young leaves. The basal portion is sold for fresh consumption as a salad and has the same flavor as the heart but a different texture.

The trunk of adult stems has a fibrous ring that has great strength and elasticity. The wood is used to make weapons – bows, arrows and spears – as well as in construction and flooring and handicrafts. The sap from apex and immature and unexpanded fronds, either unfermented or fermented, was used to prepare nutritional and intoxicating drinks. The young inflorescences were also eaten roasted, without opening the protective spathe. Infusions of the roots were used in medicine as a vermicide.

Wine Palm, Buriti

Mauritia flexuosa L. f. for some authors has a related species, *M. vinifera* Mart. (Martin *et al.*, 1987), while other authors include a third species, *M. martina*

Spruce (Cavalcante, 1974). Apparently none of these is a valid name and *M. flexuosa* includes all three as ecological variants (Meerow, 2008).

This palm is called wine palm and mirity palm in English, while in Spanish it is: *aguaje, achual, caranday-guazu, auashi, bimón, buritisol, maritî, muritî, moriti, ideuí, cananguche, chomiya, moriche, morete*; in Portuguese: *burití, burití do brejo, coqueiro burití, buritizeiro, murití, muritim, muritizeiro, palmeira dos brejos, mirity, carandaguaçu, carandaiguaçu*; in Guyana *ita* and in French *palmier báche*.

Origin and distribution

Wine palm is native to Amazonia and most probably originated in the basins of the Marañón, Huallaga and Ucayali rivers in Peru, although it is found in the Amazonian areas of Brazil, Colombia, Ecuador and Guyana and Venezuela (Flores-Paytán, 1997). Most of its production comes from wild plants that grow in swampy areas called *aguajales* in Peru, where practically no other crop would grow except some palms like *Euterpe oleracea* or those of the *Oenocarpus-Jessenia* complex that tolerate flooded areas and can be found growing in association with this palm. In other parts of the world, limited introductions have been made mainly to botanical gardens, experiment stations or plant collections.

Ecology

Soil

It grows in humid lowland soils of Amazonia that are permanently or seasonally flooded (León, 1987), preferably on swampy acid histosols. It adapts to drier areas with proper or bad drainage that are sandy or humic hydromorphic with high content of organic matter (Flores-Paytán, 1997).

Rainfall

The palm grows naturally in places with an average minimum rainfall of 900–1,000 to a maximum of 3,500–4,000 mm. It can grow in drier areas with irrigation but it is normally not cultivated because it is not economic.

Temperature

The species is found in places with annual mean maximum temperatures of 25°C and mean minimum temperatures of 17°C, normally at altitudes of 50–900 m.

Light and photoperiod

Buriti palm needs full sunshine for proper development and production. Only at young stages does the plant grow under partial shade; in association with other species, the canopy of this palm is normally the tallest.

Wind

Normally in its habitat, winds are not frequent and the strong stem and its good anchorage prevent any wind problems.

General characteristics

Plant

The round, straight and smooth stem can attain 30–35 m with a diameter at chest height of 30–60 cm. Some adult specimens 3–5 m tall have been identified (Villachica *et al.*, 1996). The primary roots can grow 60 cm in depth and can then expand horizontally up to 40 m; they have secondary roots or neumatophores that allow them to breath under flooded conditions. The fan-shaped leaves can measure up to 1.0 m in radius (León, 1987) with a 3–6 m channel-shaped petiole and 20–30 of them are grouped at the top of the stem forming a spherical canopy (Fig. 8.3). The underside of the leaves is light green and the upper side dark green (Flores-Paytán, 1997).

Flowers, pollination and fruit set

This palm is polygamous-dioecious meaning it can have male, female or hermaphrodite flowers. The inflorescences are inter-foliar, similar in size and from 1.8 to 3.0 m long with a 60–100 cm peduncle. Male flowers are larger (8–10 by 6–7 mm) than female flowers (2–3 mm). Two to eight inflorescences can appear on a plant in a year.

Fruit

The oval fruit (Fig. 8.3) is an elliptical drupe that appears in large bunches and is 4–7 cm long × 4–5 cm in diameter, and has a hard and shiny epicarp formed by brownish-red to dark red scales in a pineapple peel type of arrangement; the thin mesocarp (2–6 mm) is edible, soft, floury and yellow, orange or reddish-orange in color, representing about 8–15% of the total fruit weight. The endocarp is thin and leathery. The seed (sometimes two) is large, solid and sub-globose.

Cultivar development

Cytogenetics, genetics and breeding

In the past, three species of *Mauritia* were considered to exist, *M. flexuosa*, *M. vinifera* and *M. martina*. They are now considered to be the same species but different ecotypes with differences in plant form and aspect, fruit aspect and number of seeds per fruit. There are no formal collections being evaluated. Fortunately, there are very large wild populations that can be used as germplasm sources.

Cultivars

There are no named cultivars, although there are different types, especially in regard to plant size and fruit type. In the Peruvian Amazon, plants with fruit

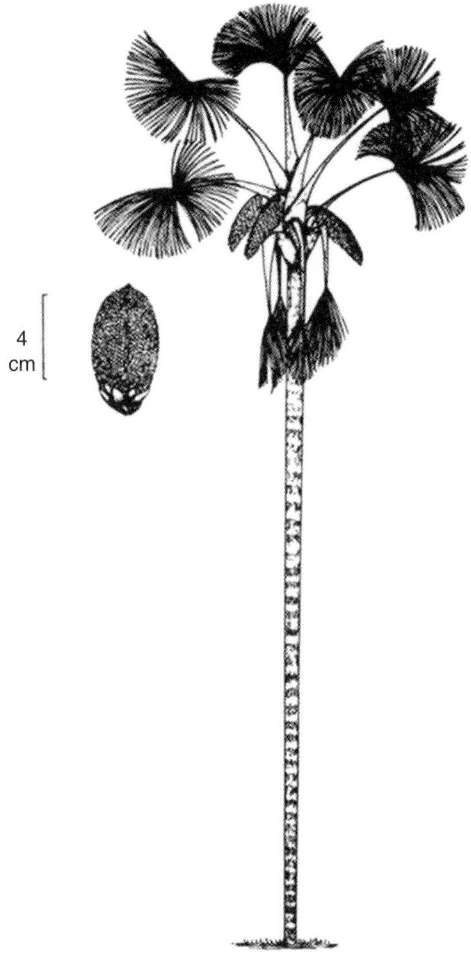

Fig. 8.3. Buriti (*Mauritia flexuosa*) showing solitary stem, fronds, fruit bunches (from Branner, J.C. (1901–1902) The palm trees of Brazil. *Popular Science Monthly* 60, 387–412).

of different pulp colors and thicknesses exist; those having fruit with thick pulp are called *aguaje carnoso* (meaty aguaje). Plants producing fruit with reddish pulp are called *shambo* and others with yellow pulp are called *ponguete* (Villachica *et al.*, 1996).

Cultural practices

Propagation

Seed is the only propagation method used. Seeds are short lived with about 30 days' viability. Cleaned seeds should be put directly in nursery bags filled with a mixture of sand, soil and decomposed manure or sawdust. Germination is slow, starting after about 80–90 days and ending at about 5 months with final germination percentages of 50–60%. Seeds can be superficially dried and put in tightly closed plastic bags in a warm room until they show signs of germination and then can be transplanted into nursery bags (25 × 30 cm); in this way no special care has to be taken for almost 3 months. The seedlings will be ready for transplanting in about 8–12 months after sowing.

Orchard establishment

Most of the plants used are wild but a few plantings have been done. For this the forest has to be cleared of trees and large shrubs and if the soil has no special problem holes can be dug. In soils that have been used for agriculture, it will be necessary to check if there is a plowpan or a hardpan in which case sub-soiling will be necessary. Holes are normally 40 × 40 × 40 cm. Planting distances are normally 8 × 8 or 7 × 9 m. It is recommended to plant two plants leaving 1 m between them in order to eliminate the male plants at first flowering to leave only 3–5% of male plants for pollination purposes. In lands flooded permanently or seasonally, plants can sometimes have problems because neumatophoric roots do not get access to the air and the plants die.

In most instances this palm is harvested in the wild state and people use some agroforestry type of management, thinning out excess plants and associating annual or permanent crops in the space between plants, using species like cashew, papaya, banana, plantain, pineapple, rice, corn and cassava among others.

Irrigation

This is not normally done because the plant grows in humid areas with abundant rainfall but in dry areas plants will need to be irrigated.

Pruning

No special pruning is needed except removing the old leaves. Thinning out plants that are growing too close to each other in order to allow them space for proper growth is another common practice.

Fertilization

Few studies have been done on fertilization. Fertilization recommendations are based on nutrient extraction by the crop, which amounts to 50–60 kg N, 10–15 kg Ca, 2–3 kg P and 0.5 0.8 kg Fe ha^{-1} $year^{-1}$. All residues of the palm and the associated crops as well as weeds are left for recycling (Villachica *et al.*, 1996).

Pest management

No important insects or diseases have been reported for this species. Villachica *et al.* (1996) report the presence of the Lepidopterae *Opsiphanes* sp., *Prenes* sp., and *Brassolis* sp., and larvae from the Gelichiidae and Oechphoridae families that skeletonize the leaves, cementing two adjacent leaflets together so they can feed in this internal channel. *Rhynchophorus palmarum* is normally found in stems that are felled when eliminating male trees; this is a way of obtaining larvae of this insect, which is a very popular food for the native people.

Weed control

Nothing is done regarding weed control since most plants are in the wild. In cultivated plants, no reports are available but using a machete or a hoe are probably the most used methods of weeding.

Orchard protection

No information is available since few cultivated plantations exist. The plant is very solid so that wind protection is probably not necessary unless very strong winds are a common occurrence.

Harvesting and postharvest handling

Yields

The plant starts producing after 7–8 years and can produce for 40–50 years (Vargas *et al.*, 1999). Fruit weight varies from 40 to 90 g and a bunch can weigh 30–40 kg. Yields can be 19–20 t ha^{-1} $year^{-1}$ in cultivated plantations and from wild trees 6–9 t ha^{-1} $year^{-1}$ (Flores-Paytán, 1997) while Villachica *et al.* (1996) indicate 15–25 t ha^{-1} $year^{-1}$, of which only 12–13% is usable pulp.

Harvesting

The harvest season south of the equator is in the first yearly semester and north of the equator in the second semester, although there is always some fruit to gather throughout the year. The bunch is harvested when the lower fruit start turning dark red before the majority of fruit on the bunch changes color to avoid dehiscing fruit falling to the ground.

Harvesting from young plants is easy since the fruit bunches are close to the ground. Since bunches arise between the leaves, on older plants fruit become difficult to reach and harvest. In the wild, people will cut down the palm giving rise to an increased percentage of male plants in wild populations that have been exploited by man. A system using wood triangles that are tied to the stem and serve as stair steps has been developed in Peru, so that people are able to climb the trees and not have to cut them down (Villachica *et al.*, 1996). Long poles with a knife or a hook at the end are used to cut the bunch or to pull the bunch or portions of it and then pick the fallen fruit from the ground.

Postharvest treatment

Harvested fruit can be held for 1 week after which rapid decomposition of the pulp occurs. During this week the pulp should be collected by submerging the fruit in hot water (60–70°C) for 10–15 minutes, eliminating the scales and removing the pulp by hand. Pulp that will not be used fresh has to be frozen or dried into a powder-like product.

Utilization

The paste-like pulp can be used fresh or can be frozen or dried for later use to prepare a refreshing drink called *aguajina* in Peru. *Aguajina* is a mixture of the pulp with water and sugar that is also used to prepare ice cream and shakes. The pulp is used to extract oil, prepare sweets and an alcoholic drink by fermentation called *masato*, and can be made into flour that has multiple uses in food preparation and to cure diarrhea. The pith of the stem is used to extract edible starch similar to sago palm in the Orient. The stump when felled oozes a sweet liquid used to prepare an alcoholic drink. The young inflorescence can be cut to obtain a sweet sap (toddy), as done with coconut. This sweet drink can be immediately consumed, fermented into an alcoholic beverage or evaporated in a pan to obtain palm sugar.

The fruit pulp is considered most nutritive, as it is rich in oils, and contains 2–3% protein, 18.7% carbohydrates, calcium, iron, phosphorus, and vitamins A, B1 and B2 (Table 8.1). In relation to vitamin A, the pulp contains a very high amount of retinol, 1,062 mg per 100 g, or provitamin A, 5,000 IU g^{-1} of oil, enough to prevent vitamin A deficiency in the native population.

The fruit endocarp is used to make handicrafts. The trunk is used to make canoes, the leaves for roofing, hats, baskets, hammocks and ropes. The petioles are used for making bottle stoppers or mattress or pillow fillings.

Pataua, Ungurahui

Oenocarpus bataua Mart. is a single-stemmed palm with an edible fruit pulp that grows in the wild with very little planted areas. Synonyms are: *Jessenia bataua* ssp. *bataua*; *J. weberbaueri* Burret. According to Jones (1995) *O. bataua* has two subspecies, *bataua* and *oligocarpus*.

Common names for this palm include pataua (English), *ungurahui*, *sacumama* (Peru), *seje*, *patabá*, *milpesos* (Colombia), *majo* (Bolivia), *chapil* (Ecuador), *palma seje* (Venezuela), *aricaguá*, *batauá*, *sagua*, *aricagua*, *colaboca* (Spanish), *batauá*, *patauá*, *patuá* (Portuguese) and *komboe* (Dutch).

Important genera and species

Its related species include: *O. bacaba* Mart. with a single stem and common in the states of Pará and Amazonas in Brazil; *O. mapora* Karsten, a clumping palm; and *O. minor* ssp. *minor* and *O. distichus*, both with a single stem and the latter with a high ornamental value. Meerow (2008) adds *O. balickii* F. Kahn, similar to *O. mapora*, but does not mention *O. minor*. Cavalcante (1974) considered another species *O. multicaulis* Spruce. There is also a hybrid of *O. bacaba* × *O. minor* (Donadio *et al.*, 2002) and another of *O. bacaba* × *O. bataua* (Meerow, 2008).

Origin and distribution

It is considered to be a native to tropical America, especially the northern part. It is found growing wild in the Amazon basin in Bolivia, Peru, Brazil, Ecuador, Colombia, Venezuela and Guyana (Flores-Paytán, 1997) and even Panama and Chocó, the Pacific part of Colombia. It is cultivated on a very small scale because of its use for refreshments and high quality oil (León, 1987).

Ecology

Soil

The palm grows in non-flooded soils with good drainage or in permanently or seasonally flooded soils with poor drainage where populations up to 1,000 plants ha^{-1} can be found, many time-associated with the *Jessenia* palm. It adapts to a wide range of soils from those rich in organic matter content to sandy nutrient-poor spodosols.

Rainfall

Growth requires a minimum of 1,000–1,700 mm of rainfall to a maximum of 4,000 mm rain with a preference for higher precipitation regimes.

Temperature

The annual mean maximum is 25–26°C and the mean minimum 17.2°C. It can be seen from sea level to 900 m indicating that it does not like the colder climates at higher altitudes.

Light and photoperiod

For proper growth and production the plant needs plenty of sunlight; it can grow under partial shade but will produce less. In well-drained areas, the population is lower due to the shading produced by other species that grow under these conditions. Normally young plants need partial shade while older plants prefer full sunlight.

Wind

Wind is normally not a problem in its native areas.

General characteristics

Plant

This palm has a single slender stem that can reach 14–25 m and a diameter at chest height of 15–25 cm with some spines that abscise (Fig. 8.4). The stem presents 5 cm wide rings every 20–30 cm in the lower parts of the stem that get closer together higher up the stem. The canopy is 6 m high and 8 m wide and has 7–16 leaves produced year round in a spiral arrangement. The pinnate leaves can be 3–8 m long and have a basal sheath (León, 1987). The petioles are 0.2–2.0 m long and the leaf has 80–100 leaflets per side in a single plane, with the medium portion of the leaf having longer and wider leaflets that are dark green on the upper side and whitish-green on the underside.

Flowers, pollination and fruit set

The creamy-brown to yellowish unisexual flowers come in intra-foliar panicles with a horse-tail appearance. The panicles arise in each leaf axil and are initially protected by deciduous bracts. Flowers are arranged in groups of three with two male flowers at the sides and one female flower between them. The 6 mm male flowers have 6–12 stamens and the female globular flowers have an ovoid

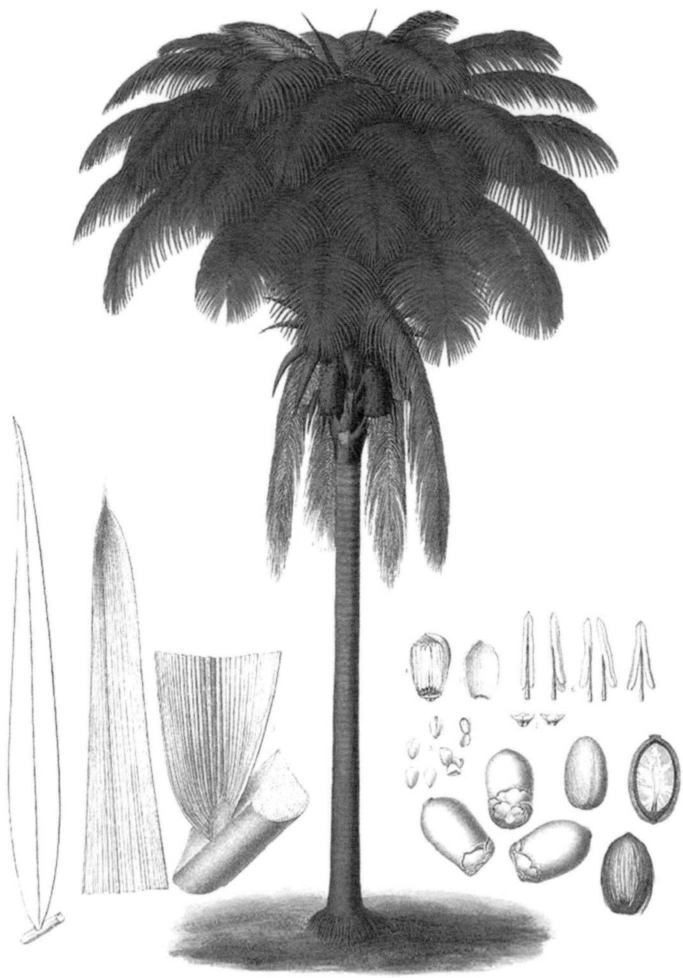

Fig. 8.4. Pataua or ungurahui (*Oenocarpus bataua*) showing solitary stem, leaves, fruit bunches and fruit (from Karsten, H. (1858–1869) *Florae Columbiae, Terrarumque Adiacemtum, Specimina Selecta*. 1: t. 98, Berolini. Image courtesy Missouri Botanical Garden, http://www.botanicus.org).

pistil. Each leaf axil produces one inflorescence and two to three, sometimes even five, are produced per stem during a year. Flowering occurs mainly from June to August.

Fruit

The fruit is an ovoid or ellipsoidal drupe 2.5–3.6 cm long and 1.7–2.3 cm wide (Fig. 8.4) weighing 1.5–2.0 g, sometimes 4.0 g. There are around 500–4,000 fruit per bunch weighing 2–15 kg. The fruit surface is smooth, dark red to an almost black-violet at maturity covered by a whitish waxy bloom. The 1.5–2.0 mm thick mesocarp is fleshy and high in oil content, and white to violet in color.

The hard endocarp is covered by large dark fibers. The fruit matures 6–8 months after flowering.

Cultivar development

Cytogenetics, genetics and breeding

There is a large genetic variability expressed by the existence of two subspecies, *bataua* and *oligocarpa*, with wide variation in fruit sizes, color at maturity and forms and oil content. EMBRAPA Centro de Pesquisa Agropecuária do Trópico Úmido (CPATU) in Belém has a collection of this palm that includes around 300 introductions of the *Jessenia*/*Bataua* complex (Villachica *et al.*, 1996).

Cultivars

There are no defined cultivars, in spite of the large variability found for this species that could allow for improvement.

Cultural practices

Propagation

Seed propagation is the only method used. Seeds have a high germination percentage and kept in a moist environment in the fruit can remain viable for 1.5 months. They are prepared for sowing by soaking the mature fruit for about 1 h in warm water, then the pulp is removed by hand and the washed seeds are allowed to superficially dry in the shade (Flores-Paytán, 1997). Decomposed sawdust can be used in the germination bed or any suitable mix that is retentive and has good drainage. Germination starts 1.5–3 months after sowing and when plants are about 10 cm they are transferred to nursery bags with 2 kg capacity until they attain 30–40 cm.

Mycorrhizal association has proven very effective in this palm so that forest soil where there are wild plants should be used for preparing the substrate for germination beds and nursery bags. The seedlings are kept under partial shade in the nursery stage.

Orchard establishment

Fields should be prepared in the usual manner. Recommended spacing is 7 × 7 m when planted as a single crop, while 6 × 12 m or 12 × 12 m should be used if planted associated with rice, corn, cassava, pineapple, cacao, avocado and other species. Plantings should be done at the start of the rainy season and plants acclimated to 50% shade in the nursery. Partial shade is required in the field during early growth by planting pigeon peas, cassava or other species around the planting hole in advance, otherwise cut palm leaves are arranged to provide partial shade during initial weeks after transplanting. Production should start after 5–6 years under cultivation and in the natural forest it should take more time due to shading of the associated tree vegetation.

Irrigation

No irrigation is normally done for a cultivated crop since it is grown in very rainy areas.

Pruning

Old leaves can be removed under cultivation as long as it is not too expensive otherwise nothing needs to be done.

Fertilization

No information is available.

Pest management

There are no mentions of specific diseases and insects.

Weed control

A leguminous cover crop can be useful. When interplanted with other crops all plant residues and pruning should be recycled in the field to improve ungurahui growth. A contact herbicide can be used for the area around the plant or manual weeding can be done.

Orchard protection

Since very little is cultivated there is no recommendation in this aspect.

Harvesting and postharvest handling

Yields

Each palm can produce two fruiting bunches per year, each with around 15 kg of fruit resulting in roughly 24 kg of clean fruit per tree. Pulp yields are between 35 and 47% of fruit weight so that a 200 tree per ha plantation can yield around 4.8 t fruit that would translate into 1.9 t of pulp ha^{-1}.

Oil content of the mesocarp is around 19–20% and 14–15% for the epicarp resulting in 6.5–8.0 % for the whole fruit and seed. In a dense wild population, it is estimated that 1.6–3.5 t ha^{-1} of fruit can be obtained; this translates into 112–260 l oil ha^{-1}. When interplanted with 140 plants ha^{-1} (12 × 6 m) the estimated fruit yield would be 3.36 t ha^{-1} or 260–320 l of oil while a monoculture with 200 plants ha^{-1} could yield between 312 and 384 l of oil (Flores-Paytán, 1997).

Harvesting

Harvest is frequently done by picking the fruit from the ground after it falls. In some cases, people climb the stems to get the bunch. A pole with a hook could be used. In extreme cases, the stem is cut down with loss of a productive stem and a reduction in the number of producing plants.

Postharvest treatment

Harvested fruit should be left two days to ripen completely. The pulp has to be processed or consumed during the first week after harvest otherwise the oils become rancid.

Table 8.2. Comparative composition of fatty acids in oils of ungurahui (*Oenocarpus bataua*) and olive (*Olea europaea*) (from Flores-Paytán, 1997).

Fatty acid	Ungurahui oil (%)	Olive oil (%)
Palmitic	13.2	11.2
Palmitoleic	0.6	1.5
Stearic	3.6	2.0
Oleic	77.7	76.0
Linoleic	2.7	8.5
Linolenic	0.6	0.5
Other	1.6	0

Utilization

The pulp is diluted in water to make a drink called *chapo* in Peru that is mixed with sugar or flour. It is also used to make ice cream and wine by fermentation. However, the main product is the oil that has a very similar appearance and fatty acid content to olive oil with around 80% of non-saturated and 2–4% of saturated fatty acids (Table 8.2). The petioles are used to make darts and the stems are used to make arrow tips and bows. In folk medicine, the oil is used as a laxative and as a liniment as well as for treating tuberculosis, asthma and other respiratory problems and to revitalize the hair.

References and Further Reading

Acai

Aguiar, F., Menezes, V. and Rogez, H. (2013) Spontaneous postharvest fermentation of açai (*Euterpe oleracea*) fruit. *Postharvest Biology and Technology* 86, 294–299.

Altman, R.F.A. (1956) O caroco de açaí (*Euterpe oleracea* Mart.). *Boletim Técnico do Instituto Agronômico do Norte (IAN)* 31, 109–111.

Anderson, A.B. (1988) Use and management of native forests dominated by acai palm (*Euterpe oleracea*) in the Amazon estuary. *Advances in Economic Botany* 6, 144–154.

Anonymous (2011) *Tabela Brasileira de Composição de Alimentos*. NEPA – UNICAMP 4ta ed. revisada e ampliada. NEPA UNICAMP, Campinas, Brazil.

Araujo, E.F. and Silva, R.F. de (1994) Evaluation of quality in acai seeds stored in different packages and environmental conditions. *Revista Brasileira de Sementes* 16, 76–79.

Bichara, C.M.G. and Rogez, H. (2011) Açai (*Euterpe oleracea* Martius). In: Yahia, E.M. (ed.) *Postharvest Biology and Technology of Tropical and Subtropical Fruits, Volume 2: Açaí to citrus*. Woodhead Publishing, Oxford, UK, pp. 1–26.

Bovi, M.L.A. and de Castro, A. (1993) Assai. In: Clay, J.W. and Clement, C.R. (eds) *Selected Species and Strategies to Enhance Income Generation from Amazonian Forests*. Food and Agriculture Organization (FAO). Miscellaneous Working Paper 93/6. FAO, Rome. Available at: http//www.fao.org/docrep/v0784e/v0784e0b.htm assai (accessed 10 November 2013).

Bovi, M.L.A., Godoy, G. Jr. and Saes, L.A. (1987) Híbridos interespecíficos de palmiteiro (*E. oleracea* x *E. edulis*). *Bragantia* 46, 343–363.

Bovi, M.L.A., Spiering, S.H. and Melo, T.M. (1989) Temperaturas e substratos para germinação de sementes de palmiteiro e açaízeiro. In: *Anais do Segundo Congresso sobre Tecnologia de Sementes Florestais*. Atibaia, São Paulo, pp. 43. Miscellaneous Working Paper 93/6, Rome. Available at: http://www.fao.org/docrep/v0784e/v0784e0b.htm#assaí (accessed 23 December 2014).

Calzavara, B.B.G. (1972) *As Possibilidades do Acaizeiro no Estuario Amazonico*. Boletim da Fundacao de Ciencias Agrarias do Para. São Paulo, Brazil.

Campos, F.A.M. (1951) Valor nutritivo de frutos brasileiros. *Instituto de Nutrição, Trabalhos e Pesquisas* 6, 72–75.

Chup, E.Y. (1999) The effects of arbuscular mycorrhizal fungi inoculation on *Euterpe oleracea* Mart. (açaí) seedlings. *Pesquisa Agropecuaria Brasileira*, Brasilia, 34, 1019–1024.

Donadio, L.C., Moro, F.V. and Servidone, A.A. (2002) *Frutas Brasileiras*. Editora Novos Talentos, Jaboticabal, São Paulo, Brazil, pp. 47–52.

Henderson, A. and Galeano, G. (1996) A revision of *Euterpe*, *Prestoea*, and *Neonicholsonia* (Palmae). *Flora Neotropica Monograph* 72, 1–90.

Jardim, M.A.G. and Anderson, A.B. (1987) Manejo de populaçães nativas do açaízeiro (*Euterpe oleracea* Mart.) no estuário Amazônico: resultados preliminares. *Boletim de Pesquisas Florestal* 15, 1–18.

Ledo, A. da Silva, Lameira, O.A., Benbadis, A.K., Menezes, I.C. de, Oliveira, M.S.P. de and Filho, S.M. (2002) Somatic embryogenesis from zygotic embryos of *Euterpe oleracea* Mart. *Revista Brasileira de Fruticultura* 24, 601–603.

León, J. (1987) *Botánica de los Cultivos Tropicales*. Instituto Interamericano de Cooperación para la Agricultura, San José, Costa Rica, pp. 81–82.

Lichtenthäler, R., Rodrigues, R.B., Maia, J.G.S., Papagiannopoulos, M., Fabricius, H. and Marx, F. (2005) Total oxidant scavenging capacities of *Euterpe oleracea* Mart. (Acai) fruits. *International Journal of Food Sciences and Nutrition* 56, 53–64.

Martin, F.W., Campbell, C.W. and Ruberthé, R.M. (1987) *Perennial Edible Fruits of the Tropics: An Inventory.* Agricultural Handbook No. 642, USDA, ARS, Washington, DC, p. 48.

Meerow, A.W. (2008) Assaí, *Euterpe oleracea* Mart. Arecaceae. In: Janick, J. and Paull, R.E. (eds) *Encyclopedia of Fruit and Nuts*. CAB International, Wallingford, UK, pp. 128–130.

Melo, C.F.M., Barbosa, W.C. and Alves, S.M. (1988) Obtenção de açaí desidratado. *Boletim Pesquisa* 92, 1–13.

Mota, S. (1946) Pesquisas sobre o valor alimentar do açaí. *Anais Associação Química Brasileira* 5, 35–38.

Nogueira, O.L., Carvalho, C.J.R. de, Ller, C.H.M., Galvão, E.U.P., Silva, H.M. de, Rodrigues, J.E.L.F, Oliveira, M.S.P. de, Carvalho, J.E.U. de, Rocha Neto, O.G. da, Nascimento, W.M.O. do and Calzavara, B.B.G. (1995) *A Cultura do Açaí*. Coleção Plantar 26, Embrapa-SPI, Brasilia, Brazil.

Oliveira, P.A.A.C. de, Silva, I.G. da, Souza, M.L. de, Furtado, C.M. and Silva, R.F. da (2011) In natura açaí beverage: quality, pasteurization and acidification. *Food Science and Technology* (Campinas) 31, 502–507. Available at: http://www.scielo.br/scielo.php?script=sci_arttext&pid=S0101-20612011000200035&lng=en&nrm=iso (accessed 19 July 2013).

Rogez, H. (2000) *Açai: Preparo, Composição e Melhoramento da Conservação*. EDUFPA, Belém, Brazil, p. 313.

Rogez, H., Pompeu, D.R., Akwie, S.N.T. and Larondelle, Y. (2011) Sigmoidal kinetics of anthocyanin accumulation during fruit ripening: a comparison between açai fruits (*Euterpe oleracea*) and other anthocyanin-rich fruits. *Journal of Food Composition and Analysis* 24, 796–800.

Strudwick, J. and Sobel, G.L. (1988) Uses of *Euterpe oleracea* Mart. in the Amazon estuary, Brazil. *Advances in Economic Botany* 6, 225–253.

Tonon, R.V., Brabet, C. and Hubinger, M.D. (2009) Influência da temperatura do ar de secagem e da concentração de agente carreador sobre as propriedades físico-químicas do suco de açaí em pó. *Food Science and Technology* (Campinas) 29, 444–450.

Urdaneta, H. (1981) *Planificación Silvicultural de los Bosques Ricos en Palma Manaca* (Euterpe oleracea) *en el Delta del Río Orinoco*. Universidad de los Andes, Mérida, Venezuela.

Vargas, O., Alix, C., Lobo, A.D. (authors), Duarte, O. and Sanchez, J. (technical reviewers) (1999) *Frutales y Condimentarias del Trópico Húmedo*. CURLA; PDBL; AFE/COHDEFOR; DICTA; SETCO; PROFORFITH, La Ceiba, Honduras, pp. 52–53.

Villachica, H., Carvalho, J.E.U., Muller, C.H., Diaz, S.C. and Almanza, M. (1996) Asai, *Euterpe oleracea*. In: *Frutales y Hortalizas Promisorios de la Amazonía*. Tratado de Cooperación Amazónica, Lima, Perú, pp. 33–43.

Zorzenon, F.J. and Bergmann, E.C. (1995) Occurrence of *Xyleborus ferrugineus* (Fabricius, 1801) (Coleoptera: Scolytidae) in fruits and seeds of two species of the genus *Euterpe*. *Revista de Agricultura Piracicaba* 70, 17–20.

Peach palm

Adin, A., Weber, J.C., Sotelo Montes, C., Vidaurre, H., Vosman, B. and Smulders, M.J.M. (2004) Genetic differentiation and trade among populations of peach palm (*Bactris gasipaes* Kunth) in the Peruvian Amazon – implications for genetic resource management. *Theoretical and Applied Genetics* 108, 1564–1573.

Arias, O. and Huete, F. (1983) Propagación vegetativa in vitro de pejibaye (*Bactris gasipaes* H.B.K.). *Turrialba* 33, 103–108.

Bovi, M.L.A. (1998) *Palmito Pupunha: Informações Básicas Para Cultivo*. Boletim Técnico 173, Instituto Agronômico, Campinas, São Paulo, Brazil, 50 pp.

Clement, C.R. (1996) Is Spiny better than Spineless for Pejibaye Heart of Palm Production? *Proceedings of the Interamerican Society for Tropical Horticulture* 40, 236–241.

Clement, C.R. (1999) 1492 and the loss of Amazonian crop genetic resources. I. The relation between domestication and human population decline. *Economic Botany* 53, 188–202.

Clement, C.R. (2000) *Pupunha (*Bactris gasipaes *Kunth, Palmae)*. Série Frutas Nativas, 8, Fundep, Jaboticabal, São Paulo, Brazil, 48 pp.

Clement, C.R. (2008) Peach palm *Bactris gasipaes* Kunth. In: Janick, J. and Paull, R.E. (eds) *Encyclopedia of Fruit and Nuts*. CAB International, Wallingford, UK, pp. 93–101.

Clement, C.R. and Bovi, M.L.A. (2000) Padronização de medidas de crescimento e produção em experimentos com pupunheira para palmito. *Acta Amazonica* 30, 349–362.

Clement, C.R., Weber, J.C., van Leeuwen, J., Domian, C.A., Cole, D.M., Arévalo Lopez, L.A. and Argüello, H. (2004) Why extensive research and development did not promote use of peach palm fruit in Latin America. *Agroforestry Systems* 61, 195–206.

DeFrank, J. and Clement, C.R. (1995) Weed control in Pejibaye Heart of Palm Plantations in Hawaii. *HortScience* 30(6), 1215–1216.

Donadio, L.C., Moro, F.V. and Servidone, A.A. (2002) *Frutas Brasileiras*. Editora Novos Talentos, Jaboticabal, São Paulo, Brazil, pp. 47–52.

FAO/WHO (1985) *Codex Standard for Canned Palmito* (Codex Stan 144-1985). Food and Agriculture Organization/World Health Organization, Rome.

Fernandes, D.N. and Sanford, R.L. (1995) Effects of recent land use practices on soil nutrients and succession under tropical wet forest in Costa Rica. *Conservation Biology* 9, 915–922.

Ferreira, E. (1999) The phylogeny of pupunha (*Bactris gasipaes* Kunth, Palmae) and allied species. In: Henderson, A. and Borchsenius, F. (eds) *Evolution, Variation and Classification of Palms*. Memoirs of the New York Botanical Garden 83, New York Botanical Garden Press, Bronx, New York, pp. 225–236.

Henderson, A. (2000) Bactris *(Palmae)*. Flora Neotropica Monograph 79, New York Botanical Garden, New York, 181 pp.

Mora Urpí, J. (1983) El pejibaye (*Bactris gasipaes* H.B.K.): origen, biología floral y manejo agronómico. In: *Palmeras Poco Conocidas de América Tropical*. Turrialba, Costa Rica, CATIE/FAO, pp. 118–160.

Mora Urpí, J. and Gainza Echeverria, J. (eds) (1999) *Palmito de Pejibaye (*Bactris gasipaes *Kunth): Su Cultivo e Industrializacíon*. Editorial Universidad de Costa Rica, San José, Costa Rica, 260 pp.

Mora Urpí, J. and Solis, E. (1980) Polinización en *Bactris gasipaes* H.B.K. *Revista de Biología Tropical* 29, 139–142.

Mora Urpí, J., Weber, J.C. and Clement, C.R. (1997) *Peach palm,* Bactris gasipaes *Kunth. Promoting the Conservation and Use of Underutilized and Neglected Crops. 20*. Institute of Plant Genetics and Crop Plant Research – (IPK), Gatersleben and International Plant Genetic Resources Institute – (IPGRI), Rome, 83 pp.

Mora Urpí, J., Bogantes Arias, A. and Arroyo Oquendo, C. (1999) Cultivares de pejibaye para palmito. In: Mora Urpí, J. and Gainza Echeverria, J. (eds) *Palmito de Pejibaye (*Bactris gasipaes *Kunth): Su Cultivo e Industrializacíon*. Editorial Universidad de Costa Rica, San José, Costa Rica, pp. 41–47.

Morcote-Rios, G. and Bernal, R. (2001) Remains of palms (Palmae) at archaeological sites in the New World: A review. *Botanical Review* 67, 309–350.

NAS (1975) *Underexploited Tropical Plants with Promising Economic Value*. National Academy of Sciences, Washington, DC, 188 pp.

Patiño, V.M. (1963) *Plantas Cultivadas y Animales Domésticos en América Equinoccial. Tomo 1. Frutales*. Imprenta Departamental, Cali, Colombia, 547 pp.

Patiño, V.M. (1992) An ethnobotanical sketch of the palm *Bactris* (*Guilielma*) *gasipaes*. *Principes* 36, 143–147.

Prance, G.T. (1984) The pejibaye, *Guilielma gasipaes* (H.B.K.) Bailey, and the papaya, *Carica papaya* L. In: Stone, D. (ed.) *Pre-Columbian Plant Migration*. Peabody Museum of Archeology and Ethnology, Paper No. 76, Harvard University Press, Cambridge, pp. 85–104.

Rodrigues, D.P., Astolfi Filho, S. and Clement, C.R. (2004) Molecular marker-mediated validation of morphologically defined landraces of pejibaye (*Bactris gasipaes*) and their phylogenetic relationships. *Genetic Resources and Crop Evolution* 51, 871–882.

Rojas-Vargas, S., Ramírez, P. and Mora Urpí, J. (1999) Polimorfismo isoenzimático en cuatro razas y un híbrido de *Bactris gasipaes* (Palmae). *Revista de Biología Tropical* 47, 755–761.

Romero Castañeda, R. (1961) *Frutas Silvestres de Colombia, Vol. 1*. Editorial San Juan Eudes, Bogotá, Colombia, pp. 15–18.

Villachica, H., Carvalho, J.E.U., Muller, C.H., Diaz, S.C. and Almanza, M. (1996) Pijuayo. In: *Frutales y Hortalizas Promisorios de la Amazonía*. Tratado de Cooperación Amazónica, Lima, Perú, pp. 215–226.

Yuyama, L.K.O., Aguiar, J.P.L., Yuyama, K., Macedo, S.H.M., Fávaro, D.I.T., Afonso, C. and Vasconcellos, M.B.A. (1999) Determinação de elementos essenciais e não essenciais em palmito de pupunheira. *Horticultura Brasileira* 17, 91–95.

Yuyama, L.K.O., Aguiar, J.P.L., Yuyama, K., Clement, C.R., Macedo, S.H.M., Fávaro, D.I.T., Afonso, C., Vasconcellos, M.B.A., Vannucchi, H., Pimentel, S.A. and Badolato, E.S.G. (2003) Chemical composition of the fruit mesocarp of three peach palm (*Bactris gasipaes*) populations grown in central Amazonia, Brazil. *International Journal of Food Sciences and Nutrition* 54, 49–56.

Buriti

Cavalcante, P.B. (1974) *Frutas Comestiveis da Amazonía II*. Publicaçoes Avulsas No. 27, Conselho Nacional de Pesquisas, Museu Paraense Emilio Goeldi, Belém, Pará, Brazil, pp. 37–39.

Donadio, L.C., Moro, F.V. and Servidone, A.A. (2002) *Frutas Brasileiras*. Editora Novos Talentos, Jaboticabal, São Paulo, Brazil, pp. 90–93.

Flores-Paytán, S. (1997) *Cultivo de Frutales Nativos Amazónicos, Manual para el Extensionista*. Tratado de Cooperación Amazónica, Secretaría Pro-Tempore, Lima, Perú, pp. 7–13.

Gomes, L.R.P., Lopes, M.T.G., Bentes, J.L.D.S., Barros, W.S., Neto, C., de Queiroz, P. and Contim, L.A.S. (2011) Genetic diversity in natural populations of Buriti (*Mauritia flexuosa* L. f.). *Crop Breeding and Applied Biotechnology* 11, 216–223.

Holm, J.A., Miller, C.J. and Cropper, W.P. (2008) Population dynamics of the dioecious Amazonian palm *Mauritia flexuosa*: simulation analysis of sustainable harvesting. *Biotropica*, 40, 550–558.

León, J. (1987) *Botánica de los Cultivos Tropicales*. Instituto Interamericano de Cooperación para la Agricultura, San José, Costa Rica, p. 62.

Martin, F.W., Campbell, C.W. and Ruberthé, R.M. (1987) *Perennial Edible Fruits of the Tropics: An Inventory*. Agricutural Handbook No. 642, USDA, ARS, Washington, DC, p. 48.

Meerow, A.W. (2008) *Mauritia flexuosa* Buriti. In: Janick, J. and Paull, R.E. (eds) *Encyclopedia of Fruit and Nuts*. CAB International, Wallingford, UK, pp. 132–134.

Padoch, C. (1988) Aguaje (*Mauritia flexuosa* Lf) in the economy of Iquitos, Peru. *Advances in Economic Botany 6*, 214–224.

Rosa, R.K. and Koptur, S. (2013) New findings on the pollination biology of *Mauritia flexuosa* (Arecaceae) in Roraima, Brazil: Linking dioecy, wind, and habitat. *American Journal of Botany* 100, 613–621.

Vargas, O., Alix, C., Lobo, A.D. (authors), Duarte, O. and Sanchez, J. (technical reviewers) (1999) *Frutales y Condimentarias del Trópico Húmedo*. CURLA; PDBL; AFE/COHDEFOR; DICTA; SETCO; PROFORFITH, La Ceiba, Honduras, pp. 54–55.

Villachica, H., Carvalho, J.E.U., Muller, C.H., Diaz, S.C. and Almanza, M. (1996) Aguaje. In: *Frutales y Hortalizas Promisorios de la Amazonía*. Tratado de Cooperación Amazónica, Lima, Perú, pp. 3–11.

Pataua, Ungurahui

Cavalcante, P.B. (1974) *Frutas Comestiveis da Amazonía II*. Publicaçoes Avulsas No. 27, Conselho Nacional de Pesquisas, Museu Paraense Emilio Goeldi, Belém, Pará, Brazil, pp. 37, 41–43.

Cifuentes, I., Moreno, F. and Arango, D.A. (2010) Reproductive phenology and fruit productivity of *Oenocarpus bataua* (Mart.) in flooded forests in the Chocó Biogeographic region, Colombia. *Biota Neotrop*. 10(4). Available at: http://www.biotaneotropica.org.br/v10n4/en/abstract?article+bn01910042010 (accessed 1 November 2013),.

Collazos, M. and Mejía, M. (1988) Fenología y poscosecha de mil pesos *Jessenia bataua* (Mart) Burret. *Acta Agronómica* 38, 53–63.

Donadio, L.C., Moro, F.V. and Servidone, A.A. (2002) *Frutas Brasileiras*. Editora Novos Talentos, Jaboticabal, São Paulo, Brazil, pp. 67–70.

Flores-Paytán, S. (1997) *Cultivo de Frutales Nativos Amazónicos, Manual para el Extensionista*. Tratado de Cooperación Amazónica, Secretaría Pro-Tempore, Lima, Perú, pp. 279–285.

Gonzáles Coral, A. and Melecio Torres-Reyna, G. (2011) *Manual de Producción de Plantones de Oenocarpus bataua* C. Martius *"Ungurahui"*. Instituto de Investigaciones de la Amazonía Peruana – IIAP, Iquitos – Perú. Available at: http://www.iiap.org.pe/cdpublicaciones2011/documentos/pdf/libros/25.pdf (accessed 1 November 2013).

Jones, D.L. (1995) *Palms throughout the World*. Smithsonian Institution Press, Washington, DC, pp. 276–278.

León, J. (1987) *Botánica de los Cultivos Tropicales*. Instituto Interamericano de Cooperación para la Agricultura, San José, Costa Rica, p. 58.

Meerow, A.W. (2008) *Oenocarpus bataua* Patuá. In: Janick, J. and Paull, R.E. (eds) *Encyclopedia of Fruit and Nuts*. CAB International, Wallingford, UK, pp. 135–138.

Montúfar, R., Laffargue, A., Pintaud, J.C., Hamon, S., Avallone, S. and Dussert, S. (2010) *Oenocarpus bataua* Mart. (Arecaceae): Rediscovering a source of high oleic vegetable oil from Amazonia. *Journal of the American Oil Chemists' Society* 87(2), 167–172.

Nuñez-Avellaneda, L.A. and Rojas-Robles, R. (2008) Biología reproductiva y ecología de la polinización de la Palma Milipesos *Oenocarpus bataua* en los Andes Colombianos. *Caldasia* 30, 101–125. Available at: http://www.scielo.org.co/scielo.php?pid=S0366-52322008000100005&script=sci_arttext (accessed 1 November 2013).

Vallejo-Rendón, D. (2002) *Oenocarpus bataua,* seje; Colombia Amazónica, Separata especies promisorias 1. Corporación Colombiana para la Amazonia – Araracuara – COA.

Villachica, H., Carvalho, J.E.U., Muller, C.H., Diaz, S.C. and Almanza, M. (1996) Ungurahui. In: *Frutales y Hortalizas Promisorios de la Amazonía*. Tratado de Cooperación Amazónica, Lima, Perú, pp. 281–287.

9 Other Families

RUTACEAE

The Rutaceae, commonly called the citrus family, has generally hermaphroditic flowers divided into four or five parts with noticeable scent. They are mostly trees and shrubs with very variable fruits including berries, drupes, hesperidiums, samaras, capsules and follicles. The family is closely related to the Sapindaceae that has litchi, longan and rambutan. A non-citrus tropical fruit genus in this family is *Casimiroa* with ten accepted species.

White Sapote

Casimiroa edulis Llav. et Lex. is commonly called white sapote and Mexican apple (South Africa) in English; *matasano*, *sapote blanco* and *zapote* in Spanish; *chalarina* in Peru; and *matasano*, *pomme mexicaine* and *sapote blanche* in French (Martin *et al.*, 1987). The meaning of *matasano* in American Spanish is "healthy person killer" because of some toxins in the fruit, though no deaths have been reported. Alkaloids that show antihypertensive activity have been isolated that could have sedative-like effects and be sleep inducers (Awaad *et al.*, 2007).

Important genera and species

Other species in the genus *Casimiroa* include three species native to Mexico: *C. pubescens*, *C. pringlei* and *C. watsonii*. Another species, *C. sapota*, also commonly called matasano, is similar to *C. edulis* (McCain 1993; Crane, 2008). The taxonomy of *C. edulis* (white sapote) is not clear with some taxonomists referring to it as *C. tetrameria* Mill., with *C. edulis* being a variant (George *et al.*, 1988).

C. tetrameria has the lower foliage covered with a dense velvety pubescence and the leaves are larger than those of white sapote; the fruit are also larger and ovoid to ellipsoid.

Origin and distribution

The plant is native to Central America and the central highlands of Mexico (Martin *et al.*, 1987). The species is distributed in many cool tropical and subtropical areas of the world but has not become commercially important and normally is grown on a small scale, mainly as a backyard tree, and very few commercial orchards exist (McCain, 1993). There are some commercial orchards in New Zealand and on a smaller scale in South Africa (Yahia and Gutierrez-Orozco, 2011). It is sold mainly in local or small town markets or at roadsides. It can be found in Mexico, Central America, California, Florida, Hawaii, Egypt, South Africa, New Zealand and Australia, and seldom in Peru.

Ecology

Soil

White sapote trees are tolerant of most well-drained soil conditions including high pH and calcareous soils. They adapt well to arid soil conditions (Yahia and Gutierrez-Orozco, 2011).

Rainfall

It grows with medium rainfall greater than 2,000 mm and tolerates heavy rainfall in well-drained soils. Trees are drought tolerant and will withstand seasonal dry periods although leaf abscission may occur (Martin *et al.*, 1987). Drought does reduce fruit size and yield (Batten, 1984).

Temperature

The tree is adapted to cool tropical and subtropical areas (Martin *et al.*, 1987). In the tropics, it grows at altitudes above 600–900 m and occasionally up to 3,000 m. Trees do not do well in the hot humid tropics. Young trees are quite cold tolerant and can withstand temperatures of −2°C to −3°C and mature trees −5°C to −6°C (Campbell *et al.*, 1977), although frost will defoliate the tree. White sapote will produce quality fruit in areas where citrus do not have enough heat units for acceptable flavor development (Batten, 1984).

Light and photoperiod

The tree prefers to grow under full sun and in the areas where it is grown, as far north as California, it flowers normally, indicating no photoperiodic response.

Wind

Young trees are reported to be susceptible to wind damage and should be staked or protected by wind barriers (George *et al.*, 1988). Fruit scarring can be severe when exposed to wind.

General characteristics

Tree

This evergreen has an erect or spreading open canopy that can reach 4.5–18 m in height with an extensive spreading root system. The trunk is dull, brownish-gray and has corky lenticels. The alternate leaves are digitally compound with three to seven leaflets, usually five, that are lanceolate or elliptic or ovate, 6–13 cm long and 2.5–5.0 cm wide (Fig. 9.1); they are smooth or hairy on the lower surface (Sauls and Campbell, 1983).

Fig. 9.1. White sapote (*Casimiroa edulis*) showing leaf and fruit (used with permission from León, J. (2000) *Botánica de los Cultivos Tropicales*. Editorial Agroamérica, Instituto Interamericano de Cooperación para la Agricultura (IICA), San José, Costa Rica).

Flowers

The small flowers are inconspicuous, greenish-yellow and appear on new wood with all leaves completely developed on terminal and axillary panicles of 15–100 or more flowers. The flowers have five sepals and petals, five stamens and a three- to five-lobed stigma. The perfect flowers are sometimes functionally unisexual, especially females.

Flowers most commonly occur during the autumn and winter and the fruit are harvested during spring and summer. Flowering may be induced by drought stress and cincturing (George *et al.*, 1988).

Pollination and fruit set

Male flowers produce variable amounts of pollen. Male sterility and/or self-incompatibility can occur and limit fruit production in some cases. According to George and Nissen (1991) some cultivars like 'Yellow', 'Golden Globe', 'Reinikie Commercial' and 'Blumenthal' require cross-pollination while others like 'Vernon' and 'Ortego' do not.

Bees and ants are very attracted to the flowers (Yahia and Gutierrez-Orozco, 2011) and along with other insects are the pollinators (McCain, 1993).

Fruit

The fruit appear singly or in groups and weigh from 70 to 650 g (Fig. 9.1). The fruit are sub-globose to oblate, up to 12 cm long and 6–12 cm in diameter. The inedible peel is thin, very tender, greenish-yellow or yellow-orange with many tiny, conspicuous, yellow oil glands. Fruit that have a green skin at maturity have white flesh while fruit with yellow to yellow-orange skin have flesh of the same color (Yahia and Gutierrez-Orozco, 2011). The flesh is soft and very sweet with one to five, usually two or three, large, hard ovoid seeds enclosed in a hard endocarp (Batten, 1984). Some fruit can have a bitter or medicinal aftertaste due possibly to alkaloids.

Fruit are generally available during late spring to early autumn although some cultivars, depending upon climate, may have from two up to four or five crops per year (Morton, 1987; George and Nissen, 1991). The fruit matures in about 120 days and there are reports that fruit thinning increases fruit size (Martin *et al.*, 1987; McCain, 1993).

Cultivar development

White sapote cultivars, selections and seedlings are under evaluation at a number of institutions in California, Florida, Australia, New Zealand and Israel (George *et al.*, 1988; Nerd *et al.*, 1990; Mizrahi and Nerd, 1996).

Cultivars

A number of cultivars have been selected, in Florida: 'Dade', 'Golden', 'Page' and 'Smathers'; in New Zealand 'Fernie'; in California: 'Coleman', 'Gillespie', 'Harvey', 'Maechtlen', 'Pike', 'Suebelle' and 'Yellow' (Mowry *et al.*, 1967; Morley-Bunker, 1986; Morton, 1987; George *et al.*, 1988; McCain, 1993). Other varieties include 'Golden Globe', 'Reinikie Commercial', 'Blumenthal', 'Vernon' and 'Ortego' (Crane, 2008).

Cultural practices

Propagation

SEXUAL White sapote trees may be propagated by seed. The seeds are sown within 3–8 weeks of extraction from the fruit as viability declines with time.

Germination will begin in 3–4 weeks. Seeds should be used only to propagate rootstocks. Trees grown from seed take 5–8 years to fruit (Vargas *et al.*, 1999).

ASEXUAL Asexual propagation is the recommended method to maintain superior cultivars. Air layering, budding and grafting can be used (Martin *et al.*, 1987). Vigorously growing seedling rootstocks are utilized to graft or bud superior cultivars with 'Pike' seedlings preferred by some nurseries. However, overgrowth of the 'Pike' rootstock has been observed (George *et al.*, 1988). Veneer grafting as well as patch and "T" budding are normally used. Scions from matured shoots and rootstocks with 1 cm diameter at 20–25 cm from the soil are used (Vargas *et al.*, 1999). Asexually propagated trees start producing after 3–4 years.

Orchard establishment

The field for a commercial planting is prepared as for any other fruit trees. Tree spacing depends upon climate, cultivar and horticultural expertise. Nursery plants 12–18 months old are normally used. Generally plant spacing ranges from 7 by 7 m to 10 by 10 m. Closer distances can be used initially and as canopy closure starts to occur, every other tree can be reduced in size by pruning and finally eliminated to achieve the final spacing.

Irrigation

No detailed irrigation recommendations are available. Young trees require frequent watering to induce vigorous growth. Drought stress during flowering and/or fruit development appears to reduce crop yields and quality, so that mature trees, although drought tolerant, require adequate soil moisture during this period for better fruit size and yields.

Pruning

Newly planted trees may be cut back to 1 m to encourage low branching to form a good tree structure with three to four scaffold branches. The tree has a tendency of producing branches at very acute angles and that can be very long and willowy. These willowy branches require periodic pruning to reduce their length to obtain more uniform branching and a well-balanced tree (Batten, 1984). Light annual pruning is recommended to produce more fruiting laterals and internal leaf coverage to protect fruit from sunburn (Crane, 2008).

Fertilization

Fertilizer practices are based on observation and experience and include application of all essential elements. Young trees should be fertilized every 2–3 months with a low analysis NPK material, increasing the rate as trees mature (Sauls and Campbell, 1983).

Pest management

The fruit as many others are attacked by anthracnose and scab (Crane, 2008). It is also susceptible to attacks by various fruit flies. Other insects such as mealy bugs, fruit spotting bug and scales have been reported but their significance is unknown.

Weed control

As with most crops weeds will have to be controlled, especially in young orchards where canopies are not producing enough shade to control weed growth.

Orchard protection

Trees will have to be protected from strong winds by planting in a safe area or installing wind barriers perpendicular to the main direction of the winds.

Harvesting and postharvest handling

Large mature trees have been reported to produce up to 1,000 fruit or about 110–140 kg (Vargas *et al.*, 1999).

Harvesting

White sapote fruit should be hand-harvested by cutting the peduncle leaving a short piece that will fall once the fruit ripens fully (Yahia and Gutierrez-Orozco, 2011). Fruit maturity may be determined by fruit size, characteristic for a particular cultivar, and development of light green or greenish-yellow or yellow color. The fruit must be handled carefully because the peel is very thin and easily damaged (Crane, 2008). Harvest season is from June to October in Mexico, while in Florida fruit mature in November to December. There are varieties that produce almost year round.

Postharvest treatment

White sapote is a climacteric fruit (Yahia and Gutierrez-Orozco, 2011). Fruit harvested when the peel color has changed to light green or greenish-yellow ripen in 3–5 days at ambient temperatures. Once the fruit softens it is very difficult to handle (Batten, 1984). Fruit destined for distant markets are usually picked less mature, stored, shipped and then allowed to ripen at room temperature at their destination. Depending upon the maturity, white sapote fruit may be stored for 2–3 weeks at 19–21°C and 85–95% relative humidity or for 3–6 weeks at 5°C (McGregor, 1987).

Chilling injury has been reported in fruit stored for 63 days at 1°C (Yahia and Gutierrez-Orozco, 2011). Packing should be done in cushioned boxes or crates. Overripe fruit have a pungent unpleasant flavor (Morton, 1987).

Utilization

White sapote is commonly eaten fresh, used in fruit salads, served with cream and sugar, and sometimes made into beverages or preserves (Martin *et al.*, 1987). Fruit can be frozen and will retain their flavor once thawed (Batten, 1984). The fruit are a moderately good source of phosphorus, vitamin C and vitamin A (Table 9.1).

Uses of the pulp, leaves, bark and seeds abound in traditional medicine. The pulp is said to induce sleep and reduce rheumatic pain. The blood pressure

Table 9.1. Composition of white sapote, black sapote, nance, ceriman, South American sapote, genipap, borojo, purui grande, and Brazil nut per 100 g edible portion of fruit.

Proximate	White sapote[a]	Black sapote[a]	Nance[a]	Ceriman[b]	South American sapote[b]	Genipap[b]	Borojo[c]	Purui grande[d]	Brazil nut[e]
Moisture (g)	82.0	82	82.8	77.88	85.3	67.6	69.41	64.7	3
Energy (kcal)	65	66	66	73.7		103	108	93	751.6
Protein (g)	1.4	0.7	0.9	1.81	0.13	5.2	0.78	1.1	16.4
Lipid (g)	0.4	1.2	14.4	0.2	0.1	0.3	0.06	0	69.3
Fiber(g)	1.7	1.6	2.2	16.19	0.5	9.4	3.5	8.3	4.6
Carbohydrates (g)	15.7	15.0	14.4	0.57			29.03	24.7	3.2
Ash (g)	0.5	1.1	0.6	0.85	0.38	1.2		1.2	
Minerals									
Calcium (mg)	8.0	18	33		18.4	40	17.7	25	0.24
Iron (mg)	0.2	1.2	2.0		0.44	3.6	0.69	1.5	
Phosphorus (mg)	18	26	17		28.5	58	18.2	160	0.66
Vitamins									
Thiamine (mg)	0.04	0.02	0.02		0.03	0.04		0.03	150
Riboflavin (mg)	0.07	0.03	0.04		0.023	0.04		0.12	Trace
Niacin (mg)	0.5	0.2	0.4		0.33	0.5		2.3	
Vitamin A activity (mcg)	15	0.19	20						Trace
Ascorbic acid (mg)	23	29	84		9.7	33		3	

[a]Leung and Flores, 1966.
[b]Morton, 1987.
[c]Ocampo *et al.*, 2012.
[d]Mejía, 1984.
[e]Villachica *et al.*, 1996.

lowering properties of white sapote have been confirmed in some studies with extracts from the leaves, bark and seeds. The seeds are reported to be poisonous, but also contain anti-mutagenic activity; several physiologically active compounds that have effects in different medicinal uses have been identified, which could suggest more interest in cultivating this tree in the future.

BOMBACACEAE

Bombacaceae is no longer recognized at the rank of family but as a subfamily (Bombacoideae) within the family Malvaceae. *Pachira aquatica*, Malabar chestnut, is also in this new taxonomic group. Another important edible tropical species formerly in this family was the durian, which has been transferred to the new family Durionaceae. The family Malvaceae has nine subfamilies; in the subfamily Bombacoideae is the genus *Quararibea* with 98 species.

South American Sapote or Chupa-chupa

The South American sapote (*Quararibea cordata* Humb. & Bonpl., Vischer) (synonym *Matisia cordata* H.B.K.) is not a relative of the other sapotes. Some authors have referred to it as "South American sapote" or "chupa-chupa", the name used in Colombia and Peru. In these countries, it is also called *zapote chupa-chupa* or *sapote chupa-chupa, zapote chupa, sapotillo* and *sapote de monte*. In Brazil, it is called *sapota, sapote-do-Peru* or *sapota-do-Solimões*. In French, it is *sapote du Pérou* and in Portuguese *sapota do Peru*.

Origin and distribution

Chupa-chupa originated and was domesticated by the indigenous people in the western Amazon (Donadio, 1983). The main area of origin seems to be where Peru, Colombia and Brazil meet. In Brazil, it grows wild in the western Amazonian area near the border with Peru and is cultivated in most towns along the rivers of western Brazil as far south as the state of Cruzeiro do Sul. It has been introduced to the coastal areas of Ecuador as well as the Cauca and Magdalena valleys in Colombia. It is grown as a backyard or home garden tree with harvested fruit being sent to some of the large cities of Colombia, Ecuador and western Brazil and Iquitos in Peru. In 1964, seeds were introduced into South Florida and started bearing 7 years later (Whitman, 1976).

Ecology

Soil

This species requires good, well-drained soils, rich in organic matter and mineral nutrients for good growth. On the poor Amazonian oxisols growth is very poor, even with fertilization. It tolerates the dry, oolitic limestone of South Florida with proper fertilization. Soils with occasional flooding are tolerated.

Rainfall

Fairly well distributed rain in excess of 2,000 mm with a short dry period is best for growth, although it grows in areas with 1,500–4,000 mm rainfall. It does not tolerate drought (Clement, 1982).

Temperature

This tropical species can grow in the subtropics and up to 1,200–2,000 m in the tropics, as long as winter is not severe. It grows in South Florida but it will be defoliated by cold weather and is very susceptible to frost injury (Martin *et al.*, 1987).

Light, photoperiod and wind

The tree should be grown in full sun and has no photoperiodic responses. No data is available on wind responses and it probably needs shelter in more exposed areas.

General characteristics

Tree

Wild trees can attain 30–45 m in height but if cultivated it normally does not exceed 12–15 m. The tree is very fast-growing and the canopy is open with a pagoda form, with five branches appearing in several whorls along the main stem. The stem is fairly straight and can be devoid of branches for over half of its length. The foliage is dense, semi-deciduous and concentrates at the end of the branches with almost no foliage inside the canopy. The alternate, glabrous, heart-shaped leaves can be 50 cm long and 30–40 cm wide (Fig. 9.2), especially in young individuals or unproductive branches, but are much smaller in branches that produce flowers and fruit. The leaf is dark green above and lighter green below with prominent coarse veins in a palmate distribution and this tree makes a good ornamental plant (Santos, 2008).

Flowers

The tree has a cauliflorous bloom of hermaphrodite flowers (2.5 cm diameter and 5–7 cm long) that come in groups of three to six, arising in the lesser branches and on the trunk. Each flower has five sepals that remain attached to the ripe fruit and five petals that are slightly orange or yellow or pinkish-white with five conspicuous stamens and a pistil (Fig. 9.2). Flowering occurs in northern Brazil during the dry season from August to November while in central Brazil it occurs June to August and in Florida during the cold season from January to February.

Pollination and fruit set

There is good evidence that hummingbirds are the main pollinating agent, as well as bees and wasps. In the afternoon, some trees become self-compatible.

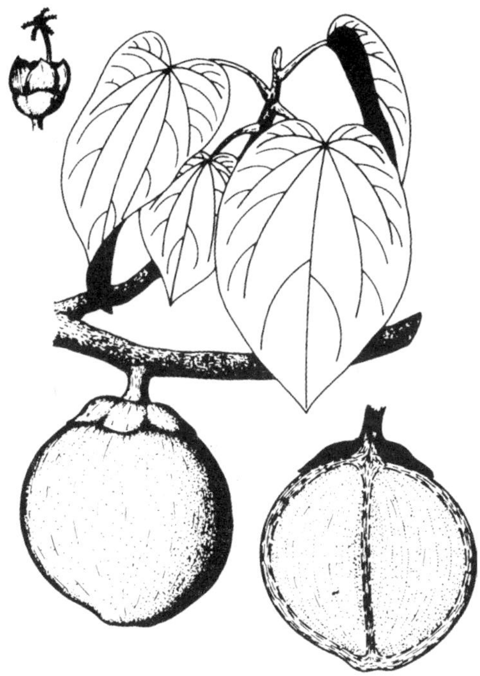

Fig. 9.2. South American sapote (*Quararibea cordata*) showing leaves, female flower and fruit (used with permission from León, J. (2000) *Botánica de los Cultivos Tropicales*. Editorial Agroamérica, Instituto Interamericano de Cooperación para la Agricultura (IICA), San José, Costa Rica).

Fruit

This large fruit is an ovoid or rounded berry that appears singly or in clusters on the old branches (Fig. 9.2). It is 7–15 cm long and 5–15 cm in diameter, sometimes 20 cm, and weighs an average of 400 g with a range from 200 to 1,400 g (Villachica *et al.*, 1996). The calyx is persistent and contracts during the final stages of ripening leaving a perimeter of lighter-colored skin that was previously covered. The rind is elastic, hairy, 1–1.5 cm thick, brownish-green and surrounds a very juicy, fibrous, orange-yellow and sweet pulp. The fruit has five sections and each one has a seed except when some abort leaving only the seed coat in the pulp. The cuneiform seeds are 4 cm by 2.5 cm, weigh ~11 g and have attached fibers like mango seeds. It takes about 240 days from anthesis to fruit ripening in the tropics, and up to 270–300 days in the subtropics.

Cultivar development

Some selection has been carried out for desirable types for introduction into new areas. There are a few accessions at EMBRAPA Centro de Pesquisa Agropecuária do Trópico Úmido (CPATU) in Belém, Brazil as well as at the Peruvian Instituto Nacional de Innovación Agraria (INIA) Research Station at Pucallpa. Manaus and Belém are potential sources for selection material (Villachica *et al.*, 1996). Apparently, the plants from the Peruvian Amazon bear higher quality fruit.

Cultivars

'Portoviejo' is a variety originated in Portoviejo, Ecuador with hermaphrodite flowers and introduced to Colombia in 1983. It has a sweet flavor and aroma, and an intense orange pulp and a high content of vitamin C (Rios-Castaño *et al.*, 1997).

Cultural practices

Propagation

SEXUAL Seed has been the normal propagation method. The fresh seeds are cleaned from most of the pulp and sown in a horizontal position. The seeds are recalcitrant and should be sown as soon as possible after extraction.

ASEXUAL Propagation by grafting on seedlings of chupa-chupa has been tried successfully. Side veneer and cleft grafting are the recommended methods (Donadio *et al.*, 2002). Apparently grafted plants do not produce earlier than seedlings, so people prefer to use the latter.

Orchard establishment

There are no commercial plantations but land preparation would include plowing and harrowing to bring the soil to a good tilth before transplanting. Spacing of 9–10 m by 9–10 m is recommended for seedling trees and 8 by 8 m for grafted or budded plants (Villachica *et al.*, 1996).

Irrigation

The tree is normally grown in areas with plenty of rainfall. In places with a dry period of more than 1–2 months, it is advisable to irrigate.

Pruning

No published results are known but it will probably benefit from topping the small trees to induce low whorls of scaffold branches and prevent the tree from becoming too tall and devoid of lateral branches on the lower stem. In later years, topping of long branches and elimination of undesirable material that is damaged or diseased or growing in the wrong direction would be advisable.

Fertilization

Little experience exists. Liming could be important in acid soils.

Pest management

No major diseases have been reported. There are some reports about *Pestalotia* attacking the foliage. The fruit is very susceptible to attack by fruit flies. White fly (*Aleurodicus dispersus*) and the Cuban May beetle (*Phyllophaga bruneri*) are reported as pests.

Harvesting and postharvest handling

The tree can be very productive yielding up to 6,000 fruit year^{-1}, although 400–1,300 is more common with an average fruit weight of around 300–400 g. In Belém, Brazil fruit is collected between September and February (Villachica *et al.*, 1996). The fruit is ready for harvest when it shows a clear color ring around the sepals. The strong peduncle will have to be cut with a curved knife and the fruit caught before they fall to the ground. Due to its thick elastic rind the fruit travels very well and can stay in good condition for a week to 10 days without refrigeration if collected at the proper stage (Clement, 1982).

Utilization

The fruit contains 82.4% pulp, 14% peel and 3.6% seeds. The fruit is normally eaten out of hand and sometimes juiced (Santos, 2008). Nectar can be prepared by extracting the pulp and diluting it in three times its volume of hot water at 75°C for 5 min; the pH is lowered to 3.5 with citric acid then the nectar is pasteurized (Villachica *et al.*, 1996). Experiments in preparing jam and jelly have not been encouraging. The nutritional value of the pulp is not very high, except for carotene content (Table 9.1).

EBENACEAE

The pantropical family Ebenaceae has more than 750 species of trees and shrubs and includes the fruit persimmon and the dense black wood ebony. Not all species in this family bear edible fruit, many fruits being very high in tannins when unripe. One of the largest genera in this family is *Diospyros* with more than 500 species mostly from the tropics. The genus name was originally applied to the persimmon (*D. lotus*) from the Caucasus. The more important persimmon in commerce is *Diospyros kaki* Thunb., the Japanese or Asian persimmon. Mabolo or velvet apple *D. blancoi* is a tropical fruit tree from the Philippines, Taiwan and Indonesia. The precious wood ebony is obtained from several species in the genus: *D. ebenum*, native to southern India and Sri Lanka; *D. crassiflora* native to western Africa; and *D. celebica*, native to Indonesia.

Black Sapote

Black sapote, *Diospyros digyna* Jacq. (Ebenaceae), is called the black persimmon or chocolate pudding tree (English), *zapote negro* (Philippines), *ebano* (Portuguese), *barbacoa, bois de´ebene, sapote noir* (French) or in Spanish *sapote negro, zapote prieto, zapote de mico, matasano de mico, ébano*. It is not related to other sapotes (*Pouteria sapota* or *Casimiroa edulis*). The synonyms are *D. nigra* (J.F. Gmelin) Perrettet, *D. obtusifolia* Humb. & Bonpl. ex. Willd. and *D. ebenaster* Hiern. *D. ebenaster* is a wild species of the West Indies, now renamed *D. revoluta* Poir., that has smaller and thicker leaves and smaller fruit than the black sapote and is apparently poisonous.

Origin and distribution

Probably a native to the forested lowlands of Mexico and Central America, it was taken by the Spanish to the Philippines and is now naturalized in the Moluccas and Sulawesi. The species has become indigenous to primary and secondary low- and medium-altitude forests in the Philippines. It is cultivated in Mexico and Guatemala and on a small scale in Hawaii, Florida and California but is a minor fruit elsewhere. Black sapote grows best in coastal areas from Jalisco to Chiapas, Veracruz and Yucatan. The tree has been cultivated in South Florida, in home gardens and small commercial orchards.

Ecology

Soil

The tree is adapted to a wide range of soil types and can withstand salinity. It does best in a light, deep well-drained soil, rich in organic matter, but can grow in shallow soils as well. It is highly adaptive to calcareous soils and shows only minimal chlorosis from micronutrient deficiency.

Rainfall

It grows in climates where rains are fairly high and well distributed with a short dry season during the year. It is sensitive to drought but can withstand sporadic floods. In arid regions of Israel and Africa, it grows well, with high production, under irrigation.

Temperature

The tree adapts itself to tropical and subtropical climates but prefers hot, low to middle elevation tropical areas without being truly tropical, since it can grow as far north as West Palm Beach in Florida, if protected from frost during the first years. Established trees can stand occasional short exposures to −2.2°C or −1.1°C but −5°C will kill small or weak trees. It tolerates 40–42°C.

Light and photoperiod

Not much is known about this but the plant flowers normally in the areas where it is grown. It seems to prefer full sun for best growth.

Wind

This fairly sturdy tree can withstand moderate winds.

General characteristics

Tree

This broad-topped, slow-growing tree (12–25 m) with an 8–10 m diameter canopy has a furrowed trunk (75 cm diameter) and black bark. The leathery, alternate, evergreen glossy leaves are 10–30 cm long and 6–10 cm wide, elliptic-oblong to lanceolate and tapered at both ends or rounded at the base and acute at the apex (Fig. 9.3). The canopy is very attractive and produces good shade.

Flowers

The tree is dioecious (occasionally polygamous) with small axillary flowers that occur on 1–2-year-old shoots. Some flowers have both male and female organs (Fig. 9.3); others that are solely male have a gardenia-like scent. The axillary male flowers are borne in clusters (three to seven) while the hermaphrodite flowers often are borne singly in the leaf axils. The white, tubular-lobed flowers (1.0–2.0 cm wide) have a persistent green calyx and resemble those of *D. kaki*. Petals fall 3–4 days after flower opening, while the calyx is persistent and stays with the fruit. The ovary is eight-, ten- or 12-celled with one ovule per cell. The juvenile stage can be as short as 3–4 years with the flowers borne on the new shoots. Self-incompatibility has been reported in solitary trees. In the Philippines, trees flower in March during the dry season. In Central America, the tree flowers during the last quarter of the year at the start of the dry season (Heinze, 1983). The presence of a male tree increases production. No information is available on natural pollination or fruit set.

Fruit

The fruit (Fig. 9.3) can be round, flattened, oblate or conical (5–15 cm diameter) and are dark, shiny and bright green, with a persistent four- to six-lobed calyx

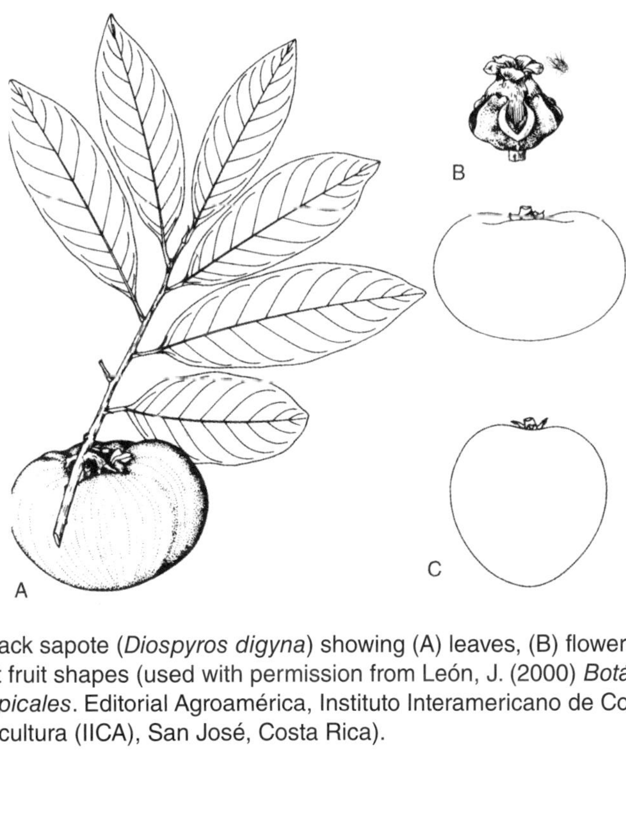

Fig. 9.3. Black sapote (*Diospyros digyna*) showing (A) leaves, (B) flower and (C) different fruit shapes (used with permission from León, J. (2000) *Botánica de los Cultivos Tropicales*. Editorial Agroamérica, Instituto Interamericano de Cooperación para la Agricultura (IICA), San José, Costa Rica).

(4–5 cm across). Immature fruit have a golden-yellow bronze color and are not edible, being astringent. The smooth, thin skin ripens to an olive- to muddy-green. The soft pulp is somewhat jelly-like, sweet and mild in flavor and brown to very dark brown, almost black, and has no aroma. The fruit has one to ten, exceptionally 15, flat, smooth, brown seeds (2–2.5 cm long) or is seedless. In Mexico, the fruit is available from August to January, while in Florida they ripen between December and February, although in some trees this occurs in June to August (Morton, 1987). Fruit takes about 4 months to mature from anthesis.

Cultivar development

Trees vary in shape and hairiness of leaves and the size, shape, seediness, flesh color and sweetness of the fruit. Selections have been made in the Philippines, Mexico, Australia and Florida. 'Manilla' and 'Valesca' are old, mainly seedless cultivars from the Philippines. Clonal selections in Australia have led to cultivars including 'Bernecker' and 'Maher'. In Florida, a selection made by the University of Florida named 'Merida' is propagated by grafting.

Cultural practices

Propagation

SEXUAL It is usually grown from seeds that germinate in 15–30 days, and the seeds stay viable for several months after drying. It is best to sow the seed directly into large bags at a depth of 1–1.5 cm. Seedlings may need to grow for up to 2 years before transplanting or grafting because of initial slow growth. Seedlings will flower after 6–10 years.

ASEXUAL Asexual propagation guarantees shorter trees and assures pistillate or female trees with good fruit quality will be obtained. The tree can be air layered and shield budded using mature but still green scions to get large 3.0–3.5 cm shields that can be budded. Cleft grafting and veneer grafting using scions 6 months of age are successful but the best grafting method is the lateral graft, leaving at least two leaves in the rootstock (Ledesma and Campbell, 2001). The scions should have two half leaves and the buds should be woody, but with no new growth. Grafted plants should be put under 70% shade and covered with a plastic bag for 3–4 weeks. The best time in Florida is September–October with 80% success. Seedling rootstocks are used for grafting and budding. Grafted or budded plants will flower in 3–4 years.

Orchard establishment

This would follow the normal procedure for field preparation. Planting distances of 8–12 m are recommended.

Irrigation

In its native area, it grows without irrigation but any dry period longer than 1–2 months justifies irrigating. In Australia and Israel, the plants have to be irrigated.

Pruning

As in most trees, the ideal for fruit production is to cut the main stem at a height of about 90–100 cm in order to get three to four main branches arising below this point. Interlacing and dead branches as well as branches hanging down to the ground need to be removed.

Fertilization

No specific studies are available but rates could initially be established using similar fruit tree species as a reference.

Pest management

No serious diseases are reported. The fruit is a fruit-fly host. Mulching with plastic, organic matter or cutting of the weeds is advisable when the tree is young.

Orchard protection

If strong winds prevail then proper protection with wind barriers is needed.

Harvesting and postharvest handling

Harvest

Due to the green fruit skin and that of the foliage, it is difficult to detect fruit on the tree or to notice the slight color change that takes place as the fruit ripens. Fruit must be harvested when fully grown but still unripe. This difficulty leads to fruit being lost due to natural dehiscence and this has also limited its use as a street tree. In Mexico, the production reached 700 t in 2001 (Yahia and Gutierrez-Orozco, 2011).

Postharvest handling

The fruit is very soft when fully ripe and can be held only for a few days in cold storage. The fruit ripens in 10 days from a bright green mature stage to an olive- to muddy-green. Those picked firm but olive-green ripen in 2–6 days. Once the fruit ripens it becomes very soft and can be kept for only a few days and handling becomes difficult. George (1984) mentions a report where green-mature fruit can be held at 10°C for several months and then ripened within 48 h after warming at 29.4°C.

Utilization

The pulp is eaten fresh when fully ripe and soft, although it does not have as much flavor or aroma. The pulp is very good when mixed with lime, orange or pineapple juice. Some people mix or blend it with milk or milk and nutmeg. It is also used for ice cream, cake fillings and desserts. Wine, brandy and liqueurs can be mixed with the pulp and served as a dessert. In Mexico, they prepare a dessert by mixing the pulp with wine and cinnamon. The brown to dark brown pulp is one of the reasons this fruit has not become more popular. It has a high content of calcium and phosphorus (Table 9.1) and more vitamin C content than oranges, with 7% sugars (Geilfus, 1989).

The green fruit is astringent and bitter, and used as a fish poison in the Philippines. The yellowish to deep yellow wood with black markings near the heart is suitable for cabinetwork and handcrafts. Leaf decoctions and other preparations are used against leprosy, ringworm and itching skin; also as an astringent and febrifuge. The tree is grown widely as an ornamental.

MALPIGHIACEAE

The tropical and subtropical family Malpighiaceae has 77 genera with about 1,300 species mostly in the Americas including the Caribbean. This family of trees, shrubs and vines of the New World has unicellular, two-branched hairs, simple opposite leaves, and bilaterally symmetrical flowers with two large oil glands on the abaxial surface of the sepals and five clawed petals (Davis and Anderson, 2010). The family has dry or fleshy, indehiscent or dehiscent fruits. One of these fleshy fruit is the acerola (*Malpighia emarginata* DC.).

The genus *Byrsonima* has about 135 species found from southern Mexico, south-eastern Florida and the Caribbean to south-eastern Brazil. All have entire leaves, yellow flowers, and fleshy, edible fruit. They are primarily pollinated by various bees that specialize in collecting the floral oils produced by the sepals. The fruit are important in the diet of certain wildlife, such as the golden conure, a parakeet, while the leaves are food for some of the American moth-butterflies (Hedylidae) and the Tehuantepec jackrabbit, an endangered species in Mexico.

Byrsonima spicata, the maricao, probably the second best known species, is the namesake tree and floral emblem of Maricao, Puerto Rico, and figures on the coat of arms of the city. In Brazil, other species called murici are: *Byrsonima verbascifolia*, *Byrsonima coccolobifolia*, *Byrsonima basifolia*, *Byrsonima amazonica*, *Byrsonima lancifolia*, *Byrsonima crispa* and *Byrsonima orbygniana*; these are used for different purposes (Donadio *et al.*, 2002).

Golden Spoon or Nance

Nance, *Byrsonima crassifolia* (L.) Kunth. (Malpighiaceae) is the accepted name and the basionym is *Malpighia crassifolia* L. There are a number of synonyms (Bayuelo-Jiménez, 2008) including *B. coriacea* (Sw.) DC., *B. cotinifolia* Kunth., *B. cubensis* A.Juss., *B. cumingiana* A. Juss., *B. fagifolia* Nied., *B. fendleri* Turcz., *B. ferruginea* Kunth., *B. jamaicensis* Urb. & Nied., *B. karwinskiana* A. Juss., *B. lanceolata* DC., *B. laurifolia* Kunth., *B. montana* Kunth., *B. moritziana* Turcz., *B. moureila* (Aubl.) Loudon, *B. panamensis* Beurl., *B. pulchra* DC., *B. rhopalifolia* Kunth. and *B. spruceane* Nied.

This evergreen tree has one of the most popular fruits in Central America, southern Mexico and parts of Brazil. There are numerous common names such as *nance*, *nanche*, *nanchi*, *changungo*, *chengua*, *chi*, *huizaa*, *mami-hfia*, *nanantze*, *nance agrio*, *nancis*, *nanche amarillo*, *nanche dulce*, *nandzin*, *nantzincuahuiti* and *nanzinxocotl* in Mexico; *nance verde* in El Salvador; *crabo* and *craboo* in Belize; *nancito, nance, crabo* in Honduras; *doncela, maricao* in the Dominican Republic; *maricao cimaroon, maricao verde* and *peralejo blanco* in Puerto Rico; *peralejo de sabana* in Cuba; *tapal* in Guatemala; *indano* in Peru; *chaparro, maache, chaparro manteca, mantequera, nanzi, noro, peraleja hembra, yaca or yuco* in Colombia; *chaparro de chinche, manero manteco* or *manteco sabanero* in Venezuela; *murici, mixiri, murici-do-campo* and *murici-da-praia* in Brazil; *sabana kwari moeleidan, sabana mango* and *hori* in Surinam; *huria* in Guyana; *savanna serrette* in Trinidad; and golden spoon in the former British West Indies and the USA.

Origin and distribution

This species is native to tropical America and the West Indies. It is cultivated in the subtropical and tropical regions of Mexico, Central and South America. In most areas, it is grown as part of a subsistence farming system. It is little known outside of the Americas.

Commercial production occurs in Mexico, most of Central America and Brazil. In many places, nance trees are left when forests are cleared and maintained in a state of semi-wild cultivation, while in other cases seedlings are borne randomly, distributed by birds or other animals.

Ecology

Soil

The tree adapts to rocky, sandy and alkaline soils, with only moderate permeability, to soils with good drainage, fertility and organic matter. It grows in open slopes with stony soils of the tropical broad-leaved deciduous forests or in soils formed by metamorphic rocks. It is also frequently found in degraded soils. Soils with poor drainage tend to cause root rots (Vargas *et al.*, 1999).

Rainfall

The ideal precipitation is between 1,500 and 2,000 mm, but it does grow successfully in the tropics and subtropics in areas subject to rainfall ranging from 800 to 1,500 mm year^{-1}. It prefers an alternation of dry and wet seasons with the wet season lasting up to 6–8 months. Flowering occurs during the rainy season.

Temperature

The tree tolerates high temperatures with an optimum of 22–28°C. It is frequently found in the tropical lowlands, but it can also be seen in a wide range of environments from the Caribbean coasts to the semidesert regions of north-eastern Brazil, humid tropical lowlands and the middle elevations of Mexico, Central and South America. The tree can be seen from sea level to 1,800 m altitude, where winter temperatures can go down to 12°C, but normally it is found below 1,000 m. The tree does not tolerate frost (Bayuelo-Jiménez, 2008).

Light

The plant seems to do best under full sun although sometimes during its first year in the field temporary shade can be used. Production of flowers seems to depend more on soil moisture than photoperiod.

Wind

The tree is fairly resistant to strong winds.

General characteristics

Tree

The tree or large bush is slow-growing and reaches up to 5–10 m and exceptionally to 20 m. It has low branches with a round or spreading to narrow canopy. The trunk is short or tall, crooked or straight and its bark peels off in scales. The young branches are densely covered by rust-like pubescence (trichomes). The leaves are opposite, obovate to elliptic, 4–15 cm long and 4–7 cm wide (Fig. 9.4). The leaf base is acute or somewhat obtusely acute and the apices are acute to shortly acuminate, and margins are revolute. The leaves are leathery and glabrous on the upper surface, and reddish-tomentose on the underside. In young leaves, the mid-vein is prominent.

Fig. 9.4. Nance (*Byrsonima crassifolia*) showing flower, leaves and fruit (used with permisión from León, J. (2000) *Botánica de los Cultivos Tropicales*. Editorial Agroamérica, Instituto Interamericano de Cooperación para la Agricultura (IICA), San José, Costa Rica).

Flowers

Flowers are 1–1.5 cm in diameter and borne on compact terminal racemes that are 8–15 cm long (Fig. 9.4). Each flower is actinomorphic, with yellow or orange petals, green sepals and six to ten sessile glands. The separate petals are 5 mm long.

Nance has a long period of flowering with inflorescence development following a sigmoid curve. Four phases of development have been described, from the bud stage to anthesis taking about 2 months, flower opening 1–2 days, petals changing from yellow to orange (3–4 days) and petals falling over 5–6 days after flower opening (Duarte and Vernon, 2002). This goes on as long as there are flowers on the inflorescence. In Mexico, the trees flower from November through July, while in Brazil they flower from December to April. The flowers exude oil whose major components are mono- and diglycerides and free carboxylic acids instead of nectar (Vinson *et al.*, 1997).

Pollination and fruit set

These oil-producing flowers are visited and pollinated by oil-collecting bees. The most common insects visiting the flowers were the stingless bee, honey bees and sporadically wasps. Natural fruit set is around 44–46% resulting from an average of 23 flowers and ten fruit per inflorescence (Fig. 9.4). Protecting the flowers from insects with wire mesh reduces fruit set to 0.5–1.2%, indicating an almost absolute dependency on insect pollination (Duarte and Vernon, 2002).

Fruit

The fruit is a globose trilocular drupe, 1.5–3 cm in diameter (Fig. 9.4) and slightly flattened at the poles. The mesocarp is creamy to yellow, with a delicate sweet and sour flavor that can be astringent if eaten too soon. The pulp encloses one to three viable seeds fused in a woody round endocarp. The fruit growth curve is a double sigmoid that takes 130–135 days to reach maturity (Duarte and Vernon, 2002). The skin changes as the fruit approaches maturity from green to paler green, yellowish-green, yellow, orange, red, or blue when ripe. Fruit abscise when fully ripe and are very perishable.

Main cultivars and breeding

The chromosome number of *B. crassifolia* is $2n = 20$. Local cultivars show considerable variation in plant growth, fruit production, size and quality. Consistent attempts to improve cultivars have apparently not been made, presumably because cultivated plants display a reasonable pattern of productivity and fruit quality. Some selections made on fruit size, color and flavor exist but sexual propagation does not favor keeping their genotype, making asexual propagation necessary.

Several named cultivars are grown for commercial or home use in Michoacán and Veracruz, Mexico and in Central America. Some cultivars from Central America are 'Grande amarillo ácido', 'Anaranjado dulce', 'Amarillo acanalado dulce', 'Amarillo pequeño ácido' and 'Grande morado dulce' (Barbeau, 1990). In Mexico, they have selected 'Nanche dulce', 'Nanche agrio', 'Amarillo jaspeado', 'Amarillo canelo', 'Amarillo rojo', 'Sangunga', 'Cónico', 'Ácido pequeño', 'Púrpura', 'Mejorado' and 'Dulce-ácido'. Brazil has also selected types by fruit color, size and flavor. All of these cultivars typically have a superior flavor to the wild types, and are commonly consumed as a fresh fruit (Bayuelo-Jiménez, 2008).

Cultural practices

Propagation

SEXUAL Propagation is generally by seeds that are dispersed by animals, especially cattle that feed on fruit. Some seeds germinate readily in 20–45 days in a well-drained potting soil while other seed lots may take several months to germinate. There is apparently a dormancy component that can be overcome by soaking the seeds for 24 h in gibberellic acid (Alix and Duarte, 1999). Drying the seeds in the sun appears to be better than in the shade. Germination is epigeal and tree growth is slow. The juvenile period is 2–4 years from seed.

ASEXUAL Due to variability of seedling populations in fruit size, color, flavor and yield, asexual propagation is used. Rooting of hardwood cuttings has been tried with little success with leafy cuttings being more successful. Leafy terminal cuttings treated with rooting hormone and held in sealed polyethylene chambers under 70% shade resulted in 66% rooting (Duarte *et al.*, 2007). For grafting, cleft, splice and side graft are used with cleft giving about 80% success. Air layers can also be used with about 85% success rate (Duarte and Escobar, 2007).

Orchard establishment

The field should be prepared as for any other fruit orchard. Planting distances vary with soil fertility, pruning system and if the plant is a seedling or comes from vegetative propagation. Seedlings in rich soil are planted at 5 × 5 m to 7 × 7 m, so that plant densities vary from 204 to 400 trees ha^{-1}, depending on growing conditions and cultivar. Where the soil is less fertile and the cultivar is compact, the distance can be reduced to 4 m. Initial shade, mulch and a cover crop are beneficial.

Irrigation

Nance normally grows in places with a regime of rain interrupted by a dry season. This makes production occur after the beginning of the rainy season. However, controlled irrigation during the dry season can induce off-season fruit production. Out of season production is economic if the return compensates for the extra investments and yearly expenses in irrigation. Irrigation can extend the flowering and harvest seasons in areas where temperatures are favorable for growth or result in more than one flowering period per year.

Pruning

Some pruning is practiced to open up the compact canopy especially in the warmer areas with ample rain to reduce excessive tree growth. Formation pruning involves topping after reaching 1 m in height to induce lateral growth. Of these laterals, three that are better distributed in the horizontal plane and spread over 20–25 cm are selected. At later stages pruning should include elimination of diseased or damaged branches, those growing too low or into the canopy and those that are overcrowding the interior of the canopy. Trees should be kept at less than 4 m height.

Fertilization

At planting time, 5 kg aged manure and 200 g superphosphate should be added to each hole. Applications of a balanced fertilizer are recommended when the tree is fruiting. No specific fertilization recommendations exist; it is suggested to initially follow formulas developed for its close relative the acerola until more experience is obtained and adjustments can be made.

Pest management

DISEASE No serious diseases have been reported for this species. Leaf spot resulting from attack by the fungus *Meliola* sp., known as "fumagina" (sooty mold), occurs in Mexico and Brazil (Villachica *et al.*, 1996). The fungus *Gloeosporium* produces brown spots in the foliage and fruit. Under certain climatic conditions anthracnose caused by *Colletotrichum gloeosporioides* can be severe and even mummify the fruit (Vargas *et al.*, 1999).

PESTS Fruit flies can lead to substantial fruit losses and greatly limit fruit export to countries such as the USA and Japan without an approved disinfestation treatment. In Brazil, nance leaves, branches and fruit are attacked by various insects. The beetle *Macraspis festiva* eats the ripe fruit. Other beetles such as *Costalimaita ferruginea*, *Oncideres dejeani* and *Orthezia insignis* cut the branches.

Weed control

The initial months are the most critical for weed control and small plants should be protected from weed competition.

Harvesting and postharvest handling

Economic production starts in the fourth year, with yields of 3–5 t ha^{-1} of fruit in Mexico and 4–6 t ha^{-1} in Brazil. Yields can fluctuate from around 15–20 kg per tree per year to 20–30 kg per tree per year (Barbeau, 1990; Vargas *et al.*, 1999).

Harvesting

When the calyx abscises and the fruit is slightly soft when pressed between the thumb and the index fingers, the fruit is ripe. Fully mature and firm fruit are harvested after they fall but they can be picked when they start to change color. Three or four pickings will have to be made every week to avoid damage from insects and animals. At abscission, the fruit is too astringent for eating and by the next day it normally has softened to the touch and loses its astringency so it can be eaten or processed. It is possible to harvest by hand with the calyx still attached at the start of the skin color change. This fruit is then stored until the calyx abscises and the fruit is not astringent.

In Mexico, fruit are marketed from June to October while in Brazil fruit are available from December to April. In Central America, fruit is mostly available in the third trimester of the year, although with irrigation fruit can be harvested at the beginning of the year.

Postharvest treatment

Nance is a climacteric fruit that can be stored for a few days at about 12°C. The fruit can be stored for several months in good condition by submerging in water. In the markets of the lowland tropics, it is common to find nance in sealed glass containers filled with water where it can stay without spoiling for 2–3 weeks (Duarte, 2011).

Utilization

The fruit was eaten by Amerindians in pre-Columbian times from Mexico to Brazil. During the harvest season, it is consumed either fresh or processed, sold in small town markets and on roadsides, and part of the harvest reaches large city markets. There are no clear statistics about production, but the volume produced and marketed is not very great. The widespread appreciation of this fruit gives it a potential to be marketed as a fresh fruit. In addition, with its adaptation to varied climates, nance could become an important processed fruit for the lowland tropics if a market can be developed.

The pulp constitutes about 64% of the fruit, while 25% is peel that is normally consumed with the pulp, and 11% is the seed (Donadio *et al.*, 2002). The seed is high in oil and protein and the fruit is a good source of ascorbic acid (Table 9.1). The peel can have as much as 20% oil. Fruit total soluble solids range from 4.4 to 13.8%, acidity from 1.1 to 2.5% and there are 20 mg pectin per 100 g. When mature, the fruit is yellow or orange or green and has a strong odor resembling a fruity, rancid cheese or certain soaps, which is due to its 2.23% oil content. The aroma is due to ethyl butanoate (fruity, sweet), ethyl hexanoate (fruity), l-octen-3-ol (mushroom-like), butyric acid (rancid cheese), hexanoic acid (pungent cheese) and phenylethyl alcohol (floral). The pulp flavor components are mainly ethyl, methyl and phenylethyl esters and carboxylic acids, terpenoids, lactones and some sulfur-containing substances (Rezende and Fraga, 2003).

The fruit is normally consumed raw out of hand or as a drink prepared in the blender mixing the pulp, water and sugar. The pulp can be obtained by parboiling the fruit and then separating the stones or waiting until the fruit is very soft and extracting the stones or putting softened fruit in the blender at slow speed. Fermented drinks are also prepared putting the fruit in glass jars and adding rum or sugarcane alcohol (Duarte, 2011). The fruit can be preserved in liquors and vinegars; it can also be candied by boiling in syrup until it becomes solid. The pulp is also used for preparing yogurts and ice creams and a good tasting jam can also be prepared with it. The cooked fruit can be added to stews, soups and tamales. Frozen pulp can be held for several months.

Other uses of this plant include the green fruit skin as a light brown dye. The bark contains 17.5–28.3% tannin and 2.7% oxalic acid and is employed in tanning and dyeing. A strong fiber is extracted from the bark. Fresh branches are cut into small pieces and thrown into streams to stupefy fish; or they are crushed at the edge of shallow waters so that the juice spills into the water for the same effect (Morton, 1987). The hard, flexible wood is used for firewood, charcoal, furniture, tool handles and building construction (Bye and Linares, 1990). The tree is well suited to restoration of infertile and burned-over land. It is a highly drought-tolerant and fire-resistant species (León, 1983).

In medicine, a bark infusion is taken to control diarrhea and fever. It apparently tightens the teeth when there are problems with the gums. It is also used to treat snake bites, skin rashes, wounds and other problems as well as stomach ache and infections. The seeds are used to treat dysentery. The bark and leaves also have microbial biocides, and antifungal and anti-nematodal properties (Bayuelo-Jiménez, 2008).

RUBIACEAE

Rubiaceae is the fourth largest angiosperm family, often called the coffee family, and includes coffee (*Coffea* spp.), quinine (*Cinchona*), noni (*Morinda citrifolia*) and gardenia (*Gardenia* spp.). The greatest diversity in this family of mostly shrubs is found in the humid tropics, and it is recognized by its opposite simple leaves, stipules and an inferior ovary. The *Alibertia* group in the subfamily Ixoroideae (tribe Gardenieae) has two strongly supported subclades based upon molecular evidence and contains about 11 genera and 120 species. One group comprises several *Alibertia* species, including the type species (*A. edulis*) and the species in the genus *Borojoa*. The division is generally supported morphologically by fruit and corolla size, number of corolla lobes, and pollen aperture (Persson, 2000). Some authors have placed the *Borojoa* species in the genus *Alibertia*. The genus *Genipa* though in the same group is regarded as a separate genus from *Alibertia*, based on fruit size and corolla tube length.

Borojo

Borojo, *Borojoa patinoi* Cuatrec., is found in the humid forests of the western Pacific coast of Colombia into Panama and Ecuador. Eleven species have been described for genus *Borojoa* (Ricker *et al.*, 1997). A related species, *B. sorbilis* Cuatrec., is similar and regarded as a synonym by some, though it is found in Bolivia, Brazil and Peru on non-flooded soils. The name borojo is from the native Emberá language of Colombia and Panama, "boro" meaning head and "ne-jo" meaning fruit (Ricker *et al.*, 1997). Besides *B. patinoi* and *B. sorbilis*, there are other wild *Borojoa* and *Alibertia* species that bear edible fruit, occasionally seen in the markets. This is unlike most species in the Rubiaceae.

Origin and distribution

This species is cultivated in Colombia where more than 3,000 ha of cultivated orchards exist (Vargas *et al.*, 1999). Outside of Colombia the species is still little known, with some specimens being grown in Central America and other places.

Ecology

Soil

It grows in soils that are well drained and nutrient poor, such as ultisols and oxisols with high clay contents. Sometimes it is found in poor soils that are periodically flooded (Romero Castañeda, 1961).

Rainfall

This understory species is found in the humid rainforest with yearly rainfall from 1,800 up to 4,000 mm.

Temperature

Along the Pacific coast of Colombia, it is found up to 1,200 m, but it grows best in areas of less than 400 m altitude with an average temperature of 26°C and relative humidity of 85%.

Light and photoperiod

It is an understory tree that grows shorter in full sun (3.5 m). Plantings do better under semi-shade like coffee. It does not seem to respond to photoperiod since it grows well in Central America at 12°N and at the equator.

Wind

No specific information is available for this species under cultivation. It seems to be able to stand winds fairly well since the plant is low branched and not too tall.

General characteristics

Tree

Borojo is a species that reaches a height of 3–7 m with soft wood and a thin bark that peels easily. It tends to branch out from close to the ground. A single-stemmed plant will have a stem diameter at 5 years of 2–18 cm. The roots are fibrous and shallow with possible mycorrhizal associations. The large stipulate opposite leaves are lanceolate to elliptic, slightly acuminated, dark green and up to 30 cm long and 15 cm wide (Fig. 9.5) with a smooth texture (Romero Castañeda, 1961). The tree grows normally as an understory plant though orchards have been developed in full sun. The shrub grows 2–3 m in the first 3 years and about 5 m after 7 years and 7 m after 25 years.

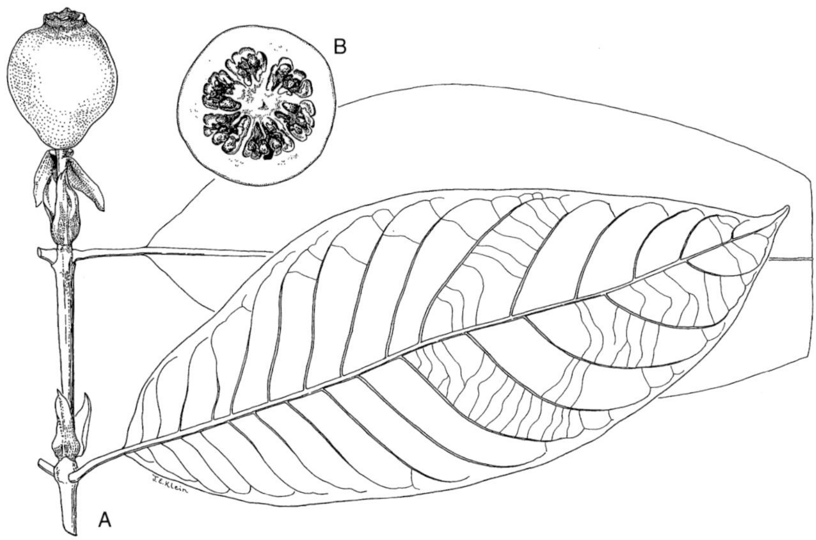

Fig. 9.5. Borojo (*Borojoa patinoi*) showing (A) leaf and fruit with (B) cross-section of fruit (used with permission from Missouri Botanical Garden from Robert E. Woodson, Jr., Robert W. Schery and John D. Dwyer (1980) *Flora of Panama*. Part IX. Family 179. Rubiaceae Part 1. *Annals of the Missouri Botanical Garden*, Vol. 67, No. 1, pp. 1–256).

Flowers

In this dioecious plant, the male flowers are white, fragrant and occur in terminal clusters. The solitary female flowers occur at the tips of branches. The corolla of male flowers is 2.5 cm long and has five petals. The calyx is pyramidal, greenish, glabrous and very short. The first flowers are produced when the tree is about 3 m tall at 1.5 years of age. Pollination is carried out by insects, hummingbirds and bats.

Fruit

The large globose-pyriform fruit, a berry, is 8–13 cm in diameter (200–1,100 g) when mature and presents persistent bracts (Fig. 9.5). The skin is thick and smooth pale green when immature and turns to dull reddish to brown when ripe with an intense and characteristic scent, which some regard as oily. The fruit pulp (90% of fruit) is white and hard becoming fleshy brown as it ripens and encloses about 300 seeds (6 by 10 mm).

Main cultivars and breeding

The Instituto Nacional Autónomo de Investigaciones Agropecuarias (INIAP) in Ecuador and Instituto Colombiano Agropecuario (ICA) in Colombia have some ecotypes grown in adaptive testing in the Amazon region. The plots of the farmers in the area of Chocó, Colombia are a good source of germplasm for this species. No named cultivars are known.

Cultural practices

Propagation

SEXUAL Propagation is by seeds that germinate in 20–45 days. Seeds can be stored for up to 3 months (Vargas *et al.*, 1999).

ASEXUAL Limited reports suggest it can be propagated by cuttings, air layering, or grafting with limited success. *In vitro* propagation of selected male and female clones is possible (Medina and Sepúlveda, 2006). Vegetative propagation would be advantageous for this species to be sure about the sex of the plants and so be able to plant 95% female and 5% male trees.

Orchard establishment

Indigenous peoples in the Chocó region cultivate the tree near river banks subject to occasional flooding. It is also planted on terraces and on small hills. Often it is planted with other fruit species in mixed orchards. At least 5% of the trees should be male plants, evenly distributed, for good pollination throughout the orchard.

Normally 1-year-old plants 30–40 cm tall are transplanted at a spacing of 5 × 5 or 6 × 6 m. The use of shade is recommended in some cases, provided by native trees or plants like *Gliricidia sepium*, and even a second shade floor could be used consisting of forest trees with commercial value (Vargas *et al.*, 1999). About 5% male plants should be uniformly interplanted.

Pruning

Some pruning to keep tree height and remove undesirable growth should be practiced, but there are no specific recommendations.

Fertilization

In Colombia, it is recommended to fertilize plants in production with 250–500 g per tree, with a 15–15–15 formula.

Diseases, pests and weeds

A number of insect pests have been reported. Leaf cutting ants, leaf scales and the larvae of an unidentified moth are among them. Anthraconose (*Colletotrichum* spp.) is a problem on the fruit.

Harvesting and postharvest handling

Harvest

The tree starts to fruit in about 3–4 years from seed and in 1.5 years from vegetative propagation and bears about 20 fruit per tree in the fifth year. The estimated yield for a plantation of 625 trees per ha can be of 30,000 fruit with 15–20 t ha^{-1}. Production can vary from year to year because of alternate bearing.

Maturity is judged when the fruit abscises. Mature fruit tends to soften and is often sold in bags or fruit nets. Fruit development takes 8–12 months.

Two peaks of fruit production occur: April through June and October through December in its native habitat (Ricker *et al.*, 1997).

Postharvest handling

The non-climacteric fruit is regarded as mature when it falls to the ground and is collected. The fruit can be harvested unripe and then ripened though the quality is lower than tree ripened fruit. Fruit should be washed and handled very carefully to take it to the market.

Utilization

Besides fruit being collected from the wild, it is also cultivated in small orchards and fruit is either sold in the local or national markets in Colombia (Ricker *et al.*, 1997). Its presence in the markets and the cultivated acreage is increasing, and it seems to be a promising crop for the region. Borojo is highly valued on local markets; 300 t are sold on the Cali market.

The fruit is consumed fresh and as a juice (*jugo del amor*), jelly, sauces, candies and wine. The juice is mixed with milk and alcoholic drinks and the pulp used for ice cream. The pulp is 88% of the fruit, the remaining 12% is seed and peel. The fruit pulp is about 30% total soluble solids, is very acidic, and is high in phosphorus and a good level of carbohydrate and calcium (Table 9.1).

The seeds are about 36.0% moisture, 1% fat, 11.0% protein, 1% ash, 13.0% carbohydrates and 39.0% crude fiber, and a high content of water soluble B vitamins.

The fruit is famous in western Colombia for its supposed aphrodisiac and health properties (Romero Castañeda, 1961). It is also reported to be beneficial to treat wounds, kidney problems, high blood pressure and cancer.

Purui Grande

Purui grande *Borojoa sorbilis* (Ducke) Cutrec. (Rubiaceae) is very similar to *B. patinoi*, the native to the Pacific coast of Colombia, also used for fruit production. The genus *Borojoa*, as mentioned above, has close similarity to the genus *Alibertia*. The synonyms are *Alibertia sorbilis* Ducke and *Thieleodoxa sorbilis* Ducke. The common names in Spanish are *borojó* and *parvi grande*, and in Portuguese *purui grande*.

Origin and distribution

This species is found at the other side of the Andes in the Amazonia area that is shared by Brazil, Bolivia and Peru. Very little distribution and cultivation occur elsewhere; some people regard it as the same species as *B. patinoi*, which developed on the other side of the Andes.

Ecology

Soils

It grows well in the acid non-flooded soils of western Amazonia, but it adapts to other areas closer to the Andes and to places like Belém do Pará (Villachica *et al.*, 1996). Good drainage is important.

Climate

The species is an understory small tree of the high tropical forest of the southwestern Amazon basin. The plant needs from 2,000 to 2,800 mm of rainfall and a mean temperature of 26°C to grow well. It has not been recorded above 300 m. No photoperiodic response is known. In its native habitat few wind problems exist.

General characteristics

Tree

This large shrub or small tree (4–5 m) has a bole up to 10 cm and opposite branches. The light brown bark is thin and sheds in large pieces. The opposite single leaves are elliptic-ovate to oblong-ovate (25–45 cm by 15–21 cm broad) and occur on 1.5–3 cm petioles and have a cordate base.

Flowers, flower induction, pollination and fruit set

The inflorescences of this dioecious species are terminal. The white male flowers arise in the leaf axils on short pedicels. The corolla of the male flower is 25–30 mm long. It has five oblong lobes 8–10 mm long and five stamens inserted in the tube on 12 mm anthers. The solitary sub-sessile female flowers are on short pedicels. The corolla is 25 mm long with six lobes. The numerous ovules are smooth and the style is longer than the corolla tube. In Tefé, Amazonia, it flowers from September to December and has fruit January to June. The juvenile period is unknown.

Fruit

The fruit is a glabrous berry (4.5–5 cm long × 6–10 cm diameter). The skin is smooth, hard and leathery (3–6 mm thick). The mesocarp is dark brown when ripe and encloses numerous flattened lens-shaped 8–20 mm long seeds.

Cultivar development

Some diversity exists in the fruit from globular to pear-shaped. The great similarity between *B. sorbilis* and *B. patinoi* raises the possibility that both are different varieties of the same species. The Instituto Nacional de Pesquisas da Amazônia (INPA) in Brazil has three accessions of wild ecotypes. However, no selected cultivars exist.

Cultural practices

Propagation

SEXUAL The tree is normally propagated by seed. The seeds are recalcitrant, being viable for about 7 months if kept cool. The percentage of germination of fresh seeds is around 80%. The seed should be taken from mature fruit of selected trees, washed with water and dried in the shade for at least 2 days. Then it should be germinated in a substrate of decomposed sawdust, sand or top soil kept wet and under about 75% shade with emergence after about 25 days and up to 55 days. After germination when the cotyledons have emerged and the hypocotyl is straight and the seedlings look like a soldier or a match they are transplanted into nursery bags and shade is gradually reduced to 50%. The weight of 1,000 seeds is about 220 g (Villachica *et al.*, 1996).

ASEXUAL Vegetative propagation is desirable for this dioecious species to be sure about the sex of the plants and to obtain 95% females. Rooting cuttings, grafting and air layering are possible. Root cuttings between 2 and 5 cm in diameter and 30 cm or longer from female plants are used. The most suitable substrate is a mixture of sand, moss and organic matter (2:1:1) or rotted sawdust. The nursery should be shaded and the relative humidity greater than 85% (Villachica *et al.*, 1996). Grafting is also possible, as is air layering.

Orchard establishment

The trees grow slowly in the nursery and normally are ready to be transplanted after about 1 year when they reach between 30 and 40 cm in height. Planting distances are sometimes 4–6 m between plants and 6 m between rows with 625 trees per ha. Seedling growth is slow and may only achieve 1–2.5 m in 3 years. At least 5% of the trees in an orchard should be male plants, evenly distributed to ensure good pollination.

Pruning

The plant should be pruned following the main harvest to maintain the height at 3.0 m to facilitate cultural operations and harvesting.

Irrigation practices

No information exists since no commercial plantations have been installed.

Fertilization

No specific recommendations exist. Initial trials should use similar amounts to those used for guava or related crops.

Diseases, pests and weeds

The species has no known diseases, but physiological problems including iron and boron deficiencies do occur on calcareous soils. Sunburn causes black spots to appear on the fruit skin that subsequently crack.

Harvest and postharvest handling

Harvesting

Fruiting from seed begins in the fifth to sixth year and the annual yield is between 5 and 6 kg of fruit per plant. The fruit is collected at full maturity from the wild and from cultivated trees, and normally it is picked from the ground. At full ripe stage, the fruit are very perishable necessitating daily collections of fallen fruit to prevent deterioration of fruit on the ground and loss to animals. The leathery skin provides some protection from mechanical injury.

Postharvest handling

The fruit is apparently non-climacteric therefore it should be harvested as close as possible to the time it will fall. After collection, the fallen fruit are washed and kept in a shady, well-ventilated location. Less than fully ripe fruit can be harvested and further ripened at 95–100% RH and greater than 20°C. Weight loss during ripening can be a problem.

Utilization

The fruit mesocarp is consumed fresh and used to make juices, jams, jellies, ice cream, candies and wine. The pulp is pasty and subacid with a flavor somewhat like tamarind. The edible flesh is about 30% of the fruit and 60–70% is water (Table 9.1).

Genipap

Genipa americana L. has numerous synonyms, including *Gardenia genipa* Sw., *G. americana* var. *caruto* (Kunth) Schum., *G. barbata* Presl, *G. caruto* Kunth, *G. codonocalyx* Standl., *G. cymosa* Spruce, *G. excelsa* K. Krause, *G. grandifolia* Pers., *G. nervosa* Spruce, *G. oblongifolia* Ruiz & Pav., *G. pubescens* DC., *G.spruceana* Steyer., and *G. venosa* Standl. It is commonly known as genipap, marmalade box (English), *genipayer* (French), *jagua, caruto, majagua* (Colombia), *caruto, xagua* (Venezuela), *bi, bi grande, ñandipa, danipa, bitu* (Bolivia), *huito, yagua-yagua, jagua* (Peru), *genipa, genipa tree* (Guyana), *taproepa, tapoeripa* (Suriname), *guaytil blanco, jagua blanco* (Panama), *irayol* (Guatemala), *tiñedientes* (El Salvador), *tapaculo, ygualti* (Nicaragua), *guaitil* (Costa Rica), *jenipapo, genipapo* (Brasil), and *jagua azul, maluco* (Mexico) (Cavalcante, 1991; Donadio *et al.*, 1998).

Origin and distribution

Genipap is still found wild in northern South America. It has been widely distributed as a cultivated fruit since pre-Columbian times. It is now distributed throughout the humid neotropics from southern Florida and Mexico, Central America and the Caribbean islands to northern Argentina and south-east Brazil (de Freitas and Salles, 2008). It is also found in the pre-Andean area below 700 m of Bolivia (Vásquez and Coimbra, 2002). Although widely distributed, it is not produced commercially. Prior to the European arrival, the most important product extracted from this plant was a juice that was used by Native Americans to paint their bodies a dark bluish-black color and was believed to have magical and medicinal properties.

Ecology

Soil

The trees grow well in a variety of humid area soils. Fertile deep soils on mild slopes, with a pH between 4.6 and 6.5 and with good water-holding capacity are preferred. It is also found in transition forests with heavy soils that do not flood. In the Amazon basin, it grows along both clear water and sediment-rich rivers, but not along black-water rivers.

Rainfall

The plant is adapted to continuously humid habitats; in some parts of its natural distribution there are dry seasons that last up to 3 or 4 months. During the dry season, the tree sheds its leaves. The species seems to grow better in areas with rainfall between 1,200 and 4,000 mm.

Temperature

The best growth occurs in areas with an average temperature between 18°C and 28°C. This species grows naturally between sea level and 1,500 m altitude. It is sensitive to low temperatures and frost.

Light and photoperiod

The plant grows best in full sun with abundant water, characteristic of semi-deciduous rainforests located in swampy areas. In the forest succession, the species is considered to be late secondary, with some climax characteristics.

General characteristics

Plant

The stoutish tree may reach 10–20 m in height and 30–80 cm bole diameter, with abundant branching but sparse foliage. The trunk is generally straight due to apical dominance, and has a somewhat thick, smooth bark, of a green-grayish color. The leafy canopy has generally horizontal branches (Vásquez and Coimbra, 2002), although the lower branches may be hanging. The leaves are simple, opposite, with a dark green glossy upper side, obovoid to oblong in shape, 10–35 cm in length, acuminate or rounded at the apex and completely glabrous on both sides (Fig. 9.6). The petiole is short. They have a marked central vein and lateral veins are alternate and run parallel.

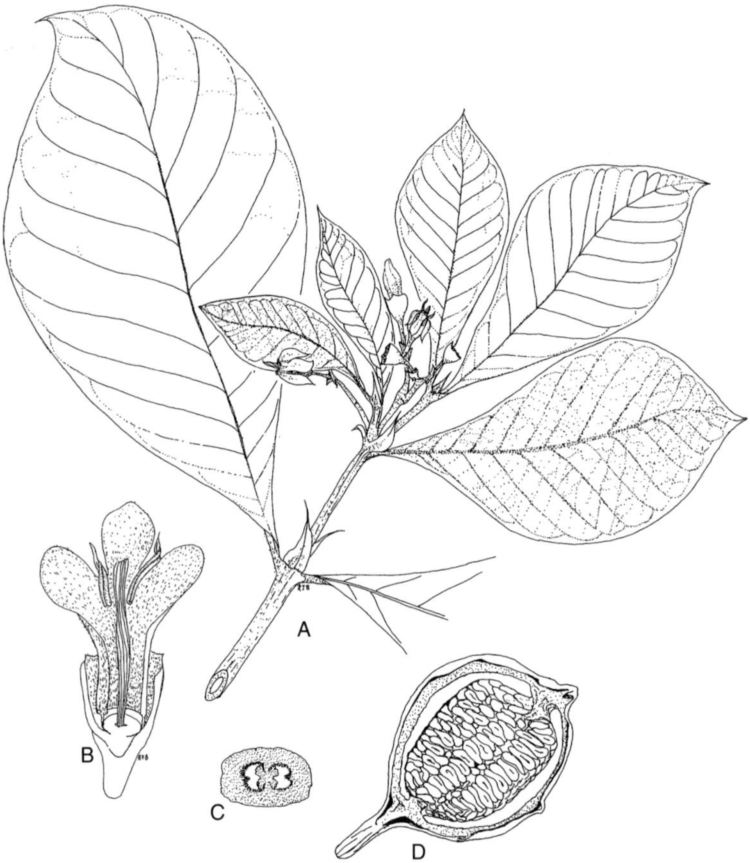

Fig. 9.6. Genipap (*Genipa americana*) showing (A) stem with flower, (B) flower, (C) fruit and (D) cross-section of fruit (used with permission from Missouri Botanical Garden from Robert E. Woodson, Jr., Robert W. Schery and John D. Dwyer (1980) *Flora of Panama*. Part IX. Family 179. Rubiaceae Part 1. *Annals of the Missouri Botanical Garden*, Vol. 67, No. 1, pp. 1–256).

Flowers

The flowers are large, hermaphrodite, shaped like long pipes (Fig. 9.6), and arranged in small cyclical sub-cymose inflorescences that are resinous and congregate in axillary terminal groups, sometimes with few or even only one flower. The 2–4 cm petals are joined at their bases, are white and as soon as they open they become yellowish, contrasting with the dark green leaves; there are five stamens and the flowers are slightly aromatic.

Pollination and fruit set

The species is monoecious, but its reproductive system is a mixture of outcrossing and selfing. Pollination is entomophilic, accomplished by several species of large bees, including *Bombus morio* and *Epicharis rustica flava* (Crestana, 1993). In Amazonia, it blooms in October, at the end of the dry season, or, in a few years, until December, at the beginning of the rainy season. Fruit ripen mainly in the rainy season, from January to March, becoming dispersed by gravity, animals and water.

Fruit

The fruit is a globose, ovoid berry, 7–10 cm diameter × 6–15 cm length, weighing 200–500 g (Fig. 9.6). When immature, the fruit has a gray color that becomes tan, russet-brown or red-brown when ripe and is slightly aromatic. The pericarp is yellowish-gray, spongy and 1.5 cm thick. The grayish pulp is succulent, fibrous, slightly acid and astringent and surrounds as many as 480 seeds per fruit (mean 300). The seeds are cream colored, flattened and measure 0.7 × 0.8 cm.

Cultivar development

Since the species is not extensively cultivated no named cultivars have been reported, neither is there a breeding program. Researchers in Puerto Rico do distinguish differences between trees (Kennard and Winters, 1960).

Cultural practices

Propagation

SEXUAL The species is generally propagated by seeds. The seeds are semi-recalcitrant and should be sown fresh. Germination of fresh seed is epigeal, fast and uniform, occurring between 25 and 30 days, and can attain 90%. Stored seed after de-pulping and drying to 10% humidity may last 3–4 months with a reasonable germination rate remaining while with 5% moisture the seed dies. When the plant has two leaves, it is transplanted into bags. The growth of the seedlings is slow, taking up to 9 months to attain an appropriate size for planting out (40–50 cm).

ASEXUAL Vegetative propagation is by grafting or patch budding and taking away the petiole can result in about 60% success rate. The vegetative propagation assures greater precocity and maintenance of the mother plant's characteristics (Villachica *et al.*, 1996).

Orchard establishment

The soil should be prepared by plowing and discing. The spacing for seedling plants is 10 × 10 m, which allows interplanting with annual or semi-perennial crops in the first years. The 60 × 60 × 60 cm planting pits should be prepared with manure and 200 g of superphosphate.

Irrigation

It grows normally in areas with enough rainfall and it is not grown intensively, in which case probably supplemental irrigation would be needed in case the rainy season is too short or during longer dry periods.

Pruning

Since new flowers and fruit develop on new branches, the plant should be pruned after fruiting by eliminating older branches (Donadio *et al.*, 1998). This will stimulate new shoots that will flower next season.

Fertilization

Though no chemical fertilizer trials have been reported, 10 kg of manure per plant per year is recommended.

Pest management

A variety of insects are reported to occur on the tree, but none cause serious damage. The plant is attacked by leaf cutting ants, trunk borers and foliage-eating caterpillars. The wood is susceptible to rot and to termite attack.

Weed control

Due to slow initial growth, the planting area must be kept clean of weeds in at least the first 3 years.

Harvesting and postharvest handling

Harvesting

The fruit are highly perishable, spoiling in a few days after harvest. Fruit should be harvested from the tree before they are fully ripe and abscise. A mature 15–20-year-old tree can produce 400–600 fruit (Jøker *et al.*, 2003).

Postharvest handling

Harvested mature fruit is left to soften (Kennard and Winters, 1960). Abscised fruit can suffer mechanical injury when they fall to the ground that reduces their postharvest life and quality. It is reported that they can be stored up to 4 weeks at 10°C. Fruit can be held up to 15 days at 10–15°C and used immediately. Below 10°C the epicarp is "burned" and turns black due to chilling injury.

Utilization

The fruit is not very appealing to the eye for those unfamiliar with the fruit. It has a sour, strongly flavored pulp that is consumed fresh or, more commonly, used to make juices, wines, liquors, syrups, jellies and crystallized candies. Sometimes it is macerated in water before preparing drinks, which causes it to have a strong pungent aroma but it will keep for a while. Its juice is often mixed with cane alcohol to prepare beverages called *huitochado* (in Spanish) or *jenipapada* (in Brazilian Portuguese). The fruit contains reasonable concentrations of two B vitamins, riboflavin and thiamine (Table 9.1) (Villachica *et al.*, 1996).

The immature fruit supply a clear yellowish juice that gradually darkens to a deep blue-black. This juice was used by Native Americans to paint their bodies, hair, clothes and other objects. Today it is used in the same way by indigenous and traditional handicraft artisans. The Native Americans also used the pulp as a dental anaesthetic and insect repellent. It has bactericidal and germicidal properties because of various phenolic substances present in the fruit.

The wood is white and easily worked, and occasionally used in civil construction, small boats, carpentry and handicrafts. It has an average density, 0.68 g cm^{-3}, is flexible, compact and has reasonable durability.

LECYTHIDACEAE

The family has about 20 genera and 300 species of woody plants that are native mainly to South America with one genus *Foetidia* in Madagascar and another subfamily in Asia through to northern Australia. In South America, they are recognized for their showy flowers and large woody fruit. The species dominate the Amazonian forests, occur at all levels in the canopy and have coevolved adaptations for pollination by bees and bats and dispersal by animals, wind, and water. The Amazon basin is regarded as the center of diversity for the Lecythidaceae.

The Brazil nut is the only species of the genus *Bertholletia*. The genus name was given by the Prussian botanist-explorer Alexander von Humbolt after his friend the French chemist Claude Louis Berthollet. Other American genera of this family that produce nuts belong to the genus *Lecythis*, which includes species such as *L. zabucajo* Aulbet (the paradise nut) and the so-called monkey pot nuts that include *L. ollaria* Loefling, *L. pisonis* Camb. and *L. minor* Jacquin. Other species of the family are the red mahot *Eschweilera corrugata* Miers; membrillo (*Gustavia superba* (Kunth) Berg) native to Colombia; and the anchovy pear (*Grias cauliflora* L.) native to Colombia and the West Indies. From Fiji to Vanuatu and the Solomon Islands, the cut-nut (*Barringtonia edulis* (Miers) Seeman) is very popular with some related species such as *B. procera*, *B. novae-hiberniae* and *B. seaturae* growing in the same area, plus *B. racemosa* that grows in the Philippines.

Brazil Nut

The taxonomic literature indicates there may be two species of *Bertholletia*; however, it appears *Bertholletia excelsa* Humb. & Bonpl. and *Bertholletia nobilis* Miers. (an earlier species name) are the same plant (Bailey and Bailey, 1976). This nut is commonly called Brazil nut or Para nut in English; *castaña*, *castaña de Brasil*, *nuez de Brasil*, *castaña brasilera*, *jubia*, *yubia*, *yuvia*, *coquito del Brasil* and *topa* in Spanish; *chataigne du Bresil*, *amande d'Amerique*, *noix de Bresil*, *noix d'Amerique*, *noix d'Para* in French; *Paranuss*, *Brasilianische Kastanie* in German; *para not* in Dutch; *noce del Brasilie* in Italian; and *castanha-do-Para*, *castanheira*, *tocari* and *castanhado Maranhao* in Portuguese (Fouque, 1972).

Origin and distribution

This species is indigenous to non-flooded areas of the Amazon forests of South America, especially Brazil, Bolivia and Peru (Villachica *et al.*, 1996), but it is also found in Colombia, Venezuela, Guyana, Ecuador and Surinam (Donadio *et al.*, 2002). Brazil nut has been taken to other tropical areas but has not become important except in Cote d'Ivoire. This important nut is generally gathered from the wild trees in the Amazon and exported to the USA and the EU. The major producers are Bolivia, Brazil, Peru and Cote d'lvoire. In Brazil, the main area of wild plants is the Marbá district in Brazil's northern state of Pará (Villachica *et al.*, 1996). Many thousands of trees have been cut down in this area through deforestation (Donadio, 1983). Information on the area formally planted to Brazil nut is incomplete, although Smith *et al.* (1992) reported on at least one 4,000 ha planting in Brazil.

Ecology

Soil

The tree occurs naturally in lowland rainforests above the flood level suggesting it does not tolerate continuously wet soils. It grows in heavy to medium textured soils from clayey to sandy-clay (Mueller and Calzavara, 1989; Verheij and Coronel, 1991) but it prefers deep rich oxisols or ultisols with good drainage.

Rainfall

It is found in hot humid tropical areas with an annual rainfall of 1,400–2,800 mm but with a dry period with rainfall less than 100 mm per month and where relative humidity is between 80 and 90% (Donadio *et al.*, 2002; Villachica *et al.*, 1996). This dry period seems to be necessary for good production (Smith *et al.*, 1992). Rainfall patterns and topography affect timing of flowering and harvest season.

Temperature

Brazil nut does best in hot, humid, tropical lowland conditions between 0 and 500 m elevation. Annual temperatures in its original habitat average between 24 and 27°C (Mueller and Calzavara, 1989), with minimums of 20–24°C and maximums of 30–33°C (Villachica *et al.*, 1996).

Light and photoperiod

The tree grows better if it is under partial shade during the first 6–7 years (Fouque, 1972). Total sunshine hours in its native area fluctuate between 2,000 and 2,500 h year^{-1} (Villachica *et al.*, 1996). There seems to be no problem with photoperiod, since the plant flowers normally in the places where it has been planted. The best places for production are equatorial areas with high light intensity (Donadio, 1983).

Wind

Wind is not much of a problem in the areas where the trees grows wild. It has a well-structured stem and will probably resist fairly strong winds.

General characteristics

Tree

This is a very large tree that can reach 30–60 m in height with a trunk diameter of 1.0–1.2 m, exceptionally 4.3 m (Donadio *et al.*, 2002). The cylinder-shaped trunk is smooth and has practically no lateral branches, except at the top where the canopy is formed. The canopy can reach 10–20 m. The deciduous leaves are simple, oblong or elliptic, glabrous, alternate, shiny light green on the upper surface and dull light green or yellowish in the lower surface, 17–50 by 6–15 cm (Fig. 9.7), with an acuminate apex and obtuse base, a 2–6 cm slightly winged petiole, and conspicuous nerves in the underside (Villachica *et al.*, 1996).

Flowers

Flowers are borne in upright axial or terminal panicles 20–40 cm long with few ramifications and a 16–20 cm straight rachis (Fig. 9.7). The individual sessile or sub-sessile bisexual zygomorphic flowers are 5 cm in diameter and have three ovate bracts. There are two to four light green sepals, which are bi-lobed at anthesis, four to six unequal white-yellowish thick and fleshy petals, numerous stamens (80–135) that are united on the lower side of the flower and an inferior ovary with four or five locules (Fig. 9.7). Each locule has four to six ovules. There is a certain degree of auto-incompatibility.

According to Motta Maués (2002) the main flowering period in Brazil is during the months of low rainfall (September to December) although some plants can flower from March to June. The maturation period extends throughout the following year with ripening and dissemination occurring during the rainy season. It takes 15–17 months from anthesis to fruit drop. The flower to fruit ratio is a very low 0.4%. The productive branches grow below the inflorescence of the previous year; at first the leaves appear and later the flowers at the end (Mueller and Calzavara, 1989).

Pollination and fruit set

Flowers are self-incompatible so that cross-pollination is necessary for optimum fruit production. In a study done in Brazil by Motta Maués (2002), it was established that the pollen to ovule ratio was 26,755 indicating obligate xenogamy. The main pollinators are medium- to large-sized carpenter bees and bumble bees, belonging to the Apidae and Anthophoridae families and the following species: *Xylocopa frontalis*, *X. aurulenta*, *Epicharis rustica*, *E. affinis*, *Centris similis*, *Eulaema nigrita*, *E. cingulata*, *Bombus brevivillus*, and *B. transversalis*. It was concluded that *B. excelsa* is a mellitophilous species dependent on the bee pollinator's activity to ensure fruit production. The decline of natural pollinators in commercial plantings has led to lower fruit production. Management programs of the main pollinators are needed for wide-scale cultivation of the nut. Other authors indicate that Brazil nut flowers can only be pollinated by an insect strong enough to lift the coiled hood on the flower (Fig. 9.7) and with a tongue long enough to be able to get the contents of the complex flower. There is an orchid that produces a scent that attracts small male long-tongued orchid bees (*Euglossa* spp.), and the male bee needs that scent to attract females. Without

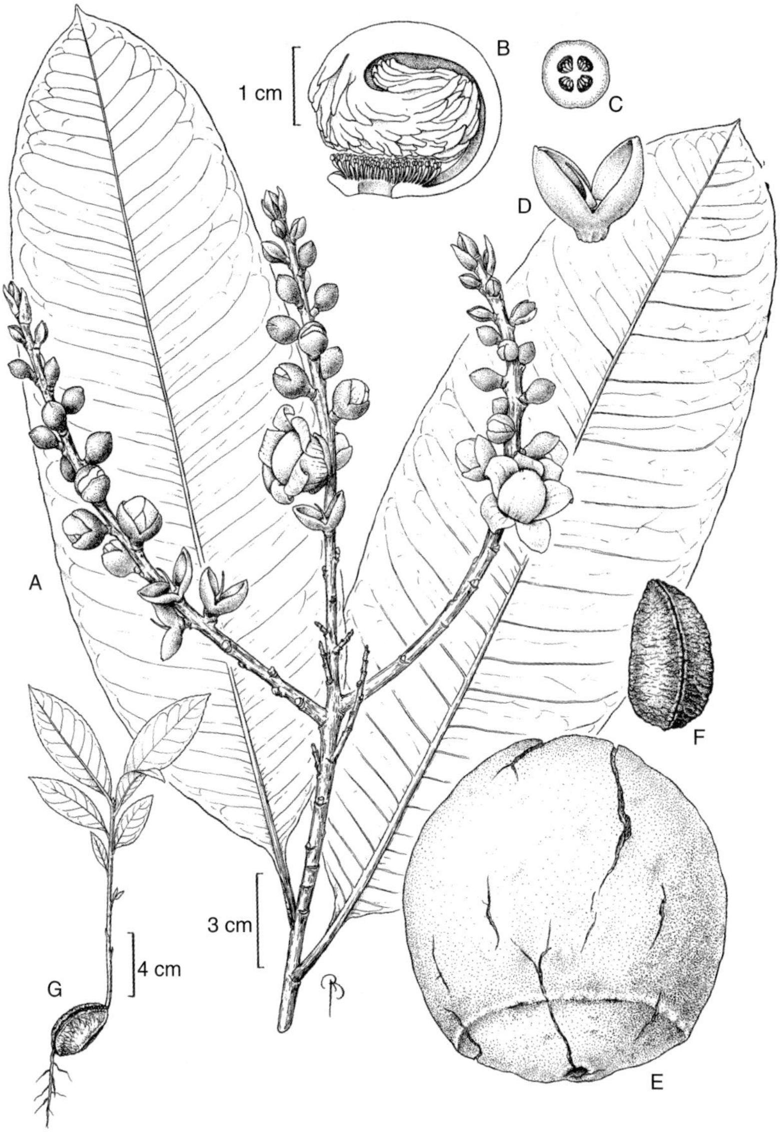

Fig. 9.7. Brazil nut (*Bertholletia excelsa*) showing (A) leaf, (B) cross-section of a flower, (C) cross-section through ovary, (D) bi-lobed bracts, (E) fruit, (F) seed or nut and (G) seedling (illustration by Bobbi Angell; reprinted by permission of artist and Dr Scott Mori, The New York Botanical Garden, copyright 1986).

this orchid, the bees would not mate and the lack of bees means flowers would not be pollinated.

Fruit

The light brown fruit has the shape of a globe or a slightly flattened sphere (Figure 9.7). It is an indehiscent woody capsule, called a pixidium, 10–25 cm in diameter (Fig. 9.8), with a thick outer shell, weighs 0.2–2.0 kg, and contains 12–18, sometimes 25, angular seeds (Donadio, 1983). Each seed is 4–7 cm long, with a brown woody shell, weighs 4–10 g and encloses an edible white, milky kernel (Fig. 9.8). Fruit take about 12–16 months to mature from anthesis. Fruit are generally gathered during the early months of the rainy season with the harvest period lasting 2–3 months. Agoutis can chew through the woody pericarp and disperse some seeds (Crane, 2008).

Cultivar development

Brazil nut has a chromosome number of $2n = 24$ (Barbeau, 1990). There is a wide genetic variability that can be seen in the size and shape of the fruit and the kernels, the time of flowering, harvest and leaf fall, and the plant shape and size.

Selection and evaluation

In Belém do Pará, EMBRAPA Centro de Pesquisa Agropecuária do Trópico Úmido (CPATU) has more than 40 accessions obtained from the Brazilian Amazon and cloning has been done. Other accessions from the Amazonas state that have

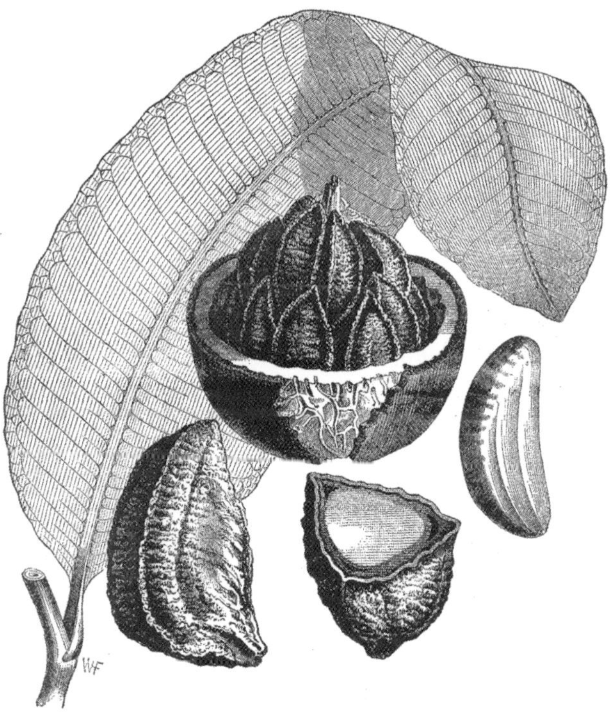

Fig. 9.8. Brazil nut (*Bertholletia excelsa*) showing fruit and seeds arranged inside the woody capsule (from Project Gutenberg eBook 11662, originally published in *Scientific American Supplement*, No. 598, June 18, 1887).

larger fruit and kernels are growing at the Aruana farm in Itacoatiara (Villachica *et al.*, 1996). The increasing economic importance and potential of Brazil nuts have led to some selection for superior types in Brazil. Other countries such as India, Malaysia and Peru are also investigating the potential of Brazil nut production using superior plant material (Smith *et at.*, 1992). Basically high-producing clones that are genetically compatible have to be selected.

Cultivars

The selected accessions will eventually become cultivars since they are being clonally propagated, but no named cultivars currently exist, with wild trees being the main source of nuts.

Cultural practices

Propagation

SEXUAL Seeds have been the most common method of propagation though they are slow and difficult to germinate and can take from 6 months to up to 2 years to start germinating. The percentage germination is also low. Apparently, the seed coat is a physical barrier for embryo expansion. Scarification with sulfuric or formic acid or acetone has had no effect on reducing this delay.

The best method for germinating the seeds is to remove the seed coats or soak in phenyl mercuric acetate, which can result in 78–80% germination after 3 months starting in 20–30 days (Donadio *et al.*, 2002). Important aspects include the use of fresh and large seeds, with white-colored kernels and without oily patches. A sample can be tested by cutting the seeds in half. To remove the seed coat, the selected seeds are soaked in water for 3 days, water is changed daily and the seeds exposed to the air for 10–15 minutes so they do not ferment. The soaked seeds are pressed with a little pressure to crack the seed coat, which is removed using special pliers. The peeled seed is soaked in a fungicide solution before being sown in raised beds filled with a 50:50 by volume mixture of aged sawdust and clean sand. The substrate should be disinfected or sterilized to prevent damping off. The seed should be sown with the widest end facing the bottom of the seedbed and covered with 1 cm of media (Mueller and Calzavara, 1989). The raised seedbed should be under semi-shade and after emergence the plantlets should be transplanted into 16–18 by 28–30 cm plastic bags filled with a mixture of 25% aged sawdust and 75% local soil.

Seeds retain their viability for 1 year. Selected seeds should be stored under shade for 2 days as exposure to the sun significantly reduces germination percentage. Seeds can be stored in a cool place and misted periodically with water to avoid excessive dehydration and sown over several months (Mueller and Calzavara, 1989). Seedlings will take 8–12 years to start producing (Martin *et al.*, 1987) and up to 20 years in some cases.

Plant regeneration from culturing immature embryos has been tried with promising results in Brazil. Embryos are separated aseptically 4–6 months after pollination when they are the size of a wheat grain and put in a Murashige &

Skoog medium supplemented with naphthalene acetic acid, benzyl amino purine, NaH_2PO_4 and activated charcoal, plants regenerate three to four weeks after (Villachica *et al.*, 1996).

ASEXUAL This is the only way to guarantee that plants with superior characteristics are obtained. These include precocity, as well as fruit and tree size. In order to achieve this, material from selected trees and, in the case of budding, the buds should be obtained from portions below last year's inflorescence and preferably from lower branches of grafted plants. Budding is performed as for rubber tree, using the shield budding method, and normally this is done in the field with seedlings that were transplanted 18–24 months before, when they are about 2 m in height. Their bark should be slippery, which indicates their cambium layer is active; this normally coincides with the presence of new leaves. The branch for obtaining the buds should be as thick as the plant to be budded and its leaves should be removed 7–10 days before taking the buds to activate them and facilitate the removal of the shield of bark containing the bud that should be budded preferably the same day. Budding is normally done at about 1 m height trying to make the superior cut of the bud shield coincide with the horizontal cut made in the rootstock to insure cambial contact. If budding is successful, 1 month later girdling is performed removing a 3–5 cm strip of bark 1 cm above the budded shield to break apical dominance and to stimulate bud growth. Budding success can be 90%. The new shoots can be plagiotropic or orthotropic; in the first case they should be straightened and tied to the stock above the girdle to grow vertically (Mueller and Calzavara, 1989). Vigorous seedlings are used as rootstocks. Budded trees begin to bear 3.5–6 years after budding.

Orchard establishment

If a commercial plantation is planned, the normal soil preparation procedure should be followed if in a previously cultivated land, otherwise the forest has to be cleared of trees and bushes and the debris eliminated before the holes are marked and prepared.

Most of the world production comes from wild trees and it has been only fairly recently that some commercial plantations have been made, especially in Brazil. The economic success of these planting has not been determined. It is important to mention that if budded plants are to be used more than one clone should be combined in the field to overcome the problem of a certain degree of auto-incompatibility of this species. Normally plants are taken to the field when 30–40 cm tall and planted in holes 40 by 40 by 40 cm. The best time to do this is at the beginning of the rainy season.

Spacing for seedling trees is from 10 to 12 m in the row and 10–15 m between rows (Paytan *et al.*, 1997) or 10 × 10 m in squares or triangles. If the crop will be combined with pasture land 10 × 20 m or 15 × 25 m should be used to allow more light for the pasture. If combined with perennial crops 10 × 25 m or 15 × 25 m should be used (Mueller and Calzavara, 1989). In many cases associated crops like citrus, cocoa, guarana, araza, cassava, pineapple or pepper are inter-planted and cultivated as long as there is sufficient light reaching their canopy. For forestry purposes, 4 × 4 m can be used and no budded plants are needed.

Irrigation

This is normally not done; possibly if large plantations succeed, irrigation will be a factor to be taken into account in drier areas.

Pruning

Training of trees begins 2 years after grafting and normally two types of pruning are recommended, the first for stem formation and the other for canopy formation. The first aims at clearing the stem base by eliminating all basal shoots below 2 m. The second aims at increasing the number of branches forming the canopy and is performed if the budding has produced a reduced number of shoots; in this case these few shoots are cut back to about 50–100 cm from the main stem and four to five leaves below this cut are eliminated to force the corresponding axillary buds to shoot (Paytan *et al.*, 1997). This pruning is done after the plagiotropic shoots coming from the budding have been re-oriented to a more vertical growth position by tying them to the part of the main stem that remains above the bud union.

Fertilization

At planting time some aged manure and 50–100 g triple superphosphate are applied in the bottom of the planting hole. After that 200–300 g triple superphosphate are applied in the second and third year respectively.

Pest management

DISEASES No major diseases have been reported. Fungal diseases that can attack Brazil nut trees include *Cercospora bertholletia* and *Phytophthora heveae* that causes budding failure. Periodic inspections are recommended as for any crop (Villachica *et al.*, 1996).

INSECTS Several insects attack Brazil nut trees including leaf cutting ants (*Atta sexdens*) and red flour beetle (*Tribolium castaneum*). Adequate measures should be taken in every case (Villachica *et al.*, 1996). Rodents can also be a problem.

Weed control

In commercial plantations, the same procedure used for other fruit trees should be followed. If there is a pasture, it will serve as a weed control and if associated crops are interplanted they should be kept clean. During the first years, the area close to the young plants should be kept clean either by manual weeding, or by the use of appropriate herbicides. Mulching around the plants using the cut weeds can be helpful.

Orchard protection

No information is available. Normally the tree is scattered among other trees and plants of the forest. In commercial plantations, probably a wind barrier would have to be installed in windy areas, especially for the initial years. At a later age, the tree is very strong and should not have problems.

Harvesting and postharvest handling

Yields

Thirty-year-old orchard trees produced 300 kg per tree per year and with 25 trees ha^{-1} about 7,500 kg ha^{-1} of nuts can be obtained (Paytan *el al.*, 1997). Other authors indicate 100–300 fruit per tree per year. In Brazil, production is expressed in liters of nuts and estimates are that a 12-year-old plant can produce 25 l nuts per year, which compares positively with 16–55 l ha^{-1} $year^{-1}$ from wild plants where normally there are three to four plants per hectare (Villachica *et al.*, 1996).

Harvest

The large, heavy Brazil nuts are collected after falling to the ground and then extracted from the exocarp with a hatchet or machete where an experienced harvester can open the fruit with a single strike and harvest between 100 and 200 l day^{-1} (Villachica *et al.*, 1996). The seeds are immediately washed and dried in the sun for a few hours before being taken to the processing plant. Harvest time coincides with the heavier rain season, which calls for good ventilation of the nuts after harvest to avoid fungal infections. Delays in drying and poor storage can cause an aflatoxin problem due to *Aspergillus* spp., resulting in the rejection of the product by the importers (Vargas *et al.*, 1999).

Postharvest handling

Once harvested the nuts are transported to urban centers that are normally far away and spoilage may occur during this journey. The dried and sorted nuts may be stored for one year if RH is kept at about 70% (Paytan *et al.*, 1997). The raw nuts are cleaned in the processing centers, where they are filled in vacuum-sealed aluminum bags, and packed in cardboard boxes. Processing yields are around 31% for shelled nuts and 64% for in-shell nuts.

Utilization

Brazil nut is eaten fresh or roasted or can be pressed to obtain "milk" and an edible oil (60–70% of the nut). The nut has a rich flavor and may be used in salads, pastries, gruels, flour and ice creams. The oil is consumed or used in cosmetics, soaps or as a lubricant.

Brazil nuts are high in energy, protein, phosphorus, potassium, iron and selenium (Tables 9.1 and 9.2). Selenium is an important anti-cancer agent, especially prostrate cancer. A single Brazil nut contains more selenium than the US Recommended Daily Allowance (RDA) for this element. Chemical analysis of Brazil nut oil has found that it primarily contains oleic, palmitic, linoleic and alpha linolenic acids along with tiny quantities of stearic and myristic acids as well as phytosterols. Proteins in Brazil nut are very high in sulfur-containing amino acids, such as cysteine (around 8%) and methionine (around 18%). They also contain high amounts of arginine, glutamic acid and glutamine.

In traditional medicine extracts from the nut are used to treat hepatitis, inflammation and hypersensitivity (Paytan *et al.*, 1997). The wood is used for

Table 9.2. Main mineral elements in Brazil nut seeds (from Villachica *et al.*, 1996).

Mineral element	ppm
Aluminum	5
Boron	2.7
Barium	1764
Calcium	1592
Cadmium	0.03
Potassium	5405
Magnesium	3370
Selenium	11
Silicon	1770
Zinc	41
Iron	93.0
Copper	18.0
Chlorine	78.0

construction and the naval industry, and to make furniture and wood chips. The hard outer shell and the seed coats are used as fuel or for handicrafts.

STERCULIACEAE

The boundaries of the families Sterculiaceae, Tiliaceae and Bombacaceae in the order Malvales have always been problematic. In more recent taxonomy revisions, these have been included in the expanded family Malvaceae as we saw for the South American sapote (*Quararibea cordata*) described above. The family Sterculiaceae is now in the subfamily Sterculioideae in the family Malvaceae. Formerly the Sterculiaceae contained 70 genera with around 1,500 species of tropical trees and shrubs. Commercially the most important products were chocolate from *Theobroma cacao* and the caffeine-containing cola nuts mainly from two species in the genus *Cola*, *C. nitida* and *C. acuminata*.

Copuazu or Cupuassu

Theobroma grandiflorum (Willd. ex Spreng.) Schum. is named cupuassu in English, *cupuaçu*, *pupu*, *pupuaçu* in Portuguese and *copuasu*, *cupuasu*, *cacao de Perú* and *cacao blanco* in Spanish, *bacau* in Colombia. It is a close relative to cacao (*Theobroma cacao* L.) and to mocambo (*Theobroma bicolor* Humb. & Bonpl.). It was also named as *Bubroma grandiflorum* Wiild. ex Spreng. and *Theobroma macrantha* Bernoulli (Villachica *et al.*, 1996).

Origin and distribution

Its origin is south and south-east Amazonia in Brazil (Andersen and Ulup-Andersen, 1988). It is found wild in the south-west of the states of Pará and

Maranhão. It has been introduced on a limited scale to other parts of the tropics but mainly in the Amazon area (Martin *et al.*, 1987). The aboriginal inhabitants of the area cultivated this fruit in pre-Columbian times, thus separation of natural occurrence from man's involvement in dispersal is difficult. At present, it is cultivated in Brazil from São Paulo state in the south to Roraima state in the north; in many cases it is planted as a backyard plant since there are not many commercial-scale plantings. It is also being cultivated on a small scale in Trinidad and Tobago, Ecuador, Guyana, Colombia, Costa Rica, Venezuela, Colombia, Martinique and Ghana.

Ecology

Soil

The plant prefers a well-drained deep fertile soil with no flooding periods. In poorer soils, it will need judicious fertilization (Andersen and Ulup-Andersen, 1988).

Rainfall

It will grow in areas with 1,800–3,200 mm rainfall. The plant is native to the humid tropical forest of the non-flooded Amazon areas where it grows under 60–95% relative humidity. It can also grow in very humid and moderately humid areas with rainfall as low as 1,800 mm as long as it is well distributed (Villachica *et al.*, 1996).

Temperature

In its native habitat, the average temperatures range from 21 to 28°C. It will grow in places with average annual temperatures of 22–23°C. In Central America, it can be cultivated up to 600 m altitude (Vargas *et al.*, 1999).

Light and photoperiod

It will tolerate some shade, since it can be found growing under the canopy of tall trees. The young plants will need shade during their first year in the field and when adult they can grow under shade or in full sun.

Wind

Since it is not a very tall plant under cultivation wind has little effect.

General characteristics

Plant

This is a tree that can reach 18–20 m although in the wild it is found as a lower plant. Cultivated specimens are 6–10 m tall. The stem is dark brown with a tricotomic branching habit, where the lower branches tend to be horizontal and upper branches ascending. The stem branches out every 1.0–1.5 m with trifurcated plagiotropic branches and it can have a diameter of 45 cm at breast height while the diameter of the canopy can reach 7 m (Donadio *et al.*, 2002). The simple petiolated oblong or oblong-ovate leaves are 25–60 cm long and 6–14 cm wide (Fig. 9.9) and delicately glabrous with a green upper side and a pale green to pale pink underside (Villachica *et al.*, 1996). Leaves in orthotropic branches are spirally arranged while those in plagiotropic branches are alternate.

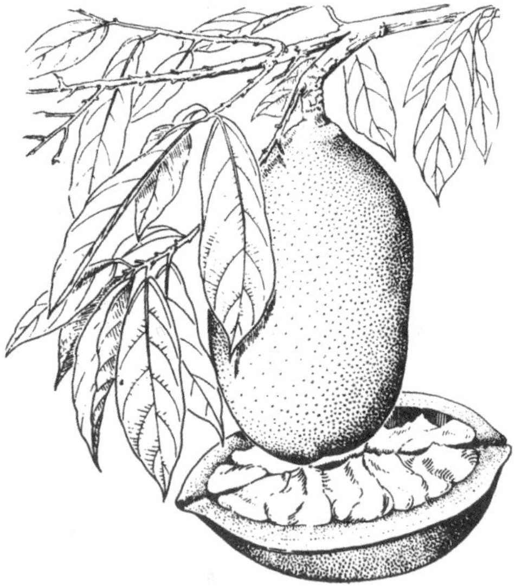

Fig. 9.9. Copuazu (*Theobroma grandiflorum*) showing leaves and fruit (used with permission from León, J. (2000) *Botánica de los Cultivos Tropicales*. Editorial Agroamérica, Instituto Interamericano de Cooperación para la Agricultura (IICA), San José, Costa Rica).

Flowers, pollination and fruit set

The hermaphrodite flowers arise in axillary inflorescences that appear in the plagiotropic branches and contain from two to five flowers, sometimes more. The peduncles have three bracteoles. The calyx has five sepals partially united or free. The corolla has five petals tunicated in their base and with the upper part colored in dark red. There are five petaloid staminodes, five stamens located in the tunic, six anthers and a pentagonal ovary with five multiovulated locules. Pollination is mainly allogamous with incompatibilities, especially in non-domesticated types. Reproductive success is low with a 7-year-old plant having 3,000 flowers only producing 17 fruit. Several species of bees pollinate the flowers but they are not abundant and thus the percentage of natural pollination is 2%, which results in many farmers doing artificial pollination. The flowers open around noon time and in the afternoon and stigmas remain receptive until next morning when artificial pollination can be done (Donadio *et al.*, 2002).

This species has very low reproductive success attributed to problems with pollination and self-incompatibility restricts its agronomic productivity. Ramos *et al.* (2005) in Brazil reported with controlled pollination between compatible and incompatible types that a late or delayed incompatibility occurred after fertilization.

Fruit

The fruit is a drupaceous berry, with an ellipsoid or oblong shape, 12–25 cm long by 10–12 cm in diameter and resembles cacao fruit (Fig. 9.9). Fruit weight can go from 0.5 to 4.0 kg with an average of 1.2 kg. The epicarp is woody and stiff, and easily broken, with a green skin covered by a dusty ferruginous layer that comes off when manipulated. The mesocarp is 6–8 mm thick, white-yellowish and harder than the endocarp, which is thin and clear and contains about 18–52 seeds in five vertical rows surrounded by a thick white-yellowish pulp that has a characteristic aroma and is slightly acid. There are seedless fruit but their acidity is lower (Cavalcante, 1991). The seeds are 2.5 by 0.9 cm.

Cultivar development

The largest genetic diversity is found in the state of Pará in Brazil, where fruit size, fruit form, pulp percentage, pulp acidity, mesocarp thickness, form and size of seeds vary. In certain areas, there is a concentration of plants with very large fruit of 3 kg and more. There are also types with a higher pulp percent. Seedless fruit are pollinated by seeded types and otherwise yield poorly. A seedless type is being planted commercially with a higher pulp yield but the flavor is inferior.

Most of the research stations in the Amazonian area of Brazil have germplasm collections. In many cases, resistance to witches' broom is being tested.

Cultivars

The first named selections are being tried in different parts of Brazil and from them more defined cultivars will become available for asexual propagation.

Cultural practices

Propagation

SEXUAL This was the usual propagation method; now it is recommended to obtain rootstocks. The best quality seeds come from large fruit from healthy plants and only the largest seeds of each fruit should be selected to get better quality plants. Seeds should be fermented for 12–24 h to get rid of the pulp or the pulp can be removed manually using shears, then by rubbing the seeds with sand and rinsing them thoroughly to get rid of the pulp residues. Sowing should not be delayed more than a couple of days to avoid dehydration. Clean seeds are sown in trays or seedbeds under 50% shade or pre-germinated in moist sand or sawdust and as soon as the radicle emerges put in a nursery bag and buried 1 cm deep with the radicle pointing down. In the seedbed or trays, germination will start in 2 weeks and will be finished in about 3–4 weeks when 90% of seeds have germinated. The seedlings should be transplanted into nursery bags and transferred under 25% shade when they are about 15 cm. The use of trays with cells makes the process much more efficient and eliminates transplant shock. Germination is hypogeal (Donadio *et al.*, 2002).

ASEXUAL This is the recommended method to propagate plants with desirable characteristics such as fruit size, yield, pulp-to-fruit ratio, plant size and seedlessness. At present grafting or budding is used when the rootstock attains 1 cm diameter at 20–30 cm from the ground. Shield or patch budding and side, cleft or splice graft can be used, and scions have to be in a semi-hardwood condition (Andersen and Ulup-Andersen, 1988). The branch from where the buds will be taken has to be defoliated and topped about 2 weeks before harvesting the buds to promote the swelling of them. In the case of grafting, upright growing branches of the mother plant tipped with mature leaves are used as scions; after cutting the scions, the upper two leaves should be reduced to 5 cm of petiole and the rest of the leaves removed, while the rootstock should be left with as many leaves as possible. The grafted plants are kept under at least 50% shade. The new growth will tend to be plagiotropic therefore orienting the branches by tying them to stakes is needed (Villachica *et al.*, 1996).

Orchard establishment

The field chosen should have good drainage and run-off. Plowing and harrowing will be needed after which the position of the holes will be marked. Holes 60 × 60 × 60 cm should be dug at 5 × 5 m or 6 × 6 m. For seedling plants, distances of 5 × 10 m or 8 × 8 m are used. If planted at 5 × 5 m the intermediate plants will have to be thinned after 7–9 years. The soil from the hole should be mixed with 10 kg aged manure and 200 g superphosphate. The plants should be protected by temporary shade for the first year (Andersen and Ulup-Andersen, 1988).

Irrigation

This is normally not done because the species is grown in areas with high and/or uniform rainfall.

Pruning

The aim is to avoid having a too tall plant, which means not leaving more than two to three main branches per plant and topping the ends of these branches to promote lateral branching and give the plant a cup form. Eliminating damaged or sick branches is also necessary, especially diseased material with witches' broom that should be eliminated during the rainiest months (Villachica *et al.*, 1996).

Fertilization

Recommendations are to apply three times a year 70, 100 and 150 g of a 10-28-20 formula during the first, second and third year respectively. The fourth year 200 g of the same formula should be applied three times (Villachica *et al.*, 1996).

Pest management

DISEASES "Witches' broom" (*Crinipellis perniciosa*) is one of the most serious and common diseases. The disease damages the branches by causing swelling and abnormal growth and excessive sprouting of the affected area resulting in bending, drying and breaking of the branches. It can also attack the fruit and flowers. Infected fruit and branches should be removed; in the case of branches, they should be cut well below the diseased part and this should be done soon after the peak of the rainy season. As the rainy season ends copper fungicides should be sprayed. Anthracnose (*Colletotrichum gloeosporioides*) attacks leaves, flowers and fruit producing black spots. *Phomopsis* fungus causes rust-colored necrotic spots in the leaves; it can also attack the fruits and branches. *Lasiodiplodia theobromae* causes internal rot in fruit it enters through wounds made by insects. *Pellicularia koleroga* can also cause damage.

INSECTS Leaf cutting ants (*Atta*) can be a problem with young plants. Crickets (*Grillus*) can also damage young plants. Aphids, mainly *Toxoptera citricidus*, can attack young growth. Chrysomelidae can also be a problem as well as some caterpillars and *Trigona* bees in certain cases. The fruit borer *Conotrachelus humeropticus* enters the fruit and causes internal rots; it is one of the worst insect problems. Rodents like to eat freshly sown seeds in the seedbed (Andersen and Ulup-Andersen, 1988; Villachica *et al.*, 1996).

Weed control

The area around the young plants should be cleaned as much as possible to insure a good initial growth; when older the area under the canopy should also be clean.

Orchard protection

Not necessary except in areas where grazing animals could be a problem. Winds should not be a problem in the areas where it is grown.

Harvesting and postharvest handling

Yields

A 5-year-old plant can produce 20–30 fruit of 1–1.5 kg. This will increase to 60–70 fruit in 7–8-year-old plants (Andersen and Ulup-Andersen, 1988). Under poor soil conditions, yields are much lower.

Harvesting

The fruit drops from the tree when ripe normally at night, leaving the peduncle on the tree. It would be ideal to find out the right stage to pick it from the plant to avoid damage to the fruit from falling to the ground. No good maturity indicators exist and harvested unripe, immature, acid fruit do not fully ripen and it is difficult to remove the pulp. In its native area, fruit mature from January to May while in Belém from October to May. Harvesting should be done daily (Vargas *et al.*, 1999).

Postharvest treatment

Cleaning the fruit from any residues is the normal procedure. Fruit should be processed at most 5–6 days after harvest.

Utilization

The juicy and sweet-acid pulp surrounding the seeds is used for many purposes in addition to as a fresh fruit (Table 9.3). The pulp can be made into a juice in a blender and this is gaining popularity in foreign markets. Other uses of the pulp are to make canned juices, sherbets, compotes, jellies, ice cream, yogurt, liquor and wine. Yields from 10 kg fruit are around 4.6 kg pulp, 3.8 kg peel and 1.6 kg seeds. There are seedless fruit that yield 67% of pulp in weight (Andersen and Ulup-Andersen, 1988).

Table 9.3. Nutritional value of 100 g copuazu pulp and seeds (from Villachica *et al*., 1996).

	Pulp	Seeds
Moisture (g)	89.9	56.6
Protein (g)	1.92	20.0
Lipid (g)	0.48	50.8
Carbohydrates (g)		15.9
Reducing sugars (g)	3.0	
Fiber (g)		9.6
pH	3.3	
Acidity (g)	2.15	
Ether extract (g)	0.53	
Ash (g)	0.67	3.7
Total solids (g)	11.00	
Brix	10.80	
Amino acids (mg)	21.9	
Pectins (mg)	390.0	
Phosphorus (mg)	310.0	
Calcium (mg)	40.0	
Vitamin C (mg)	23.1	

The seeds contain 50–60% fat that is very digestible, similar to cacao, but does not contain caffeine, only theobromine. The seeds have 20% protein and 16% carbohydrates and 9.6% fiber (Table 9.3). The seeds can be used to make a chocolate-like product called in Brazil "cupulate" that has very good flavor (Donadio *et al.*, 2002) and the fat can substitute for cacao butter; 1 kg seed produces 0.25 kg of "cupulate".

ARACEAE

This family of monocots has flowers borne on a spadix subtended by a leaf-like bract, the spathe. The family is commonly referred to as aroids or arum. In a recent taxonomic revision of the family, the genera formerly in the Lemnoideae (duckweeds) were included based upon molecular evidence. The revised Araceae contains nine subfamilies. Most aroids are grown as ornamentals (*Aglaonema, Anthuriums, Diffenbachia, Caladium, Philodendron*) with some very notable exceptions that are root crops in the genera *Colocasia* (*C. esculenta* – taro) and *Xanthosoma* spp. (cocoyam, tannier). The only species grown for its edible fruit is *Monstera deliciosa.*

Ceriman

Monstera deliciosa Liebm. is grown commercially for its compound fruit that develops from the spadix. Synonyms include *Philodendron pertusum* Kunth & Bouche., *P. anatomicum* Kunth., *Monstera borsigiana* K. Koch., *M. lennea* K. Koch., and *M. tacanaensis* Matuda.

Origin and distribution

The species is native to the wet forests of southern Mexico, Guatemala, and parts of Costa Rica and Panama. Another English name is monstera. In Mexico and Latin America, it is known as *piñanona* or *piña anona*. Other names include, in Venezuela – *ojal* or *huracán*, in Colombia – *hojadello, balazo*, in Peru – *costilla de Adán*, in Guatemala – *harpón* or *arpón común*, in Guadeloupe – *caroal, liane percee*, or *liane franche*, in Martinique – *siguine couleuvre* and in French Guiana – *arum du pays* or *arum troud*. As an ornamental and as a foliage house plant, it is known as Mexican breadfruit, hurricane plant, swiss-cheese plant, window leaf, and split leaf philodendron. A close relative *M. adansonii* from Surinam is called five-hole plant.

In 1908, the fruit was reported to be cultivated in Florida, Portugal, and Algeria (Labroy, 1908). It has been spread around the world as an ornamental though apparently is no longer cultivated for its fruit. Small quantities are seen in local markets most likely from backyard production.

Ecology

Soils

The plant grows well in almost any soil including calcareous soils that are well-drained and preferably rich in organic matter. It is intolerant of saline conditions. When cultivated in the wild it tends to be an epiphyte. As an ornamental, it will grow in the substrate used for most indoor foliage plants.

Climate

It grows in a humid climate from sea level to 1,500 m (Romero Castañeda, 1969) at temperatures from 20 to 30°C (Lim, 2012). It can withstand cold with growth ceasing below 10°C and it must be sheltered from frost. Annual rainfall needs to be above 1,000 mm. This is a strictly tropical species that can be cultivated under subtropical conditions where growth is slower. As an ornamental, it is grown indoors in many temperate climate zones with artificial room temperature.

It can stand full sun but the ideal is to grow it with some shade and as an ornamental. When grown for fruit it does best in semi-shade under high moisture conditions.

General characteristics

Plant

This fast-growing stout herbaceous vine up to 24 m long spreads over the ground forming mats and can climb trees. The rarely branched stems are cylindrical (6.25–7 cm thick), rough with leaf scars, and produce numerous tough aerial roots normally at a node (Fig. 9.10). The oval glabrous leaves are leathery on stiff flattened petioles (up to 105 cm long). Mature leaves are ovate, up to 30–90 cm long and 20–80 cm wide, deeply cut at the margins to 23 cm strips

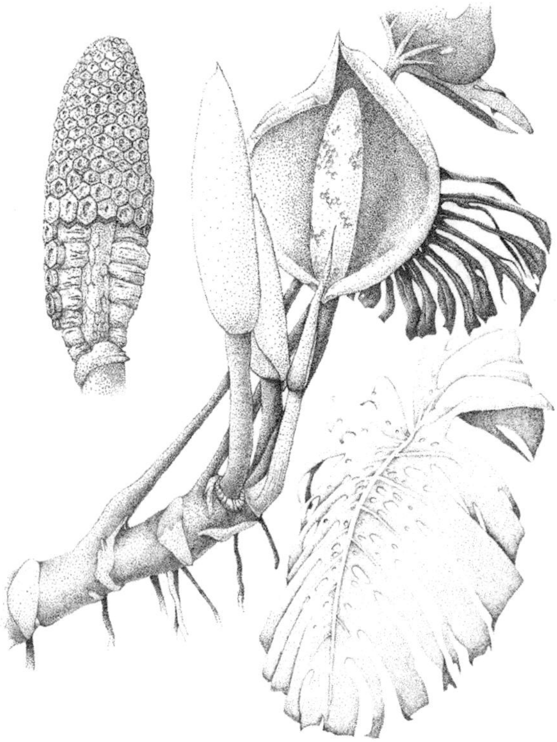

Fig. 9.10. Ceriman (*Monstera deliciosa*) showing whole plant with leaf, flower and fruit (from Alexander KhoiTu, used with permission).

and perforated on each side of the midrib with oblong holes of various sizes with a cordate base and with prominent veination especially at the lower side (Fig. 9.10). Young leaves are heart-shaped and without holes.

Flowers

Multiple inflorescences arise from the leaf axil on tough 12–20 cm cylindrical glabrous stalks, the spadix (Fig. 9.10). The spadix is cream to tan, surrounded initially by a waxy white concave spathe that develops into a green compound fruit (20–30 cm long, 5–8 cm wide). The sessile flowers with no sepals or petals have a two-carpellate and two-locular gynoecium. A thick, gum-like nectar is produced near the stigma that attracts bees. The thick hard rind (scales) corners cover individual segments of juicy ivory pulp. Between the individual segments are thin black membranes.

It flowers about 3 years after planting. Flowering and fruiting overlap as it requires 12–14 months from the opening of the inflorescence to fruit maturity. Cross-fertilization is required for fertilization and seed initiation.

Fruit

The fruit is a syncarp; a fleshy aggregate of drupes from separate ovaries (Fig. 9.10). Fruit development can take 12–14 months and does not require fertilization.

As the fruit matures, the rind takes on a lighter shade of green and progressively ripens toward the apex over 5–6 days while the pulp becomes creamy-white, sweet and fragrant (Fig. 9.10). The portion eaten is only that from which the rind can be easily removed. Occasionally, pale green pea-sized seeds occur.

Cultivar development

There are ornamental varieties including variegated forms and regional selections.

Cultural practices

Propagation

It can be raised from seeds that should be sown as soon as possible after removal from the fruit. The main propagation is by using stem cuttings (stem, internodal pieces, and apical tip with leaves) with at least two buds. Stem pieces (50–100 cm long) are also used and laid horizontally in a shallow trench. Plant suckers, with or without roots, can also be used as planting material. Suckers fruit in 2–4 years and cuttings in 4–6 years. Tissue culture can also be used.

Irrigation and fertilization

Foliage applied fertilizer encourages early growth.

Diseases, pests, and weeds

Leaf spot, anthracnose, bacterial soft rot, and root rot have been reported. Scale insects, mites, mealy bugs, and caterpillars do attack the plant. It is not regarded as a host to Caribbean fruit fly. Weeds are normally not a problem after the plant is established, since it is either climbing or its dense foliage shades out most weeds.

Harvesting and postharvest handling

The fruit is mature and ready to harvest when the fruitlet caps at the base of the spadix start to spread (Fig. 9.10) and show a creamy color between them. The fruit can be broken off the flower stem. The green caps of each floret fall as the fruit ripens from the base of the spadix upwards. The fallen caps expose the edible flesh beneath. Prior to cap dehiscence the flesh is very acrid.

There is a pronounced ethylene and respiratory climacteric during ripening concurrent to a rapid conversion of starch to sugars. The whole fruit can be ripened in plastic wrap or paper bag. The ideal is to put the fruit in a refrigerator for 24 h after harvest and then transfer it to room temperature to bring out the best flavor.

Utilization

The fully ripe fruit pulp is served as a dessert with cream, in fruit salads, sherbets or ice creams. The flavor is between pineapple, cherimoya and banana with sweet, lactone and coconut overtones, somewhat like *piña colada*. The flavor is due mainly to ethyl esters with the only terpene being linalool (Peppard, 1992). It can be stewed. Ripe fruit soluble solids are about 19% with 7–8 meq per 100 g titratable acidity; oxalic acid is the major acid component (Peters and Lee, 1977). Acridity is due to the oxalic acid raphides and associated proteins that occur in the growing fruit and floral remnants and all other parts of the plant. Aerial roots have more raphides than soil-borne roots, suggesting tissue variation. Sensitive individuals suffer throat irritation, urticaria, and anaphylaxis due to the acridity of the raphides. Ripe fruit lack acridity.

A root and leaf infusion is taken for arthritis. The inflorescences have high concentrations of the aromatic amines, tyramine, and dopamine, though not in the reproductive organs. The aerial roots are used as ropes and to fashion baskets. It is a very popular house plant that can be grown under about 50% shade.

References and Further Reading

White sapote

Awaad, A.S., Maitland, D.J. and Moneir, S.M. (2007) New alkaloids from *Casimiroa edulis* fruits and their pharmacological activity. *Chemistry of Natural Compounds* 43, 576–580.

Batten, D.J. (1984) White sapote *Casimiroa edulis*. In: Page, P.E. (comp.) *Tropical Tree Fruit for Australia*. Queensland Department of Primary Industries, Brisbane, Australia, pp. 171–174.

Bertin, R., Garcia-Argaéz, A., Martìnez-Vàzquez, M. and Froldi, G.(2011) Age-dependent vasorelaxation of *Casimiroa edulis* and *Casimiroa pubescens* extracts in rat caudal artery in vitro. *Journal of Ethnopharmacology* 137, 934–936.

Campbell, C.W., Knight, R.J. Jr. and Zareski, N.L. (1977) Freeze damage to tropical fruits in southern Florida in 1977. *Proceeding of the Florida State Horticultural Society* 90, 254–257.

Crane, J.H. (2008) *Casimiroa edulis* White sapote. In: Janick, J. and Paull, R.E. (eds) *Encyclopedia of Fruit and Nuts*. CAB International, Wallingford, UK. pp. 770–772.

George, A.P. and Nissen, R.J. (1991) *Casimiroa edulis Llave & Lex*. In: Verheij, E.W.M. and Coronel, R.E. (eds) *Edible Fruits and Nuts*. Plant Resources of South East Asia No. 2. PROSEA Foundation, Pudoc, Wageningen, the Netherlands, pp. 113–114.

George, A.P., Nissen, R.J. and Wallace, D.J. (1988) The casimiroa. *Queensland Agricultural Journal* 114, 57–62.

Martin, F.W., Campbell, C.W. and Ruberté, R.M. (1987) *Perennial Edible Fruits of the Tropics: An Inventory.* USDA-ARS, Agricultural Handbook No. 642, US Government Printing Office, Washington, DC, p. 56.

McCain, R. (1993) Goldenberry, passionfruit, and white sapote: potential fruits for cool subtropical areas. In: Janick, J. and Simon, J.E. (eds) *New Crops*. Wiley, New York, pp. 479–486.

McGregor, B.M. (1987) *Tropical Products Transport Handbook.* Agricultural Handbook No. 668. USDA, AMS, Washington, DC, p. 113.

Mizrahi, Y. and Nerd, A. (1996) New crops as a possible solution for the troubled Israeli export market. In: Janick, J. (ed.) *Progress in New Crops*. ASHS Press, Alexandria, Virginia, pp. 37–45.

Morley-Bunker, M. (1986) Miscellaneous and minor fruit crops. In: Jackson D. (ed.). *Temperate and Subtropical Fruit Production*. Butterworths of New Zealand Ltd., Wellington, New Zealand, p. 268.

Morton, J.F. (1987) *Fruits of Warm Climates*. Creative Resource Systems, Winterville, North Carolina, pp. 191–196.
Mowry, H., Toy, L.R. and Wolfe, H.S. (revised by Ruehle, G.D.) (1967) *Miscellaneous Tropical and Subtropical Florida Fruits.* Bulletin 156A. Agricultural Extension Service, IFAS, University of Florida, Gainesville, Florida, pp. 40–44.
Nerd, A., Aronson, J.A. and Mizrahi, Y. (1990) Introduction and domestication of rare and wild fruit and nut trees for the desert areas. In: Janick, J. and Simon, J.E. (eds) *Advances in New Crops*. Timber Press, Portland, Oregon, pp. 355–363.
Sauls, J. and Campbell, C.W. (1983) *White sapote*. Fruit Crops Fact Sheet, FC-59. Fla. Coop. Ext. Serv., IFAS, University of Florida, Gainesville, Florida, pp. 1–2.
Vargas, O., Alix, C., Lobo, A.D. (authors), Duarte, O. and Sanchez, J. (technical reviewers) (1999) *Frutales y Condimentarias del Trópico Húmedo*. CURLA; PDBL; AFE/COHDEFOR; DICTA; SETCO; PROFORFITH, La Ceiba, Honduras, pp. 252–253.
Vázquez-Cruz, B., Vázquez-Muñoz, M., Navarrete-Bastida, R., Trujillo-González, I.D., De Haro, R., Segura-Cobos, D. and Muñoz-López, J.L. (2009) Antihypertensive and vasorelaxant effects of the aqueous extract of *Casimiroa edulis*. *Pharmacologyonline* 3, 73–83.
Verheij, E.W.M. and Coronel, R.E. (1991) *Plant Resources of South-East Asia, No. 2 – Edible Fruits and Nuts*. Pudoc-DLO, Wageningen, Netherlands, pp.113–114.
Yahia, E.M. and Gutierrez-Orozco, F. (2011) White sapote (*Casimiroa edulis* Llave & Lex). In: Yahia, E.M. (ed.) *Postharvest Biology and Technology of Tropical and Subtropical Fruits,* Volume 4. *Mangosteen to white sapote*. Woodhead Publishing Ltd., Cambridge, UK, pp. 474–482.

South American sapote or chupa-chupa

Clement, C.R. (1980) Teste preliminar sobre a enxertia de Sapota (*Matisia cordata* H.B.K., Bombacaceae). *Revista Brasileira de Fruticultura* 2, 69–73.
Clement, C.R. (1982) Considerações sobre a Sapota da América do Sul. *Proceedings of the Tropical Region of the American Society for Horticultural Science* 25, 425–432.
Clement, C.R. (1989) A center of crop genetic diversity in western Amazonia. A new hypothesis of indigenous fruit-crop distribution. *Bioscience* 39, 624–630.
Clement, C.R. (1992) Frutas da Amazônia. *Ciência Hoje* 14, 29–37.
Donadio, L. (1983) Review of South American Fruits. *Proceedings of the Tropical Region American Society for Horticultural Science* 27A, 21–41.
FAO (1986) *Food and Fruit-bearing Forest Species. 3: Examples from Latin America*. FAO Forestry Paper 44/3, Rome, Italy, 308 pp.
Hackett, C. and Carolane, J. (1982) *Edible Horticultural Crops – A compendium of information on fruit, vegetable, spice and nut species*. Academic Press, Sydney, Australia, 673 pp.
Hodge, W.H. (1960) The South American "Sapote." *Economic Botany* 14, 203–206.
Kerr, W.E. and Clement, C.R. (1980) Práticas agrícolas de conseqüências genéticas que possibilitaram a os índios da Amazônia uma melhor adaptação às condições ecológicas da região. *Acta Amazonica* 10, 251–261.
Lehmann, J., Kern, D.C., German, L.A., Mc Cann, J., Martins, G.C. and Moreira, A. (2003) Soil fertility and production potential. In: Lehmann, J., Kern, D.C., Glaser, B. and Woods, W.I. (eds) *Amazonian Dark Earths – Origins, Properties, Management*. Kluwer Academic, Dordrecht, pp. 105–124.
Leung, W.T.W. and Flores, M. (1966) *Tabla de Composición de Alimentos para uso en América Latina*. INCAP, Editorial Interamericana.
Martin, F.W., Campbell, C.W. and Ruberté, R.M. (1987) *Perennial Edible Fruits of the Tropics: An Inventory.* Agriculture Handbook No. 642, US Department of Agriculture, p. 21.

Morton, J.F. (1987) *Chupa-Chupa*. In: *Fruits of Warm Climates*. Creative Resource Systems, Inc., Miami, Florida, pp. 291–292.
Nilsson, S. and Robyns, A. (1974) Pollen morphology and taxonomy of the genus *Quararibea* (Bombacaceae). *Bull. Jard. Bot. Nat. Belg./Bull. Nat. Plantentium Belg*. 44, 77–99.
RCF (Rainforest Conservation Fund) (2004) Agroforestry and Ethnobotany. Available at: http://www.rainforestconservation.org/data_sheets/agroforestry/Quararibea_cordata.html.
Rios-Castaño, D., Barona, P. and Evaristo, C. (1997) Zapote 'Portoviejo'. *Proceedings of the Interamerican Society for Tropical Horticulture* 41, 223–227.
Santos, R.P. (2008) *Quararibea cordata* South American sapote. In: Janick, J. and Paull, R.E. (eds) *Encyclopedia of Fruit and Nuts*. CAB International, Wallingford, UK, pp. 184–186.
Villachica, H., Carvalho, J.E.U., Muller, C.H., Diaz, S.C. and Almanza, M (1996) Zapote. In: *Frutales y Hortalizas Promisorios de la Amazonía*. Tratado de Cooperación Amazónica, Lima, Perú, pp. 301–305.
Whitman, W.F. (1976) South American sapote. *Proceedings of the Florida State Horticultural Society* 89, 226–227.

Black sapote

Geilfus, F. (1989) *El Árbol al Servicio del Agricultor* Vol. 2. *Guía de Especies*. ENDA-Caribe y CATIE, Santo Domingo, República Dominicana, 778 pp.
George, A.P. (1984) Ebenaceae. In: Page, P.E. (compiler) *Tropical Tree Fruits for Australia*. Queensland Department of Primary Industry, Information Services QI 83018, pp. 58–60.
Heinze, H. (1983) Some field observations on *Pouteria caimito, P. campechiana, Diospyros digyna* and *Solanum topiro* under humid conditions. *Proceedings of the Tropical Region of the American Society for Horticultural Science* 27, 111–116.
Howard, R.A. (1961) The correct name for *Diospyros ebenaster*. *Journal of the Arnold Arboretum, Harvard University* 42, 430–435.
Ledesma, N. and Campbell, R.J. (2001) Review of cultivars and propagation of black sapote (*Diospyros digyna* Jacq.) in Southern Florida. *Proceedings of the Interamerican Society for Tropical Horticulture* 45, 12–14.
León, J. (1987) *Botánica de los Cultivos Tropicales*. IICA, San José, Costa Rica, 445 pp.
Leung, W.T.W. and Flores, M. (1966) *Tabla de Composición de Alimentos para uso en América Latina*. INCAP, Editorial Interamericana.
Martin, F.W., Campbell, C.W. and Ruberté, R.M. (1987) *Perennial Edible Fruits of the Tropics: An Inventory*. Agriculture Handbook No. 642, ARS, USDA, Washington, DC.
Morton, J.F. (1987) *Fruits of Warm Climates*. Creative Resource Systems Inc., Winterville, North Carolina, pp. 418–419.
Ng, F.S.P. (1992) *Diospyros digyna* Jacq. In: Verheij, E.W.M. and Coronel, R.E. (eds) *Plant Resources in South-East Asia. No. 2, Edible Fruits and Nuts*. Prosea, Bogor, Indonesia, pp. 152–154.
Patiño-Rodríguez, V.M. (2002) *Historia de los Frutales Nativos del Neotrópico*. CIAT Cali, Colombia, 655 pp.
Paull, R.E. and Duarte, O. (2008) *Diospyros digyna* Black sapote. In: Janick, J. and Paull, R.E. (eds) *Encyclopedia of Fruit and Nuts*. CAB International, Wallingford, UK, pp. 324–326.
Popenoe, W. (1920) *Manual of Tropical and Subtropical Fruits*. A facsimile of the 1920 edition, Hafner Press, New York, 474 pp.
Vargas, O., Alix, C., Lobo, A.D. (authors), Duarte, O. and Sanchez, J. (technical reviewers). (1999) *Frutales y Condimentarias del Trópico Húmedo*. CURLA; PDBL; AFE/COHDEFOR; DICTA; SETCO; PROFORFITH, La Ceiba, Honduras, pp. 114–115.
Yahia, E.M. and Gutierrez-Orozco, F. (2011) Black sapote (*Diospyros digyna* Jacq.). In: Yahia, E.M. (ed.) *Postharvest Biology and Technology of Tropical and Subtropical Fruits, Volume 3. Acai to citrus*. Woodhead Publishing Ltd., Cambridge, UK, pp. 244–250.

Golden spoon or nance

Alix, C. and Duarte, O. (1999) *Propagación de Especies Frutales Tropicales.* CURLA; PDBL; AFE/COHDEFOR; DICTA; SETCO/PROFORFITH. La Ceiba, Honduras, p. 58.

Andersen, O. and Ulup-Andersen, V. (1988) *As Frutas Silvestres Brasileiras*, 3ª Ed. Publicaciones Globo Rural, São Paulo, Brazil, pp. 156–159.

Anderson, W.R. (1993) Chromosome numbers of Neotropical Malpighiaceae. *Contributions from the University of Michigan Herbarium* 19, 341–354.

Barbeau, G. (1990) *Frutas Tropicales en Nicaragua.* Dirección General de Técnicas Agropecuarias, MIDINRA, Editorial Ciencias Sociales, Managua, Nicaragua, pp. 225–226.

Bayuelo-Jiménez, J.S. (2008) *Byrsonima crassifolia* Nance. In: Janick, J. and Paull, R.E. (eds) *Encyclopedia of Fruit and Nuts.* CAB International, Wallingford, UK, pp. 459–461.

Bye, R.A. Jr. and Linares, E. (1990) Mexican market plants of 16th century. Plants recorded in Historia Natural de Nueva España. *Journal of Ethnobiology* 10, 151–158.

Cavalcante, P.B. (1974) *Frutas Comestiveis da Amazonía II.* Instituto Nacional de Pesquisas da Amazonia, Museo Paraense Emilio Goeldi, Publicaciones Avulsas No. 27, Belém, Pará, Brasil, pp. 9–10.

Davis, C.C. and Anderson, W.R. (2010) A complete generic phylogeny of Malpighiaceae inferred from nucleotide sequence data and morphology. *American Journal of Botany* 97, 2031–2048.

Donadio, L.C., Moro, F.V. and Servidone, A.A. (2002) *Frutas Brasileiras.* Editora Novos Talentos, Jaboticabal, São Paulo, Brazil, pp. 221–224.

Duarte, O. (2011) Nance (*Byrsonima crassifolia* (L.) Kunth). In: Yahia, E.M. (ed.) *Postharvest Biology and Technology of Tropical and Subtropical Fruits, Volume 4. Mangosteen to white sapote.* Woodhead Publishing Ltd., Cambridge, UK, pp. 44–50.

Duarte, O. and Escobar, O. (2007) Propagación del nance (*Byrsonima crassifolia* (L.) H.B.K.) por acodo aéreo e injerto. *Proceedings of the Interamerican Society for Tropical Horticulture* 51, 39–41.

Duarte, O. and Vernon, R. (2002) Biología floral y reproductiva del nance (*Byrsonima* sp). *Proceedings of the Interamerican Society for Tropical Horticulture* 46, 40–41.

Duarte, O., Escobar, O. and Soriano, L. (2007) Propagación del nance [*Byrsonima crassifolia* (L.) H.B.K.] por estacas terminales con hojas y estacas leñosas. *Proceedings of the Interamerican Society for Tropical Horticulture* 51, 30–34.

León, J. (1983) Underutilized fruits of Central America and northern South America. *Proceedings of the Tropical Region American Society for Horticultural Science* 27, 1–20.

Leung, W.T.W. and Flores, M. (1966) *Tabla de Composición de Alimentos para uso en América Latina.* INCAP, Editorial Interamericana.

Martin, F.W., Campbell, C.W. and Ruberté, R.M. (1987) *Perennial Edible Fruits of the Tropics: An Inventory.* Agriculture Handbook No. 642, US Department of Agriculture, p. 36.

Morton, J.F. (1987) *Fruits of Warm Climates.* Creative Resources System Inc., Winterville, North Carolina, pp. 207–209.

Rezende, C.M. and Fraga, S.R.G. (2003) Chemical and aroma determination of the pulp and seeds of murici (*Byrsonima crassifolia* L.). *Journal of Brazilian Chemistry Society* 14, 425–428.

Vargas, O., Alix, C., Lobo, A.D. (authors), Duarte, O. and Sanchez, J. (technical reviewers) (1999) *Frutales y Condimentarias del Trópico Húmedo.* CURLA; PDBL; AFE/COHDEFOR; DICTA; SETCO; PROFORFITH, La Ceiba, Honduras, p. 156–157.

Villachica, H., Carvalho, J.E.U., Muller, C.H., Diaz, S.C. and Almanza, M. (1996) Indano. In: *Frutales y Hortalizas Promisorios de la Amazonía.* Tratado de Cooperación Amazónica, Lima, Perú, pp. 163–168.

Vinson, S.B., Williams, H.J., Frankie, G.W. and Shrum, G. (1997) Floral lipid chemistry of *Byrsonima crassifolia* (Malphigiaceae) and a use of floral lipids Centris Bees (Hymenoptera: *Apidae). Biotropica* 29, 76–83.

Borojo

Arango, A.G.J. and Quijano, T.J. (1986) Estudios de los frutos de *Borojoa patinoi* (Cuatrec.). *Revista Latinoamericana de Química* 17, 167–168.

Arenas Martínez, L.E., Cuellar López, H.F. and De la Cruz Aparicio, G. (1985) El borojó, *Borojoa patinoi* Cuatr., cultivo promisorio para trópico húmedo colombiano. *Acta Agronómica* 35, 106–116.

Giraldo-Molina, C.I., Rengifo, L., Aguilar-Benítez de Lugo, E. and Alegría, A.H. (2004) Determinación del sexo en borojó (*Borojoa patinoi*, Cuatrecasas) mediante marcadores moleculares. *Revista Colombiana de Biotecnología* 6, 9–14.

Hincapié Llanos, G.A., Palacio Piedrahíta, J.C., Páez Sierra, S., Restrepo Flórez, C.E. and Vélez Acosta, L.M. (2012) Elaboración de una bebida energizante a partir de borojó (*Borojoa patinoi* Cuatrec.). *Revista Lasallista de Investigación* 9, 33–43.

Lynch, J. and Gonzalez, A. (1993) Canopy nutrient allocation in relation to incident light in the-tropical fruit tree *Borojoa patinoi* (Cuatr.). *Journal of the American Society for Horticultural Science* 118, 777–785.

Medina, M.A. and Sepúlveda, N. (2006) Micropropagación del borojó vía embriogésis somática, método para la selección clonal para el sexo. *Revista Institucional Universidad Tecnológica del Chocó* 25, 31–35.

Ocampo, R.D., Zapateiro, L.G., Gómez, J.M.F. and Torres, C.V. (2012) Caracterización bromatológica, fisicoquímica, microbiológica y reológica de la pulpa de borojó (*Borojoa patinoi* Cuatrec). *Revista Ciencia y Tecnología* 5(1), 17–24.

Persson, C. (2000) Phylogeny of the Neotropical *Alibertia* group (*Rubiaceae*), with emphasis on the genus *Alibertia*, inferred from ITS and 5S ribosomal DNA sequences. *American Journal of Botany* 87, 1018–1028.

Quito Ramón, L.S. and Torres Criollo, G.S. (2007) Estudio de prefactibilidad técnico económico de una planta para elaborar una bebida a base de noni (*Morinda citrifolia*) y borojó (*Borojoa patinoi*). Available at: http://bibdigital.epn.edu.ec/handle/15000/2769 (accessed 3 November 2014).

Ricker, M., Jessen, J.H. and Daly, D.C. (1997) The case for *Borojoa patinoi* (Rubiaceae) in the Chocó Region, Colombia. *Economic Botany* 51, 39–48.

Romero Castañeda, R. (1961) *Frutas Silvestres de Colombia,* Volumen 1. Editorial San Juan Eudes, Bogotá, Colombia, pp. 307–309.

Vargas, O., Alix, C., Lobo, A.D. (authors), Duarte, O. and Sanchez, J. (technical reviewers) (1999) *Frutales y Condimentarias del Trópico Húmedo*. CURLA; PDBL; AFE/COHDEFOR; DICTA; SETCO; PROFORFITH, La Ceiba, Honduras, pp. 244–245.

Purui grande

Braga, L.F., Sousa, M.P., Braga, J.F. and Sá, M.D. (1999) Efeito da temperatura na germinaçãode sementes de purui (*Borojoa sorbilis* (Duque) Cuatre – Rubiaceae): morfologia das sementes e das plântulas. *Revista Brasileira de Sementes,* Brasília, 21, 47–52.

Clement, C.R. and Paull, R.E. (2008) Purui grande *Borojoa sorbilis* Cuatrec. Rubiaceae In: Janick, J. and Paull, R.E. (eds). *Encyclopedia of Fruit and Nuts*. CAB International, Wallingford, UK, pp. 761–762.

Clement, C.R. and da Silva Filho, D.F. (1994) Amazonian fruits with commercial potential. *Fruit Varieties Journal* 48, 152–158.

FAO (1986) *Food and Fruit-bearing Forest Species 3: Examples from Latin America*. FAO Forestry Paper #44/3, Rome, Italy.

Fouqué, A. (1972) *Espèces Fruitières d'Amérique Tropicale*. Institut Français de Recherches Fruitierès Outre-mer (IFAC).

Mejía, M. (1984) Borojó. Fruta Ecuatorial Colombiana. *Colombia Amazónica* 1, 89–106, Bogotá, Colombia.

Ricker, M., Jessen, J.H. and Daly, D.C. (1997) The case for *Borojoa patinoi* (Rubiaceae) in the Choco region, Colombia. *Economic Botany* 51, 39–48.

Villachica, H., Carvalho, J.E.U., Muller, C.H., Diaz, S.C. and Almanza, M. (1996) Borojó. In: *Frutales y Hortalizas Promisorios de la Amazonía*. Tratado de Cooperación Amazónica, Lima, Perú, pp. 57–63.

Genipap

Cavalcante, P.B. (1991) *Frutas Comestíveis da Amazônia*, 6th edn. Edições CEJUP, Belém, Pará, Brazil, 279 pp.

Crestana, C.S.M. (1993) Biologia da reprodução de *Genipa americana L.* (Rubiaceae) na Estação Ecológica de Moji Guaçú, Estado de São Paulo. Tese (Doutorado), Universidade Estadual Paulista, Rio Claro, São Paulo, Brazil, 222 pp.

Crestana, C. de S.M. (1996) Fruit/seed dispersal of *Genipa americana* L. Rubiaceae in a riparian forest of the Moji-Guacu river, São Paulo. *Revista do Instituto Florestal* 8, 213–221.

Donadio, L.C., Nachtigal, J.C. and Sacramento, C.K. (1998) *Frutas Exóticas*. FUNEP, Jaboticabal, São Paulo, Brazil, 279 pp.

Figueiredo, R.W. de, Maia, G.A., Holanda, L.F.F. de and Monteiro, J.C.S. (1986) Physical and chemical characteristics of genipap fruit. *Pesquisa Agropecuaria Brasileira* 21, 421–428.

de Freitas, D.V. and Salles, A.E. (2008) Genipap *Genipa americana* L., Rubiaceae. In: Janick, J. and Paull, R.E. (eds) *Encyclopedia of Fruit and Nuts*. CAB International, Wallingford, UK, pp. 762–764.

Jøker, D., Salomão, A.N. and Vasquez, W. (2003) *Genipa americana L.* EMBRAPA/CATIE *Seed Leaflet* No. 67. Available at: http://curis.ku.dk/ws/files/20647055/genipaamericana_int.pdf (accessed 3 November 2014).

Kennard, W.C. and Winters, H.F. (1960) *Some Fruits and Nuts for the Tropics*. Miscellaneous Publication No. 801, Agricultural Research Service, USDA, Washington, DC, pp.70–72.

Morton, J.F. (1987) *Fruits of Warm Climates*. Creative Resources Systems Inc., Winterville, North Carolina.

Silva, A.P. da, Lima, C.L.C. de and Vieites, R.L. (1998) Chemical and physical characteristics of stored genipa (*Genipa americana*) fruits. *Scientia Agrícola* (Brazil) 55, 29–34.

Vásquez, R. and Coimbra, G. (2002) *Frutas Silvestres Comestibles de Santa Cruz*. Editorial FAN, Santa Cruz de la Sierra, Bolivia, pp. 186–187.

Villachica, H., Carvalho, J.E.U., Müller, H.C., Díaz, C.S. and Almanza, M. (1996) Huito. In: *Frutales y Hortalizas Promisorios de la Amazonia*. Tratado de Cooperación Amazónica, Lima, Perú, pp. 151–156.

Brazil nut

Bailey, L.H. and Bailey, E.Z. (1976) *Hortus Third, A Concise Dictionary of Plants Cultivated in the USA and Canada*. Macmillan Publishing Co., New York, p. 159.

Barbeau, G. (1990) *Frutas Tropicales en Nicaragua*. Dirección General de Técnicas Agropecuarias, MIDINRA, Editorial Ciencias Sociales, Managua, Nicaragua, p. 215.

Crane, J.H. (2008) *Bertholletia excelsa* Brazil nut. In: Janick, J. and Paull, R.E. (eds) *Encyclopedia of Fruit and Nuts*. CAB International, Wallingford, UK, pp. 450–452.

Donadio, L. (1983) Review of South American Fruits. *Proceedings of the Tropical Region of the American Society for Horticultural Science* 27, 21–41.

Donadio, L.C., Moro, F.V. and Servidone, A.A. (2002) *Frutas Brasileiras*. Editora Novos Talentos, Jaboticabal, São Paulo, Brazil, pp. 137–141.

Fouque, A. (1972) Famille des Lecythidacées. *Fruits* 27(II), 788–795.

Martin, F.W., Campbell, C.W. and Ruberté, R.M. (1987) *Perennial Edible Fruits of the Tropics: an Inventory.* Agriculture Handbook No. 642. US Department of Agriculture (USDA), Agriculture Research Service (ARS), US Government Printing Office, Washington, DC, p. 32.

Motta Maués, M. (2002) Reproductive phenology and pollination of the Brazil nut tree *(Bertholletia excelsa* Humb. & Bonpl. Lecythidaceae) in Eastern Amazonia. In: Kevan, P. and Imperatriz Fonseca, V.L. (eds) *Pollinating Bees - The Conservation Link Between Agriculture and Nature*. Ministry of Environment, Brasília, Brazil, pp. 245–254.

Mueller, C.H. and Calzavara, B.B.G. (1989) *Castaha-do-Brasil. Recomendaciones Básicas* No. 11. EMBRAPA, Centro de Pesquisa Agropecuária do Trópico Úmido, Belém, Pará, Brazil, 6 pp.

Paytán, S.F., Botto, M.P., Brack Egg, A., Guzmán, Y. and Kember, M. (1997) *Cultivos de Frutales Nativos Amazónicos: Manual Para el Extensionista*. Tratado de Cooperación Amazónica, Lima, Perú, pp. 63–69.

Smith, N.J.H., Williams, J.T., Plucknett, D.L. and Talbot, J.P. (1992) *Tropical Forests and Their Crops*. Comstock Publ. Assoc, Ithaca, New York, pp. 371–413.

Vargas, O., Alix, C., Lobo, A.D. (authors), Duarte, O. and Sanchez, J. (technical reviewers) (1999) *Frutales y Condimentarias del Trópico Húmedo*. CURLA; PDBL; AFE/COHDEFOR; DICTA; SETCO; PROFORFITH, La Ceiba, Honduras, pp. 148–149.

Verheij, E.W.M. and Coronel, R.E. (1991) *Edible Fruits and Nuts. Plant Resources of South East Asia No. 2*. Pudoc-DI.O, Wageningen, the Netherlands, 321 pp.

Villachica, H., Carvalho, J.E.U., Müller, H.C., Díaz, C.S. and Almanza, M. (1996) Castaña. In: *Frutales y Hortalizas Promisorios de la Amazonia*. Tratado de Cooperación Amazónica, Lima, Perú, pp. 85–95.

Copuazu or cupuassu

Alves, R.M., Artero, A.S., Sebbenn, A.M. and Figueira, A. (2003) Mating system in a natural population of *Theobroma grandiflorum* (Willd. ex Spreng.) Schum., by microsatellite markers. *Genetics and Molecular Biology* 26, 373–379.

Andersen, O. and Ulup-Andersen, V. (1988) *As Frutas Silvestres Brasileiras*. Publicaciones Globo Rural, São Paulo, Brazil, pp. 103–107.

Bayer, C., Fay, M.F., De Bruijn, A.Y., Savolainen, V., Morton, C.M., Kubitzki, K., Alverson, W.S. and Chase, M.W. (1999) Support for an expanded family concept of Malvaceae within a recircumscribed order Malvales: a combined analysis of plastid atpB and rbcL DNA sequences. *Botanical Journal of the Linnean Society* 129, 267–303.

Boulanger, R. and Crouzet, J. (2000) Free and bound flavour components of Amazonian fruits: 2. Cupuaçu volatile compounds. *Flavour and Fragance Journal* 15, 251–257.

Carpenter, D.R., Hammerstone, J.F., Romanczyk, L.J. and Aitken, W.M. (1994) Lipid composition of *Herrania* and *Theobroma* seeds. *Journal of the American Oil Chemists' Society* 71, 845–851.

Cavalcante, P.B. (1991) *Frutos Comestíveis da Amazônia*, Vol II. 6th edn. Edições CEJUP, Belém, Pará, Brazil, pp. 61–62.

Chaisieri, S., Arruda, D.H., Dimick, P.S. and Enriquez, G.A. (1989) Thermal characteristics and composition of fats from *Theobroma* species. *Turrialba* 39, 468–472.
Clement, C.R. (1999) 1942 and the loss of Amazonian crop genetic resources. I. The relation-between domestication and human population decline. *Economic Botany* 53, 188–202.
Cuatrecasas, J. (1964) Cacao and its allies: a taxonomy of the genus *Theobroma*. *Contributions from the U.S. National Herbarium* 35, 379–614.
Donadio, L.C., Moro, F.V. and Servidone, A.A. (2002) *Frutas Brasileiras*. Editora Novos Talentos, Jaboticabal, São Paulo, Brazil, pp. 149–153.
Franco, M.R.B. and Shibamoto, T. (2000) Volatile composition of some Brazilian fruits: Umbu-caja (*Spondias cytherea*), camu-camu (*Myrciaria dubia*), Araça-boi (*Eugenia stipitata*) and Cupuaçu (*Theobroma grandiflorum*). *Journal of Agricultural Food Chemistry* 48, 1263–1265.
Giacometti, D.C. (1994) Cupuaçu. In: Hernándo Bermejo, J.E. and León, J. (eds) *Neglected Crops: 1492 from a Different Perspective*. Plant Production and Protection Series No. 26. FAO, Rome, pp. 205–209.
Gilabert-Escrivá, M.V., Gonçalves, L.A.G., Figueira, A. and Silva, C.R.S. (2002) Fatty acid and triacylglycerol composition and thermal behavior of fats from seeds of Brazilian Amazonian *Theobroma* species. *Journal of the Science of Food and Agriculture* 82, 1425–1431.
Hammerstone, J.F., Romanczyk, L.J. and Aitken, W.M. (1994) Purine alkaloid distribution within *Herrania* and *Theobroma*. *Phytochemistry* 35, 1237–1240.
Martin, F.W., Campbell, C.W. and Ruberté, R.M. (1987) *Perennial Edible Fruits of the Tropics: An Inventory*. Agriculture Handbook No. 642, US Department of Agriculture, p. 64.
Ramos, A.A., Venturieri, G., Cuco, S. and Castro, N.M. (2005) The site of self-incompatibility action in cupuassu (*Theobroma grandiflorum*). *Revista Brasileira de Botanica* 28(3), 569–578.
Salvador, J.O., Muraoka, T., Rosseto, R. and Ribeiro, G.A. (1994) Sintomas de deficiencias nutricionais em cupuaçuzeiro (*Theobroma grandiflorum*) cultivado em solução nutritiva. *Scientia Agricola* 51, 407–414.
Silva, C.R.S., Figueira, A. and Souza, C.A.S. (2001) Diversidade no gênero *Theobroma*. In: Dias, L.A.S. (ed.) *Melhoramento Genético do Cacaueiro*. FUNAPE, UFG, Goiania, pp. 49–80.
Vargas, O., Alix, C., Lobo, A.D. (authors), Duarte, O. and Sanchez, J. (technical reviewers) (1999) *Frutales y Condimentarias del Trópico Húmedo*. CURLA; PDBL; AFE/COHDEFOR; DICTA; SETCO; PROFORFITH, La Ceiba, Honduras, pp. 322–323.
Velho, C.C., Whipkey, A. and Janick, J. (1990) Cupuassu: A new beverage crop for Brazil. In: Janick, J. and Simon, J.E. (eds) *Advances in New Crops*. Timber Press, Portland, Oregon, pp. 372–375.
Venturieri, G.A. (1993) *Cupuaçu: a espécie sua cultura, usos e processamento*. Belém, Clube do Cupu, 108 pp.
Vieira, M.C., Teixeira, A.A. and Silva, C.L.M. (2000) Mathematical modeling of the thermal degradation kinetics of vitamin C in cupuaçu (*Theobroma grandiflorum*) nectar. *Journal of Food Engineering* 43, 1–7.
Villachica, H., Carvalho, J.E.U., Müller, H.C., Díaz, C.S. and Almanza, M. (1996), Copuasu. In: *Frutales y Hortalizas Promisorios de la Amazonia*. Tratado de Cooperación Amazónica, Lima, Perú, pp. 103–112.

Ceriman

Abdel-Aal, A.S.I. (2011) Effect of plant density and fertilization on the production of *Monstera deliciosa* L. MS Thesis, Cairo Univ., Faculty of Agriculture, Egypt, Cairo (Egypt).

Available at: http://agris.fao.org/agris-search/search/display.do?f=2013/EG/EG1301.xml; EG2012000634 (accessed 3 November 2013).

Barabe, D. and Chretien, L. (1985) Anatomie florale de *Monstera deliciosa* (Araceae). *Canadian Journal of Botany* 63, 1423–1428.

Eyde, R.H., Nicolson, D.H. and Sherwin, P. (1967) A survey of floral anatomy in Araceae. *American Journal of Botany* 54, 478–497.

Gould, W.P. and Hallman, G.J. (2001) Laboratory and field infestation studies on *Monstera* to determine its host status in relation to the Caribbean fruit fly (Diptera: Tephritidae). *Florida Entomologist* 84, 437–438.

Hinchee, M.A.W. (1981) Morphogenesis of aerial and subterranean roots of *Monstera deliciosa*. *Botanical Gazette* 142, 347–359.

Labroy, M.O. (1908) Le ceriman de Mexique (*Monstera deliciosa* Liebm.) espece fruitiere. *Journal d'Agriculture Tropicale* 8, 169–171.

Lim, T.K. (2012) *Monstera deliciosa*. In: *Edible Medicinal and Non-Medicinal Plants*. Springer, the Netherlands, pp. 252–256.

Morton, J.F. (1987) *Fruits of Warm Climates*. Creative Resources Systems Inc., Winterville, North Carolina, pp. 15–17.

Paull, R.E. and Duarte, O. (2008) Ceriman (*Monstera deliciosa* Liebm.), Araceae. In: Janick, J. and Paull, R.E. (eds) *Encyclopedia of Fruit and Nuts*. CAB International, Wallingford, UK, pp. 77–78.

Peppard, T.L. (1992) Volatile flavor constituents of *Monstera deliciosa*. *Journal of Agriculture and Food Chemistry* 40, 257–262.

Peters, R.E. and Lee, T.H. (1977) Composition and physiology of *Monstera deliciosa* fruit and juice. *Journal of Food Science* 42, 1132–1133.

Pochet, M., Martin-Tanguy, J., Marais, A. and Martin, C. (1982) Hydroxycinnamoyl acid amines and aromatic amines in the inflorescence of some *Araceae* species. *Phytochemistry* 21, 2865–2869.

Romero Castañeda, R. (1969) *Frutas Silvestres de Colombia*. Universidad Nacional de Colombia, Bogotá, Colombia, pp. 22–24.

Xiao-xiao, L., Ling, C., Hong-xi, H. and Yu, H. (2009) Establishment of tissue culture and plant regeneration system of *Monstera deliciosa*. *Guangxi Tropical Agriculture* 2, 004.

Index

Page numbers in **bold** refer to figures and tables